NUCLEAR POWER
TECHNOLOGY

NUCLEAR POWER TECHNOLOGY

Available separately

NUCLEAR POWER TECHNOLOGY

EDITED BY
W. MARSHALL

Volume 2
Fuel Cycle

CLARENDON PRESS · OXFORD
1983

Oxford University Press, Walton Street, Oxford OX2 6DP

London Glasgow New York Toronto
Delhi Bombay Calcutta Madras Karachi
Kuala Lumpur Singapore Hong Kong Tokyo
Nairobi Dar es Salaam Cape Town
Melbourne Auckland

and associated companies in
Beirut Berlin Ibadan Mexico City Nicosia

Oxford is a trade mark of Oxford University Press

Published in the United States
by Oxford University Press, New York

British Library Cataloguing in Publication Data
Nuclear power technology.
Vol. 2: Fuel cycle
1. Atomic power
I. Marshall, W
621.48   TK9145
ISBN 0-19-851958-3

Library of Congress Cataloging in Publication Data
Nuclear power technology.
Contents: v. 1. Reactor technology – v. 2. Fuel cycle –
v. 3. Nuclear radiation.   1. Atomic power.
I. Marshall, W. (Walter), 1932–
TK9145.N83 1983   621.48   83–4124
ISBN 0-19-851958-3 (v. 2)

Printed in Great Britain
by The Thetford Press,
Thetford, Norfolk

PREFACE

Some time ago the Oxford University Press asked if I would be interested in preparing a book on nuclear power oriented towards increasing public knowledge and for the public interest. The United Kingdom Atomic Energy Authority was approaching its twenty-fifth anniversary and it therefore seemed appropriate to gather a set of chapters reviewing the state of nuclear science. The obvious way of doing this was to list all the subjects involved and then persuade an expert in each subject to write a chapter. Having chosen authors of wide experience in their fields, the result is an authoritative source book on the major aspects of nuclear science and nuclear power.

To achieve some cohesion, I set up an editorial board which has reviewed all the chapters to make sure they are factually correct, and we believe that in all the papers the choice of language makes it clear what is fact and what is opinion.

The chapters do not represent an official UKAEA point of view. Each one is very much the personal opinion of the individual author, and I have not attempted to influence his perspective of the subject. The chapters are, therefore, presented in a variety of styles, but it has been our intention throughout that they should be free of unnecessary technical jargon and comprehensible to the non-specialist. While in a publication of this standard it has been necessary to treat the technical and scientific basis for some of the subjects in depth, I hope that the book will prove helpful and informative to the general reader who, conscious of the 'Nuclear Debate' taking place around him, wishes to have access to a reliable reference document.

Many of the chapters contain bibliographies so that the reader may delve further into the technical facts and figures if he so chooses. There is also a summary of each chapter and, at the end, a comprehensive glossary and index to complete the texts.

Most of the authors come from within the UKAEA. My special thanks are due to Dr S. H. U. Bowie, who was Chief Consultant Geologist to the UKAEA from 1955 to 1977, for completing the series by contributing an article.

I especially thank Dr R. H. Flowers for the part which he has played, as Deputy Editor, in bringing this work to completion and Mrs N. M. Hutchins, as secretary to the Editorial Board, for organizing and progressing the contributions.

May 1981
                                        W. MARSHALL
                                        Chairman,
                          United Kingdom Atomic Energy Authority

ACKNOWLEDGEMENTS

Acknowledgements are due to the Editorial Board set up by the Chairman of the UKAEA to review the articles during their various stages of preparation. The members of this Editorial Board were:

Dr W. Marshall, Chairman (UKAEA) and Chairman of the Editorial Board
Dr R. H. Flowers, Deputy Editor

Mr J. G. Collier	Dr R. G. Sowden
Mr J. M. Hutcheon	Mr J. G. Tyror
Dr P. M. S. Jones	Mr P. N. Vey
Dr A. B. Lidiard	Dr B. O. Wade

Acknowledgements are due also to Mr G. Gibbons who was responsible for the artwork, to Mr J. A. G. Heller, who commented extensively on the articles from the lay-reader's point of view, to Miss P. E. Barnes, who undertook the vast typing task and to Mrs N. M. Hutchins who, as secretary to the Editorial Board, organized the authors contributions and did a large amount of editorial work.

Special thanks are due to Mr S. U. Bowie, now retired from the Institute of Geological Sciences, who prepared the article on Uranium and Thorium Raw Materials.

R. H. FLOWERS
Deputy Editor

CONTENTS

CONTRIBUTORS

R. H. ALLARDICE *Deputy Director, Dounreay*

S. BOWIE *Geological Consultant to UKAEA, Crewkerne, Somerset*

R. F. BURSTALL *Technical Services and Planning Directorate, Risley*

A. A. FARMER *Central Technical Services, Risley*

D. W. HARRIS *Manager, Reprocessing Development Group, Dounreay*

A. W. HILLS *Economics and Programmes, UKAEA, London*

C. T. JOHN *Technical Services and Planning Directorate, Risley*

N. J. KEEN *Authority Fuel Processing Directorate, Harwell*

A. L. MILLS *Chemical Technology Division, Harwell*

D. O. PICKMAN *Head of Laboratories, Springfields*

K. SADDINGTON *Consultant to UKAEA, Windscale Laboratories, Seascale, Cumbria*

J. H. TAIT *Theoretical Physics Division, Harwell*

J. D. THORN *Technical Services and Planning Directorate, Risley*

SYMBOLS AND ABBREVIATIONS

Abbreviation or Symbol	Name of unit and quantity measured	Notes, Units, Value
A	ampere, electric current	
α	alpha particle (He^{++})	
amu	atomic mass unit	1.66×10^{-27} kg
bar	bar, pressure	10^5 Pa
barn	unit of cross-section	10^{-28} m^2
β	beta particle (electron)	
Bq	becquerel, radioactivity	s^{-1}
c	velocity of light	3×10^8 m s^{-1}
C	coulomb, electric charge	A s
°C	celsius temperature interval	degree celsius
Ci	curie, radioactivity	3.7×10^{10} Bq
d	day, time	
EFPH	effective full power hours	
eV	electronvolt, energy	1.59×10^{-19} J
g	gram, mass	
γ	gamma radiation (photon)	
Gy	gray, absorbed dose	J kg^{-1}
h	hour, time	
h	Planck's constant	$6.6255\ 10^{-34}$ J s
Hz	hertz, frequency	s^{-1}
J	**joule, energy**	
k	Boltzmann constant	1.3805×10^{-23} J K^{-1}
K	kelvin, absolute temperature interval	**kelvin**
m	metre, length	
M	concentration	10^3 mol m^{-3}
min	minute, time	
mol	mole, amount of substance	gram-molecule
MW(e)	unit of power station output	megawatt, electrical
MW(t)	unit of power station output	megawatt, thermal
n	neutron	
N	newton, force	kg m s^{-2}
p	proton	
Pa	pascal, pressure	1 newton m^{-2}
pH	unit of acidity	$-\log_{10} M_{H^+}$
rad	absorbed dose	10^{-2} Gy
rem	absorbed dose equivalent	10^{-2} Sv
s	second, time	
s.t.p.	standard temperature and pressure	1 bar 273.16 K
SWU	separative work unit	see Chapter 2
Sv	sievert, absorbed dose equivalent	J kg^{-1}
t	metric tonne	1000 kg
T	tesla, magnetic flux density	V s m^{-2}

tce	tonne coal equivalent	
tha	tonne heavy atoms	
V	volt, potential difference	
W	watt, power	$J\,s^{-1}$
y	year, time	

Prefixes

Multiple	Prefix	Abbreviation
10^{-18}	atto	a
10^{-15}	femto	f
10^{-12}	pico	p
10^{-9}	nano	n
10^{-6}	micro	μ
10^{-3}	milli	m
10^{-2}	centi	c
10^{3}	kilo	k
10^{6}	mega	M
10^{9}	giga	G
10^{12}	tera	T

Additional less commonly used symbols and units are explained and defined as they occur in the text.

The Elements

A complete list of the elements and their symbols, is included here for reference.

Element	Symbol	Element	Symbol	Element	Symbol
Actinium	Ac	Dysprosium	Dy	Lutetium	Lu
Aluminium	Al	Einsteinium	Es	Magnesium	Mg
Americium	Am	Erbium	Er	Manganese	Mn
Antimony	Sb	Europium	Eu	Mendelevium	Md
Argon	Ar	Fermium	Fm	Mercury	Hg
Arsenic	As	Fluorine	F	Molybdenum	Mo
Astatine	At	Francium	Fr	Neodymium	Nd
Barium	Ba	Gadolinium	Gd	Neon	Ne
Berkelium	Bk	Gallium	Ga	Neptunium	Np
Beryllium	Be	Germanium	Ge	Nickel	Ni
Bismuth	Bi	Gold	Au	Niobium	Nb
Boron	B	Hafnium	Hf	Nitrogen	N
Bromine	Br	Helium	He	Nobelium	No
Cadmium	Cd	Holmium	Ho	Osmium	Os
Caesium	Cs	Hydrogen	H	Oxygen	O
Calcium	Ca	Indium	In	Palladium	Pd
Californium	Cf	Iodine	I	Phosphorus	P
Carbon	C	Iridium	Ir	Platinum	Pt
Cerium	Ce	Iron	Fe	Plutonium	Pu
Chlorine	Cl	Krypton	Kr	Polonium	Po
Chromium	Cr	Lanthanum	La	Potassium	K
Cobalt	Co	Lawrencium	Lr	Praseodymium	Pr
Copper	Cu	Lead	Pb	Promethium	Pm
Curium	Cm	Lithium	Li	Protactinium	Pa

Element	Symbol	Element	Symbol	Element	Symbol
Radium	Ra	Sodium	Na	Titanium	Ti
Radon	Rn	Strontium	Sr	Tungsten	W
Rhenium	Re	Sulphur	S	Uranium	U
Rhodium	Rh	Tantalum	Ta	Vanadium	V
Rubidium	Rb	Technetium	Tc	Xenon	Xe
Ruthenium	Ru	Tellurium	Te	Ytterbium	Yb
Samarium	Sm	Terbium	Tb	Yttrium	Y
Scandium	Sc	Thallium	Tl	Zinc	Zn
Selenium	Se	Thorium	Th	Zirconium	Zr
Silicon	Si	Thulium	Tm		
Silver	Ag	Tin	Sn		

9

Recycling of fuel

A. A. FARMER

There are three options presently available for thermal reactor fuel cycles: (1) the once-through mode, (2) reprocessing to recover the uranium but dispose of the plutonium and fission products, and (3) reprocessing to recover both the uranium and the plutonium. The recovered plutonium can be used as fuel either in thermal or fast reactors. However, plutonium can be used more efficiently in a fast neutron flux. In a suitably designed fast reactor it is possible, by irradiation of ^{238}U to *breed* a larger quantity of fissile material than is consumed, making it possible to use the preponderant ^{238}U isotope for power production. In this way recycle of uranium and plutonium in a fast breeder reactor system can achieve around 50 times better utilization of the original uranium than a thermal reactor system can. Reprocessing and recycle of irradiated fuel thus represents a substantial contribution towards fuel conservation.

The different properties of plutonium affect the safety characteristics of thermal and fast reactors in differing ways. There is a change in isotopic composition of plutonium as recycle proceeds. The chapter describes this and gives data on the build-up of the higher actinides, americium and curium.

Attention is drawn to the problems of fabricating and handling plutonium-enriched fuel for use in thermal reactors. As an example, a plant for refabricating mixed uranium oxide–plutonium oxide LWR fuels is described. Mention is made of the problems associated with the transport of plutonium and the impact of the special requirements of plutonium handling on fuel costs. The main difference between plutonium-enriched thermal reactor fuel and plutonium fast reactor fuel is shown to lie in the concentration of plutonium in the fuel and in the complexity of the fuel assemblies.

An approach is suggested for evaluating the economics of recycle in thermal and fast reactors. The choice between reactor strategies may not depend only on financial considerations but may also reflect the longer-term strategic issue of minimizing future uranium imports, by fully utilizing the 'waste' ^{238}U isotope. This means that recycle programmes must be evaluated in the context of their impact on the nuclear generating system as a whole. It is seen that the total uranium requirements can be restricted by the introduction of fast reactors. The quantities and distribution of plutonium in alternative reactor strategies are also described.

The technology for recycle of plutonium in thermal reactors is established, though an industrial-scale capability has yet to be demonstrated. Fast reactors are well established up to the large prototype scale, and progress is being made with the establishment of commercial programmes.

Before adopting a fuel recycle policy, certain basic factors must be considered. Among these are the questions of the basic feasibility, economics, and the availability and utilization of fuel, and attention must also be given to proliferation considerations.

The chapter concludes with a suggestion as to the timing required for the availability of fast reactors as an established technology for the generation of electricity.

Contents

1 Introduction

At the present time the generation of electricity by nuclear power is almost exclusively based on the use of thermal reactors with uranium as their basic fuel. The nuclear fuel cycle begins with the mining of uranium ore and its partial purification and conversion into uranium ore concentrates. After this, for those reactors using natural uranium, the ore concentrates are converted into pure uranium metal or its oxide. These materials are then fabricated into fuel elements and used in nuclear power stations of the gas–graphite Magnox (metal fuel) or heavy water CANDU (oxide fuel) type. Other types of thermal reactor, such as advanced gas-cooled reactors (AGR), or light water reactors (LWR), require uranium as uranium oxide in which the concentration of the fissile ^{235}U isotope has been increased (enriched). Natural uranium contains about 0.7 per cent uranium-235 (^{235}U) (the remainder being mainly ^{238}U), whereas thermal reactors such as AGR and LWR require enriched uranium with some 2–4 per cent of the ^{235}U isotope.

The starting point for obtaining enriched uranium is a chemical compound of uranium—uranium hexafluoride, *Hex*—which is gaseous at temperatures above 330 K. Natural Hex is fed into an enrichment plant, from which is obtained enriched Hex or *product* at the top end and depleted Hex or *tails* at the bottom end. The ^{235}U content of the tails is normally around 0.2 to 0.3 per cent, the rest being ^{238}U, and this material can be considered merely as a waste product as far as its use in thermal reactors is concerned. The enriched Hex is converted to uranium oxide and fabricated into fuel elements for use in power-producing thermal reactors.

During the operation of a thermal reactor, most of the thermal energy produced is derived from the fission of the ^{235}U isotope. At the end of the prescribed irradiation period, the discharged uranium will have a reduced ^{235}U content. This is around 0.2 to 0.4 per cent ^{235}U for the natural-uranium reactors, and around 0.8 per cent ^{235}U for the enriched reactors, which is similar to that found in natural uranium, and thus could be recovered during a reprocessing stage and recycled. The discharged uranium from the natural-uranium reactors, having a ^{235}U content similar to enrichment plant tails, is mainly a waste product.

As explained in §2, during the operation of thermal reactors some neutrons will be captured by ^{238}U atoms to form ^{239}U which, after two further nuclear decay processes, is converted into plutonium-239 (^{239}Pu). Like ^{235}U, ^{239}Pu is fissile in thermal reactors and is therefore a valuable fuel. Indeed, some of the plutonium created during the operation of a thermal reactor is fissioned before the fuel reaches the scheduled irradiation (or burn-up) level at which it is discharged. Some of the created plutonium, however, remains in the discharged fuel and therefore constitutes a valuable resource that could be recovered by chemical reprocessing for future use.

There are three options presently available for fuel management in thermal reactors. The first is the once-through mode in which uranium fuel is discharged from the reactor at a specified burn-up level and subsequently stored with a view to ultimate disposal of the contained uranium, plutonium, and fission products. The second option is to reprocess irradiated fuel to recover the uranium, but dispose of the plutonium and fission products. The third option is to reprocess irradiated fuel to recover both the uranium and plutonium, and dispose of the waste products only.

There are several factors which affect the choice of options to be followed. In the once-through mode, the utilization of uranium is only around one-half per cent. That is to say, of the original uranium mined, only one-half per cent of the total uranium atoms (both ^{235}U and ^{238}U isotopes) are actually fissioned, the rest are wasted. In the case of the thermal reactors using enriched uranium, if the fuel is reprocessed, the recovered uranium can be used in new fuel elements, thus reducing requirements for both new uranium ore and uranium enrichment capacity by some 15 to 20 per cent. If the plutonium were also to be recycled in new fuel, a further 20 to 25 per cent saving in uranium would be obtained. The overall effect, therefore, of a single recycle of both uranium and plutonium would be to enable some 35 to 40 per cent more power to be generated from the same amount of uranium ore, and hence to increase the utilization to around 0.7 per cent. Subsequent further recycle is also possible, but as the number of plutonium atoms produced in a thermal reactor per fissile atom fissioned is substantially less than one, the process soon converges, and the maximum utilization of uranium in an enriched thermal reactor is about 1 per cent. Reprocessing of irradiated natural uranium fuel is not worthwhile just to recover the uranium. However, recovery of the plutonium is worthwhile, and its recycle in reactors such as CANDU would increase the utilization of uranium to between 1 and 2 per cent.

But, as will be explained in §2, where the characteristics of plutonium in both thermal and fast reactors are discussed, the use of plutonium as a fuel is more efficient in a fast neutron flux than in a thermal neutron flux. Moreover, the so-called *fast breeder reactor* can be so designed that it converts more fertile ^{238}U into fissile plutonium than the fissile material it uses. The use in fast reactors of plutonium and the waste depleted uranium obtained from enrichment plant tails and irradiated thermal reactor fuel can therefore, in theory, fission virtually all the mined uranium. In practice, because the fast reactor fuel must be reprocessed after achieving a specified irradiation level to recover the uranium and plutonium for recycle, some of these materials will be lost to waste residues. The practical utilization of uranium will range from around 50 to 80 per cent depending upon the fraction of the material lost to residues. But in any event, recycle of uranium and plutonium in a fast breeder reactor system will achieve at least some 50 times greater utilization of the original uranium than with thermal reactors. Reprocessing and recycle of irradiated fuel, therefore, represents an attractive and very substantial contribution towards fuel conservation. Such recycling accords with sound waste management principles. The waste is likely to be stored for a period of at least some decades under safe and secure conditions prior to disposal. Methods of long-term waste stabilization and of waste disposal are currently under active development.

Section 3 describes how the different properties of plutonium affect the safety characteristics of thermal and fast reactors in differing ways. Also included in this section is an account of the change in plutonium isotopic composition as recyle proceeds and data are given on the build-up of the higher actinides, americium and curium.

The fabrication and reprocessing of plutonium-enriched fuels are the subject of §4. Technical features of the fabrication and reprocessing of plutonium as well as uranium fuels are dealt with in Chapter 13 and Chapter 14, and this chapter merely raises a few additional points relating to the economic effects of the plutonium fuel cycle.

Discussion on economics of recycle in thermal and fast reactors is continued in §5. The point is made that the choice between one reactor strategy and another may not be entirely a matter of the economic saving that might be achieved, but may also reflect the desire to minimize future uranium imports— and hence the strategic dependence upon other countries—by fully utilizing the 'waste' uranium arising from thermal programmes. To assess such factors it is necessary to consider the effects of plutonium recycle in thermal and fast reactors on a whole nuclear generating system. Such an assessment is outlined in the second part of §5, from which it will be seen that total uranium requirements can be restricted by the introduction of fast reactors. The quantities and distribution of plutonium in alternative reactor strategies are also described.

Section 6 indicates that, although the technology for recycle of plutonium in thermal reactors is established, there is still the need for a large demonstration programme to prove a capability on an industrial scale. Fast reactors are well established up to the large prototype scale, and progress is being made with the establishment of commercial programmes.

Finally, §7 discusses some of the factors that will govern a decision in support of recycle by summarizing some of the points established in earlier sections. There is, however, new material included: in particular, reference is made to proliferation aspects, and a short discussion on the availability of nuclear fuel is introduced. The chapter concludes with a suggestion as to the timing required for the availability of fast reactors as an established technology for the generation of electricity.

2 Plutonium characteristics

During the operation of a reactor fuelled with uranium some of the isotope ^{238}U will be converted into plutonium. There are four major plutonium isotopes, ranging in atomic weight from 239 to 242. Neutron capture in ^{238}U leads to ^{239}Pu, and further neutron captures produce the other plutonium isotopes. The amount of ^{238}U converted to plutonium will increase as the time the fuel is irradiated increases. Similarly, the longer the irradiation time the greater becomes the proportion of the higher isotopes. Recycle of plutonium will also increase the proportion of the higher isotopes, though repeated recycling will tend to lead to an equilibrium plutonium composition.

Nuclides in the trans-actinium group, i.e. with atomic number greater than or equal to that of actinium (89), may in general be divided into three groups, the fissile group, the fertile group, and the group which only captures neutrons. All the nuclides are fissionable if the neutron energy is high enough. The first group however contains nuclides like ^{235}U and ^{239}Pu, for which fission is the most probable result of a neutron absorption over a wide energy range of the incident neutron. The second group contains nuclides for which a neutron capture leads to a fissile nuclide. Examples are ^{238}U and ^{240}Pu, which are converted to ^{239}Pu and ^{241}Pu, respectively. The third group contains nuclides like ^{236}U and ^{242}Pu, whose main effect in a reactor is that they capture neutrons, at a slow rate but in an unproductive manner.

2.1 Nuclide production and removal processes

Figure 9.1 illustrates the decay and neutron reaction processes of the main nuclides involved in uranium–plutonium fuel cycles. This picture is inevitably fairly complex, since some processes lead to increases in atomic number or atomic weight, while others lead to decreases, and by steps of greater than unity. For this reason, only the nuclides of major importance are shown, and it should be realized that other nuclides, and other routes of decay or production, do exist. Figure 9.1 gives the route involved in the transition from one nuclide to another, and also gives the half-life of each nuclide. Unless the mode of decay is indicated by a diagonal line (from right to left upwards for β-decay and from right to left downwards for α-decay), the mode of radioactive decay is also indicated under the half-life. In particular, it is to be noted that α-decays have not generally been indicated by diagonal lines, in order to avoid an undue cluttering of the diagram which might confuse the reader. Also, fission reactions are not shown since they do not lead to other nuclides in Fig. 9.1.

To understand the capture and decay processes possible within a reactor, Fig. 9.1 needs to be studied with some care. For example, starting with ^{238}U there are three events that can occur.

The first is associated with the natural radioactivity of this isotope: ^{238}U decays with a very long half-life of 4.5×10^9 y to ^{234}Th by emitting an alpha particle. Subsequently, a series of beta and alpha decays occur in a chain of events ending in the stable isotope of lead ^{206}Pb. Such decay chains are ignored in reactor studies.

The second event that might occur is that an incident neutron is captured by the ^{238}U nucleus and two neutrons are immediately emitted from the nucleus; that is, an (n, 2n) reaction occurs, and the ^{238}U atom is converted to

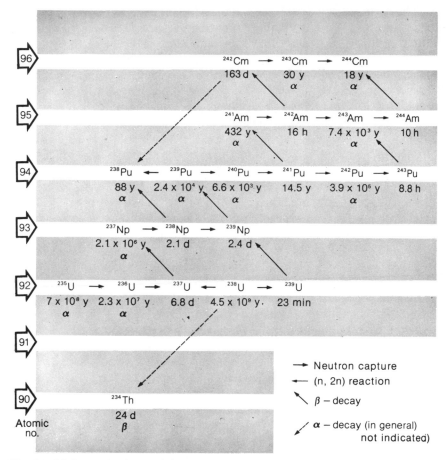

Fig. 9.1 Main production and decay processes

a ^{237}U atom. The ^{237}U atom will then decay with a half-life of 6.8 days and be transformed into ^{237}Np after the emission of a beta particle. ^{237}Np is itself alpha radioactive with a very long half-life of 2.1 × 10^6 y. Within a reactor it is more likely to capture a neutron and be converted to ^{238}Np, which itself is beta active, having a short half-life of 2.1 days to convert to ^{238}Pu. Alternatively, the ^{238}Np atom might capture a second neutron to become ^{239}Np, which is also beta active with a short half-life of 2.4 days to give ^{239}Pu.

The third and most probable event that can occur is that the ^{238}U atom will capture a neutron to become ^{239}U, which is beta active with a half-life of 23 minutes. On emitting a beta particle the ^{239}U is transformed to ^{239}Np which, as has already been shown, also decays by emitting a beta particle to give ^{239}Pu. ^{239}Pu is a member of the fissile group of nuclides and is an important secondary nuclear fuel. However, as with ^{238}U, three other events might occur in a reactor.

Firstly, ^{239}Pu is naturally radioactive with a half-life of 2.4 × 10^4 y, emitting an alpha particle to give ^{235}U. Secondly, an (n, 2n) reaction can occur to give ^{238}Pu. Thirdly, ^{239}Pu can capture a neutron to give ^{240}Pu, and subsequent neutron captures can yield ^{241}Pu, ^{242}Pu, and ^{243}Pu.

Both ^{240}Pu and ^{242}Pu are alpha radioactive and decay to give ^{236}U and ^{238}U respectively. ^{241}Pu and ^{243}Pu are both beta active and decay to ^{241}Am and ^{243}Am respectively if this should occur before neutron absorption, which would result in either fission or capture to give a higher plutonium isotope.

^{241}Am decays by emission of an alpha particle with a half-life of 432 years. Capture of a neutron can also occur to give ^{242}Am, which is beta active with a half-life of 16 hours. On decay in this manner ^{242}Cm is obtained, which may decay to ^{238}Pu by emitting an alpha-particle or capture a neutron to become ^{243}Cm and, via a further neutron capture, ^{244}Cm.

Thus it is seen that there is a very wide range of half-lives of the nuclides. Several of the uranium, neptunium, plutonium, and americium isotopes have half-lives of thousands of years and raise long-term issues of waste storage and disposal. Other isotopes decay rapidly, e.g. the two nuclides in the link between ^{238}U and ^{239}Pu, and are of limited importance. A third group of nuclides have half-lifes in an intermediate range (of the order of years), and these are often important as this is the timescale over which nuclear fuel will have to be handled.

The relative abundances of the various nuclides shown in Fig. 9.1 cannot be estimated easily from the diagram alone, as they depend upon their radioactive half-lives, the time they spend within a reactor, whether or not the fuel is reprocessed and recycled, and finally on the times irradiated fuel is stored before and after reprocessing. For example, starting with ^{238}U, it is seen that six successive neutron captures must occur before the production of ^{244}Cm. The longer the irradiation period, the greater the production of higher plutonium isotopes and the greater the possibility of the production of curium via production of americium from ^{241}Pu. However, if irradiated fuel containing plutonium is discharged and held for a considerable time before reprocessing, some of the ^{241}Pu will decay to ^{241}Am and will be removed during the subsequent reprocessing. Thus, the production of ^{242}Cm in recycled plutonium fuel can depend upon the time taken for reprocessing. Similarly, after reprocessing and refabrication of the plutonium into new fuel a long delay will allow some more of the ^{241}Pu to decay into ^{241}Am. This ^{241}Am will be present in the fuel when reloaded into a reactor. Hence the longer the delay after refabrication, the greater will be the build-up of ^{241}Am and the production of the ^{242}Cm isotope.

2.2   Plutonium isotope cross-sections

Of the four main plutonium isotopes, ^{239}Pu and ^{241}Pu are readily fissionable, and compare with ^{235}U for use in a reactor. The physical properties of a nucleus depend upon the parity of its number of protons and neutrons (i.e. whether this number is even or odd). All three of the isotopes just mentioned are of the even-proton-number, odd-neutron-number type (i.e. even-odd), which is more liable than any other combination to undergo fission on absorbing a neutron.

The concept of *cross-sections* as a measure of the probability of various neutron reactions, has been introduced in Chapter 1. The fission cross-sections of the three main fissionable isotopes are plotted in Fig. 9.2 against neutron energy over the range of 0.01 eV to 1 keV. The curves for ^{235}U and ^{239}Pu are not shown in the energy range 5–200 keV because the cross-section varies rapidly with energy, passing through a series of closely-spaced peaks (the resonances in the so-called *resolved resonance region*). At energies above the resolved resonance region the resonances still occur but they are too close to be separated experimentally. As far as the two plutonium isotopes are concerned, in addition to the resonances in the keV energy range, there is a single large resonance at about 0.3 eV. This resonance is important for thermal reactor studies, as it raises the overall fission cross-section of the plutonium isotopes, and it also has implications for the safety of such reactors.

In thermal reactors, a plot of neutron flux against energy shows a variation at low energies that exhibits the typical characteristics of Maxwell's probability function for the mean reactor temperature. This shows a few neutrons having low velocities (low energies) rising to a maximum number at the most probable velocity (energy), corresponding to the mean temperature, and falling off towards zero neutrons with high velocities (high energies). It may be noted that the variation is proportional to the reciprocal of the energy at

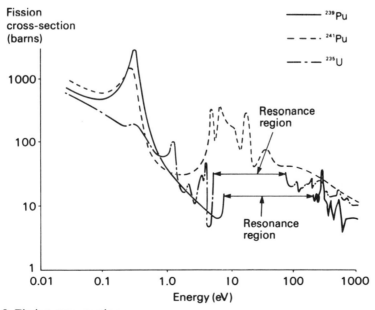

Fig. 9.2 Fission cross-sections

higher energies. There is a further small peak in the flux in the range at which neutrons are emitted in the fission process. The position of the peak of the Maxwellian curve also depends upon the degree of moderation of the neutrons and differs for different thermal reactors. For example, in a typical PWR the peak is at about 0.06 eV, which is fairly close to the plutonium cross-section resonance at 0.3 eV. If for some reason the temperature of the reactor rises, either in normal operation or because of a fault, then the neutron flux Maxwellian curve moves to a higher-energy range and the importance of the plutonium resonance rises. It cannot, however, be stated that the rate of neutron production will rise, because there are also neutron capture effects as discussed below.

The capture cross-sections of the main plutonium isotopes are plotted in Fig. 9.3. The ^{239}Pu and ^{241}Pu curves again show a high resonance at 0.3 eV. The ^{240}Pu curve shows an important resonance at about 1 eV. This latter effect is important in the consideration of the use of plutonium in thermal reactors. As the amount of ^{240}Pu isotope builds up with increasing irradiation, its cross-section is sufficiently large to increase significantly the number of neutrons captured, and so adversely to affect the overall neutron balance. This effect is somewhat mitigated in most thermal reactors by concentrating the fuel in discrete amounts, with the result that, as neutrons slow down through the resonance energies within the moderator, there is an improved chance that they will not at that stage have contact with the resonance absorber. Thus the *effective* resonance capture probability is reduced.

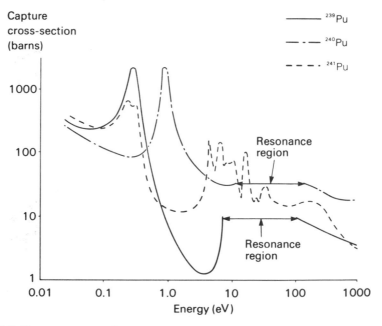

Fig. 9.3 Capture cross-sections

Throughout the energy range of Fig. 9.3 the capture cross-section of ^{240}Pu is comparable with those of ^{239}Pu and ^{241}Pu, and all three are significant in comparison with the fission cross-sections of Fig. 9.2. These facts mean that the use of plutonium in fast reactors does have advantages, a point which is discussed in more detail later.

Clearly in thermal reactors containing appreciable amounts of plutonium the low-energy resonances play a significant role, and it is important to carry out calculations in sufficient detail. In fast reactors, in which there is no specific material to slow neutrons, the low-energy resonances are not very important, as the neutron flux spectrum peaks at an energy of about 0.1 MeV, and the variation in flux with neutron energy is a flatter shape than in thermal reactors.

2.3 Thermal reactor cross-sections

As Chapter 2 indicates, it is necessary in reactor calculations to divide the energy range of Figs. 9.2 and 9.3 into intervals or groups, which are represented by average properties. The energy groups are chosen to give adequate representation of the cross-section variations of the different neutron reactions. The choice of the number of energy groups depends on the type of calculation. It is rarely possible to employ only one group, but one-group cross-sections are useful for comparison purposes. As there are a variety of thermal reactors, each with different one-group cross-sections, it is customary for these systems to quote two parameters, the cross-section at a neutron velocity of 2200 m/s (an energy of 0.025 eV), which is the most probable velocity at room temperature, and the resonance integral, which is the integral over the energies of the resonance region of the cross-section weighted by the reciprocal of the energy. Table 9.1 gives these parameters for the main nuclides.

The figures demonstrate the high fission cross-section of ^{239}Pu and ^{241}Pu in comparison with that of ^{235}U. However, the capture cross-sections of the plutonium isotopes are also high and examination of the relative capture-

Table 9.1 Cross-sections for thermal reactors (barns)

	Fission 2200 m/s value (0.025 eV)	Resonance integral	Capture 2200 m/s value (0.025 eV)	Resonance integral
^{235}U	579	275	100	144
^{236}U	—	—	5	365
^{238}U	—	—	3	275
^{239}Pu	741	301	267	200
^{240}Pu	—	—	290	8013
^{241}Pu	1009	570	368	162
^{242}Pu	—	—	19	1130
^{241}Am	3	21	832	1477

to-fission cross-section ratios indicates they are higher for ^{239}Pu than ^{235}U. This has an adverse effect on the prospects for using plutonium as a thermal reactor fuel. For example, in a typical PWR about 20 per cent of the neutrons absorbed by ^{235}U lead to ^{236}U by capture, the remainder causing fission. For ^{239}Pu the equivalent percentage (i.e. for capture producing ^{241}Pu) is up to about 36 per cent. In addition, the ^{240}Pu, if present in sufficient quantity, has a high capture cross-section and so increases further the proportion of neutrons being captured in plutonium. This adverse capture cross-section feature is partially cancelled by the high average number of neutrons produced per fission event. For ^{239}Pu this is 2.9, which compares with 2.5 for ^{235}U. Even so, the capture effect predominates, and the use of plutonium in thermal reactors is less attractive, from the physics viewpoint, than use of enriched uranium.

2.4 Fast reactor cross-sections

Table 9.2 gives one-group cross-sections (i.e. cross-sections averaged over the whole energy range) in a typical fast reactor for the main nuclides. The most noticeable difference between the thermal and fast reactor data of Tables 9.1 and 9.2 is that the latter are on average about two orders of magnitude lower than the former. This is a reflection of the general drop in cross-section with increasing neutron energy, though resonance regions are an exception.

A further important difference is that the fission cross-sections of all the nuclides of Table 9.2 are significant, while for Table 9.1 only three nuclides have significant values. This is an important advantage for fast reactors over thermal reactors. In a typical fast reactor about 15 per cent of all fissions occur in ^{238}U, which means that fewer fissions are needed in the *fissile* nuclides. In thermal reactors the percentage is much lower.

Comparison of the total cross-sections of the fertile and fissionable nuclides in Tables 9.1 and 9.2 also helps to explain why fast reactors need higher enrichments of fissile atoms than thermal reactors to function. The total cross-section of ^{238}U is lower than that of ^{239}Pu by a factor of around seven in Table 9.2. The corresponding relation for thermal reactors is over 300 if all the neutrons are assumed thermalized to a velocity of 2200 m/s. The

Table 9.2 One-group cross-sections for fast reactors (barns)

	Fission	Capture
^{235}U	2.0	0.5
^{236}U	0.1	0.6
^{238}U	0.05	0.3
^{239}Pu	1.9	0.6
^{240}Pu	0.4	0.6
^{241}Pu	2.6	0.6
^{242}Pu	0.3	0.4
^{241}Am	0.4	1.9

spread of neutron energies and the resonances will reduce the factor, but it is still an order of magnitude greater than that for fast reactors. Hence for a given mixture of, say, ^{238}U and ^{239}Pu, absorption of a neutron in plutonium rather than uranium is ten times more likely in a thermal reactor than in a fast reactor. As the fraction of all neutrons absorbed in plutonium causing fission is about the same for both thermal and fast reactors, it follows that the enrichment of plutonium in uranium required in a thermal reactor is about a tenth that required in a fast reactor. There are of course many other factors that affect the enrichment required to operate power reactors.

The last point to be observed in the comparison of the two tables is that the capture cross-sections of the fissionable plutonium isotopes are no longer high in comparison to the fission cross-sections. For ^{239}Pu only 24 per cent of the neutrons absorbed lead to ^{240}Pu by capture. The figure for ^{235}U is 20 per-cent. Hence, from a physics viewpoint, use of plutonium in a fast reactor is more attractive than use in a thermal reactor. It does not rule out use of plutonium as a thermal reactor fuel, but the element is more efficiently used in fast reactors.

2.5 Plutonium worths

When a fissile nucleus absorbs a neutron, it generally reacts in one of two ways: it may undergo fission and liberate more neutrons, or it may retain the neutron to form a less fissionable higher isotope, e.g. ^{236}U or ^{240}Pu as already seen in Fig. 9.1. The number of neutrons released per neutron absorbed by the fissile nucleus is given by:

$$\eta = \nu\sigma_f/(\sigma_c + \sigma_f),$$

where ν is the number of neutrons per fission, and σ_c and σ_f are respectively the capture and fission cross-sections of the nucleus. The quantity ν varies for the different nuclides and with the energy of the incident neutron. Likewise, σ_c and σ_f vary as already seen in Tables 9.1 and 9.2. It follows that the quantity η varies for different nuclides and with reactor type, i.e. thermal or fast. The quantity η contributes directly to the reactivity of a reactor assembly. The larger η is, the lower need be the enrichment of fissile atoms within the heavy fuel atoms for a given size of reactor to be just critical—a state that applies when the total neutrons produced by fission in fuel, less the neutrons absorbed in the fuel, just equals the neutrons lost by parasitic absorption in materials other than fuel in the reactor and by leakage from the reactor. For fuel containing several fissile nuclides it follows that each will contribute differently to the reactivity of the system according to the values of ν, σ_f, and σ_c. In particular, if plutonium is used as a fissile material, its isotopic composition has an important bearing on the enrichment required.

In thermal reactors it has already been stated that the isotopes ^{239}Pu and ^{241}Pu are fissile, and it is found that they have virtually equal reactivity worths.

In fast reactors all the plutonium isotopes are fissionable but have widely different reactivity worths. As ^{239}Pu is the most abundant isotope in any reactor-grade plutonium, it is usual to express the worths of the other isotopes relative to ^{239}Pu, which is taken as equal to 1.0. Typical values are 0.15 for ^{240}Pu and ^{242}Pu, and 1.5 for ^{241}Pu. A kilogram of plutonium can be equivalent to more or less than a kilogram of *equivalent* ^{239}Pu (kg Pu(E)), depending on the proportions of the various isotopes. The Appendix to this chapter shows that a little simple algebra enables expressions to be derived for these equivalent-^{239}Pu worths.

2.6 Delayed neutrons

While most of the neutrons are produced promptly during the fission process, a small percentage are *delayed* by a wide range of times (up to minutes). The percentage of the delayed neutrons is small, but their presence is crucial to the control of a reactor. Provided that a chain reaction cannot be sustained without their contribution, then they prevent the reactor from responding too rapidly to any change.

The delayed neutrons are produced by fission-product decay, and the relative and absolute abundances of those fission products which decay with emission of a neutron are functions of the fission type, i.e. the fissioning nuclide and the energy of the neutron causing fission. There are quite large differences between the nuclides as may be seen from the data of Table 9.3. The variation with neutron energy is much less significant, and in fact for the accuracies of the figures given in Table 9.3 there is no effective difference between delayed neutron production from neutrons in thermal and fast reactors.

Table 9.3 Delayed neutron yield (per fission)

^{235}U	0.016
^{238}U	0.044
^{239}Pu	0.006
^{241}Pu	0.015

Table 9.3 shows that fewer delayed neutrons are produced by fission in ^{239}Pu compared with the uranium isotopes. In fast reactors the relatively low yield from ^{239}Pu is balanced to some extent by the high yield from ^{238}U. In thermal reactors, however, for which fission in ^{238}U is negligible, use of plutonium as a fuel does suffer the disadvantages of the low delayed neutron yield, thus making their response to control changes rather sensitive. However, for thermal reactors using plutonium with a significant percentage of the higher isotopes the situation is mitigated by the fairly high delayed neutron yield of ^{241}Pu.

2.7 Plutonium fuel types

Early reactor types, such as the Magnox reactors, used fuel in a metallic form. The present generation of reactors, including the AGRs, use a ceramic oxide fuel. There is ample evidence that mixed uranium–plutonium oxide fuels may be fed to a variety of reactors. Advantages may be achieved with other fuels, and these are briefly reviewed below.

It is anticipated that oxide fuel may eventually be succeeded by another ceramic form, carbide fuel. This has more attractive material properties than oxide fuel: in particular, it is more dense and has a higher thermal conductivity. The advantage of an increased thermal conductivity is that for a given peak fuel temperature and a given fuel pin size more heat can be produced in each pin. The advantage of a high density is that the ratio of atoms of fuel to other atoms is high, and so there is less probability of wasteful capture of neutrons in non-fuel materials (e.g. steel). In fast reactors this results in a breeding ratio of about 1.4 for carbide fuel compared with 1.25 for oxide. (The breeding ratio is the ratio of the rate of plutonium production to consumption in the reactor overall, i.e. core plus breeder fuel elements.)

Use of metallic fuels is attractive from the physics viewpoint, as they also have a high density and thermal conductivity compared with oxide fuels. For a large power fast reactor this could result in an initial loading of plutonium about 20 per cent lower than for an oxide design, and a breeding ratio of about 1.5. The snag with the use of metallic fuels is that high irradiations involve serious swelling problems. The swelling is increased if plutonium is present in the feed fuel. While the swelling may be reduced by the addition of small amounts of other metals, use of metallic fuel for plutonium–uranium fuel is unlikely at present.

3 Use of plutonium in reactors

3.1 Thermal reactors

As explained in §2, some of the ^{238}U in the fuel is converted to ^{239}Pu during the normal operation of thermal reactors. Some of this plutonium will be fissioned before the fuel reaches its scheduled burn-up level and some will remain in the discharged fuel. The amount and quality of the plutonium remaining in such irradiated fuel varies between the different types of reactor depending on characteristics such as fuel power rating, ^{235}U enrichment of the feed fuel, and its burn-up. The main characteristics of thermal reactors operating on uranium are set out in Table 9.4, whilst Table 9.5 lists the general isotopic composition of the plutonium produced at the mean fuel burn-up shown.

From Table 9.4 it is seen that average uranium enrichment in the reject fuel is around 0.2 per cent to 0.4 per cent ^{235}U for natural uranium reactors and about 0.8 per cent for enriched reactors, which can be recovered by reprocessing the spent fuel, and recycled. The spent fuel from the natural

Table 9.4 Characteristics of thermal reactors with uranium feed

	Units	Gas-cooled reactors		Water-cooled reactors		
		Magnox	Advanced gas-cooled	Boiling-water reactor	Pressurized-water reactor	CANDU
Nominal output	MW(e)	500/600	625	1100	1100	510
Thermal efficiency	%	32	41	32.6	34.2	29
Steam cycle		Indirect	Indirect	Direct	Indirect	Indirect
Moderator		Graphite	Graphite	H_2O	H_2O	D_2O
Fuel: type		U-metal	UO_2	UO_2	UO_2	UO_2
geometry		rod	pellet	pellet	pellet	pellet
cladding		Magnox	S.S.	Zircaloy	Zircaloy	Zircaloy
mean rating	MW(t)/t h.a.	3.2	14.66	23.8	37.0	18.8
mean burn-up	MWd/t h.a.	4500	18 000	27 500	30 400	7500
initial weight	t h.a./GW(e)	975	166.5	129	79	182.7
replacement weight	t h.a. GW(e)$^{-1}$ y^{-1}	255	50.0	40.7	35.1	167.3
Enrichment: initial	% ^{235}U	0.71 (natural)	1.65	2.07	2.14	0.71 (natural)
replacement—feed	% ^{235}U	0.71 (natural)	2.22	2.60	3.00	0.71 (natural)
average replacement—reject	% ^{235}U	0.4	0.72	0.75	0.83	0.22
Uranium utilization	%	0.4	0.5	0.6	0.6	0.8
Plutonium production: total	t GW(e)$^{-1}$ y^{-1}	0.60**	0.23	0.35	0.31	0.63
fissile	t GW(e)$^{-1}$ y^{-1}	0.48	0.15	0.25	0.22	0.49
Uranium requirement† initial	t U/GW(e)	975	561	509	324	183
replacement (net)	t U GW(e)$^{-1}$ y^{-1}	255	162	165	167‡	167
Separative work requirement† initial	t SW/GW(e)	nil	253	263.5	171.6	nil
replacement (net)	t SW GW(e)$^{-1}$ y^{-1}	nil	113	120.9	129.9	nil
Coolant		CO_2	CO_2	H_2O	H_2O	D_2O
inlet temperature	°C	250	290	277	280	250
outlet temperature	°C	410	635	296	320	290
pressure	kg/cm²	24.6	42.4	72.0	155.0	11.0
Pressure containment		PCPV§	PCPV§	Steel PV	Steel PV	Zircaloy tube
Steam quality temperature at TSV*	°C	395	540	291	266	255
pressure	kg/cm²	50	163	68	55	44

† waste strip 0.25%, excluding processing losses   ‡ on a once-through basis the requirement becomes 210t U/GW(e) and uranium utilization 0.5

§ prestressed concrete pressure vessel   * turbine stop valve   ** equals about 0.80 at current operating conditions

Table 9.5 Average isotopic composition of plutonium produced in uranium-fuelled thermal reactors

Reactor type	Meanfuel burn-up (MW d/t)	Percentage of Pu Isotopes at Discharge				
		^{238}Pu	^{239}Pu	^{240}Pu	^{241}Pu	^{242}Pu
Magnox	3 000	0.1	80.0	16.9	2.7	0.3
	5 000	*	68.5	25.0	5.3	1.2
CANDU	7 500	*	66.6	26.6	5.3	1.5
AGR	18 000	0.6	53.7	30.8	9.9	5.0
BWR	27 500	2.6	59.8	23.7	10.6	3.3
	30 400	*	56.8	23.8	14.3	5.1
PWR	33 000	1.5	56.2	23.6	13.8	4.9

* information not available.

uranium reactors, being similar to enrichment plant tails, is merely a waste product. Such an operation would reduce the total uranium requirements of the enriched reactors by 15–20 per cent compared with the once-through mode. Similarly the plutonium can be recovered during reprocessing and recycled, but as explained in §2, the isotopic composition of the plutonium has an important bearing on its worth as a fissile material.

The recycling of plutonium in thermal reactors has been studied extensively in several countries, particularly for the case of light water reactors, when the saving in uranium requirement due to recycling both the recovered uranium and the plutonium is increased to about 35–40 per cent compared with once-through.

· The method of recycling plutonium in thermal reactors can vary. For countries having very large integrated electricity grids operated by a single authority, such as in England and Wales or France, particular reactors could be dedicated wholly to plutonium fuelling using the plutonium arising from those reactors fuelled with uranium. An alternative, and perhaps of more interest to countries with many individual utilities such as the USA, is for each reactor to recycle its own recovered plutonium. Further possible regimes exist, of course, and it would be possible for the plutonium-fuelled reactor to use more or less plutonium than recovered from the spent fuel of the one reactor. Studies have shown that the natural uranium saving due to plutonium recycle is about the same whether it is recycled in wholly or partially plutonium-fuelled reactors, and as most experimental and pilot-scale practical experience to date has been obtained on partial plutonium fuelling in LWRs, this will be taken as the reference case.

Next, consideration must be given to the distribution of plutonium within the fuel pins and fuel sub-assemblies. One possibility is that recycled plutonium is distributed uniformly in new replacement fuel, which can then have a uranium enrichment less than that of the uranium-only feed reactors. When equilibrium is reached—a condition that will apply when the recycled uranium and plutonium recovered from spent fuel requires the same make-up of enriched uranium as that used in the previous cycle—the whole reactor will contain mixed plutonium–uranium oxide fuel with some 2 per cent each of ^{235}U and plutonium in ^{238}U. However, plutonium has a greater associated radiation hazard than uranium and greater precautions must be taken during its handling in fuel fabrication processes which lead to high fabrication charges. To reduce this fabrication premium to a minimum it follows that mixed oxide (MOX) should be restricted to only a fraction of fuel pins by mixing plutonium recovered from irradiated fuel with natural or depleted uranium. All other fuel pins will then contain uranium oxide only with a ^{235}U enrichment the same as that used in replacement fuel.

3.1.1 *The self-generation reactor*
When the plutonium contained in the discharged fuel from a thermal reactor is mixed with natural or depleted uranium and recycled in the same reactor, thus displacing an equivalent amount of ^{235}U, the reactor is known as a self-generation reactor (SGR). However, because of the differences in neutron absorption cross-section between ^{235}U and plutonium, care must be taken when considering the physical disposition of the MOX fuel pins in the fuel sub-assemblies to avoid adverse effects on power distribution and control rod worth.

3.1.2 *The effect on power distribution*
As seen in Table 9.1 the isotopes ^{239}Pu and ^{241}Pu have fission cross-sections which are up to approximately twice that of ^{235}U in a thermal neutron spectrum. Nearly the same number of fissions are required to produce a given amount of thermal energy whether the atoms are ^{235}U, ^{239}Pu, or ^{241}Pu, and power peaking might be expected at the start of life of fuel if mixed oxide fuel were to be placed adjacent to ^{235}U-enriched fuel. But the high neutron absorption cross-section of the plutonium depresses the flux adjacent to the MOX pin. In LWR lattices the mean free path of thermal neutrons (i.e. the average distance a neutron travels between thermalization and absorption) is short, and in gas-cooled reactors each channel is effectively shielded (with respect to thermal neutrons) from all others by the graphite moderator. In neither case, therefore, can there be any net inflow of neutrons from adjacent ^{235}U-enriched fuel to mitigate the effect. Thus it is the mean moderator density immediately adjacent to the fuel that determines the number of neutrons which are thermalized and hence absorbed by the fuel. For equal fuel pin (or channel) power output—i.e. no power peaking—the flux in mixed-oxide fuel will be half that in ^{235}U-enriched fuel.

3.1.3 *Control rod worth*
Since the thermal neutron flux in plutonium fuels will be half that of ^{235}U fuel for the same power output (§3.2.1), the worth of control rods would be correspondingly reduced if they were located in this flux.

In a BWR, control is provided by cruciform plates which slide along the sides of the square-section fuel elements. To avoid these control plates being located in a depressed neutron flux, the strategy is to locate the MOX pins as an island surrounded by uranium-enriched fuel. In a PWR the fuel elements are provided with control rod clusters which displace several fuel pins. These rod clusters tend to be situated towards the centre of a fuel element, and so it may be necessary to limit the number of mixed-oxide fuel pins in the element to allow the control pins to be surrounded by ^{235}U-enriched fuel pins. A possible alternative would be to have some fuel assemblies in PWRs consisting entirely of MOX fuel rods and some with none. However, one result of such a strategy could be a substantial increase in the number of control rods within the MOX fuel assemblies to offset the reduction in worth.

3.1.4 *Long-term reactivity effects*

Long-term reactivity effects are due to the depletion of fissile material and the build-up of neutron-absorbing fission products in the fuel. The fall-off of reactivity during irradiation is less pronounced for mixed-oxide fuel than for UO$_2$ fuels for two reasons. Firstly, neutron capture in ^{239}Pu leads to ^{240}Pu, which, because of its high capture cross-section, is an excellent fertile material, having a very high probability of capturing a neutron to produce ^{241}Pu. Secondly, neutron capture in ^{235}U leads to ^{236}U, which is neither fissile nor fertile.

This lower sensitivity of MOX fuel reactivity to burn-up means that a smaller excess reactivity is needed in fresh fuel compared with UO$_2$ fuel to achieve a given burn-up. This in turn means that a lower total control rod capacity is needed to balance the lower excess reactivity at startup. Hence to some extent the reduced worth of control rods due to the lower flux level in MOX is mitigated by the reduced requirement in total capacity.

3.1.5 *Flux spatial instability*

Spatial instability of the neutron flux distribution can occur in large thermal reactors as a result of the build-up and burn-out of ^{135}Xe, which has a very large thermal cross-section. This is formed by the decay of the fission product ^{135}I (half-life 6.7 hours). ^{239}Pu has a slightly higher iodine yield (6.6 per cent) than ^{235}U (6.4 per cent), but since the thermal neutron flux in mixed-oxide fuel is smaller than in UO$_2$ fuel for a given power level, a greater percentage of the ^{135}Xe is lost by radioactivity decay and less by neutron capture. It follows that shifts in the power density have smaller local reactivity effects in mixed-oxide fuel and the reactor will tend to be more stable.

3.1.6 *Kinetics effects*

So far the introduction of plutonium has been discussd in the context of its longer-term effects on power distribution, control rod requirements, and xenon stability. Normally a reactor is controlled on short timescales by adjusting the insertion of the control rods. However, in order to ensure the safety of the reactor, it is necessary to be able to control transient behaviour over a period as short as a few seconds. The analysis of transients requires knowledge of the reactivity coefficients of major components such as the moderator, the coolant, and the fuel.

The *moderator coefficient* expresses the rate at which the reactivity of the system increases with temperature. In uranium-fuelled LWRs—and still more so in plutonium-fuelled LWRs—the coefficient is negative, an increase in moderator temperature resulting in a decrease in reactivity.

Reactivity varies linearly with the fraction of neutrons causing fission, so that the coefficient will be determined by the way in which this fraction changes with increasing moderator temperature. The process is a complex one, but may for simplicity be regarded as involving two main effects.

Firstly, there is a *temperature effect*: raising the temperature of the moderator broadens the energy distribution of neutrons in the *thermal* energy range (i.e. near to ambient moderator temperature). This is important in its effect on the relative nuclear cross-sections of core materials, in particular those which are strongly temperature dependent; in uranium-fuelled LWRs it results in an increase in the capture/fission ratio and a consequential decrease in reactivity.

Secondly, the temperature increase will result in a decrease in moderator density, and hence in the moderating power. This effect, which may be termed the *density effect*, has three main consequences:

(1) fewer neutrons will be captured in the moderator itself, meaning that more are available to react with fuel materials;

(2) there will be a reduced impedance to neutron diffusion meaning that more neutrons will be lost from the core; and

(3) neutrons will decelerate less rapidly and will spend longer in the *epithermal* energy range (the middle range of the spectrum with upper limit about 100 eV); this is of importance in relation to materials which exhibit an unusual difference between their epithermal and their thermal cross-sections.

Considering the overall impact on reactivity, it is clear that the effect of (1) will be positive and (2) negative. Although (3) is more complex, in the case of uranium fuelled LWRs it will again prove to be negative, the dominant item here being the very large epithermal capture resonance of the ^{238}U. For uranium-fuelled LWRs then, the density effect results overall in a negative moderator temperature coefficient.

The effect of moderator temperature on the reactivity of a Pu-fuelled LWR can best be understood by considering the behaviour of the individual isotopes. Both ^{240}Pu and ^{242}Pu have a large capture resonance at about 1 eV, and the flattening of the thermal neutron energy distribution results in increased resonance capture at this energy. (The effect is a complex one, and the density effect also has a role in this enhanced capture.) The fission cross-sections of the ^{239}Pu and ^{241}Pu isotopes are strongly energy dependent in the thermal range, and, as with ^{235}U, the temperature effect again results in an overall reduction in reactivity. These effects combine to produce an overall moderator temperature coefficient which is more negative than that for U-fuelled LWRs.

An increase in coolant voids in an LWR decreases the effective moderator density. The void coefficient, like the density coefficient, becomes more negative when uranium is replaced by plutonium.

As neutrons slow down, 20 to 30 per cent are captured in the absorption resonances of the fuel materials in an LWR. The figure is somewhat less for graphite-moderated reactors, which usually distribute the fuel less uniformly through the moderator so that some neutrons never encounter fuel while they are slowing down. The resonances have very high peak cross-sections but extend over a very narrow speed band. As the fuel temperature increases, the vibrational speed of the fuel atoms is increased, and the resonance is effectively broadened although its peak value is reduced. The net effect, however, is increased absorption as the fuel temperature increases. The broadening of the resonance is called Doppler broadening and the fuel temperature coefficient is called the Doppler coefficient. The effect is small in fissile isotopes since their resonances have small peak cross-sections. In an all-UO_2 core the effect is primarily due to ^{238}U, and as irradiation continues there are increased contributions from the fertile plutonium isotopes and from ^{236}U. For cores containing mixed-oxide fuel, the negative Doppler coefficient can be up to 10 per cent greater.

3.1.7 *Plutonium recycle in LWR*

Table 9.6 gives the change in plutonium composition as plutonium is recycled in the self-generation mode in BWR and PWR with initial enriched uranium fuel parameters as defined in Table 9.4. In both cases the initial reloads will comprise UO_2 fuel only, enriched with ^{235}U. In subsequent reloads the plutonium recovered from all discharged fuel is recycled in MOX fuel pins. The table indicates that, for a BWR, recycling the plutonium from 1.15 reactors the proportion of MOX in each reload batch rises from 28 per cent in the first recycle to 40 per cent in the third recycle, which is assumed to represent equilibrium conditions. The percentage of plutonium in MOX rises also, from 3.4 per cent plutonium in natural uranium in the first recycle to 4.6 per cent plutonium at equilibrium. Equilibrium is said to occur when as much plutonium is recovered from spent MOX and UO_2 fuel pins as had originally been loaded into the MOX fuel rods. For the case of a PWR recycling its own plutonium only, the fraction of MOX fuel in each reload batch increases from 18 per cent at the first recycle to just under 30 per cent at equilibrium. The proportion of plutonium in the MOX rises from 4.7 per cent to 8 per cent during the same time. It is also to be noted that the quality of the plutonium worsens with recycle in respect of its fissile content. For both the BWR and PWR studied, the fissile content of plutonium in discharged UO_2 fuel is a little over 70 per cent. In the recycle mode this percentage falls to 59 per cent and 57 per cent in BWR and PWR, respectively, at equilibrium.

It is worth pointing out that the time taken to reach equilibrium is some 17

Table 9.6 Leading fuel cycle details for Pu recycle BWR† and PWR‡

	First reload	Second reload (first recycle)	Third reload (second recycle)	Fourth reload (third recycle)	Fifth reload (fourth recycle)	Sixth reload (fifth recycle)
BWR feed						
% ^{235}U§ in UO$_2$	2.60	2.60	2.60	2.60		
% Pu* in MOX	—	3.44	4.05	4.61		
% MOX	nil	27.9	36.8	39.7		
Discharged Pu composition %:						
^{238}Pu	2.6	2.5	3.1	3.7**		
^{239}Pu	59.8	51.4	46.4	42.8		
^{240}Pu	23.7	25.7	26.2	26.6		
^{241}Pu	10.6	14.0	15.3	16.3		
^{242}Pu	3.3	6.4	9.0	10.6		
PWR feeds						
% ^{235}U§ in UO$_2$	3.0	3.0	3.0	3.0	3.0	3.0
% Pu* in MOX	—	4.72	5.83	6.89	7.51	8.05
% MOX	nil	18.4	23.4	26.5	27.8	28.8
Discharged Pu composition %:						
^{239}Pu	56.8	49.7	44.6	42.1	40.9	40.0**
^{240}Pu	23.8	27.0	38.7	29.4	29.6	29.8
^{241}Pu	14.3	16.2	17.2	17.4	17.4	17.3
^{242}Pu	5.1	7.1	9.5	11.1	12.1	12.9

  † self-generation reactor recycling plutonium from 1.15 reactors
  ‡ self-generation reactor recycling plutonium from itself only
  § based on total U only
  * based on total natural U + Pu
 ** equilibrium attained

years (BWR) to 25 years (PWR) for reactors operating at a mean load factor of 75 per cent, and that during this time the fraction of MOX in-reactor increases from 6–7 per cent at the start of the first recycle to the equilibrium values of 40 per cent and 30 per cent in BWR and PWR, respectively.

Considering the savings achieved by recycling fuel, it may be seen from Tables 9.4 and 9.6 that 35.1 tonnes of metal fuel are required to produce 1 gigawatt-year of electrical energy from a PWR. The PWR when operated on enriched uranium requires a mean feed enrichment of 3 per cent ^{235}U. At this level and assuming no recycle (as advocated by the US in INFCE) the natural uranium requirements are 210 tonnes GW(e)$^{-1}$ y^{-1}; recycle of the uranium contained in the discharged fuel (which the UK would consider to be a necessary part of the fuel cycle) would reduce the uranium requirement to 167 tonnes GW(e)$^{-1}$ y^{-1}. The recycle of plutonium in a self-generative version of PWR would eventually further reduce the uranium requirements to 128 tonnes. As pointed out before however, this figure would apply only at self generation recycle equilibrium, some 25 years after the reactor is first brought on line.

The current UKAEA philosophy is that larger reductions in uranium requirements can be made by investing the plutonium in a programme of fast reactors. Indeed, as explained later, the use of plutonium in fast reactors could lead ultimately to a self-sufficient nuclear programme with no requirement for newly mined natural uranium to fuel thermal reactors.

3.1.8 *Production of higher actinides in PWR*

It has already been indicated in §2 that the production of the higher actinides americium and curium will depend upon various factors such as fuel burn-up and rating, and whether or not plutonium is recycled. The presence of the higher actinides in discharged fuel is important because of their impact on fuel-handling. As seen from Fig. 9.1 many of the isotopes of plutonium, americium, and curium are alpha radioactive. Some, particularly the curium isotopes, have relatively short half-lives, but these are nevertheless important on the timescale of fuel reprocessing and refabrication. The alpha emissions themselves are not important from a handling point of view but they can give rise to neutron emission by (alpha, neutron) reactions with other elements present in the fuel. Secondly, ^{242}Cm, like ^{240}Pu, can undergo spontaneous fission, releasing additional neutrons. Hence the greater the concentration of the higher actinides in discharged irradiated fuel, the more intense its neutron activity and the greater the necessity to provide adequate shielding to prevent a health hazard. Table 9.7 gives the amounts of higher actinides produced in 1 GW(e) y of PWR operation. Three different whole-core fuel loadings taken to the full burn-up specified in Table 9.4 have been considered, namely:

  (1) uranium oxide only enriched with ^{235}U,
  (2) mixed oxide enriched with plutonium recovered from discharged UO_2 fuel, i.e. plutonium of composition as given in column 1 of Table 9.6, and
  (3) MOX enriched with plutonium recovered from equilibrium self-generated PWRs, i.e. plutonium of composition as given in column 6 of Table 9.6.

As already noted in §2, the production of ^{241}Am depends critically upon the production of the plutonium isotopes ^{240}Pu and ^{241}Pu, and the amount of these two isotopes in the initial fuel. This is seen clearly in Table 9.7. In the first row, plutonium is produced only during irradiation of the uranium fuel, and so a relatively modest amount of ^{241}Am is produced. In the second row, appreciable amounts of ^{240}Pu and ^{241}Pu are loaded into the reactor in fresh fuel and a marked increase in ^{241}Am production is seen. The effect is enhanced in row three due to the even larger component of the higher plutonium isotopes. Reference to Fig. 9.1 indicates that the high concentrations of ^{240}Pu, ^{241}Pu, and ^{241}Am will inevitably lead to the higher levels of ^{243}Am and the curium isotopes seen in Table 9.7.

Table 9.7 Production of the higher actinides in PWR (at one year after discharge)

Enrichment	kg GW(e)$^{-1}$ y^{-1}			
	^{241}Am	^{243}Am	^{242}Cm	^{244}Cm
^{235}U	3.6	2.87	0.094	0.68
Plutonium ex UO$_2$ fuel	30.5	36.5	1.16	9.05
Plutonium ex equilibrium SGR	82.0	77.8	1.85	11.4

3.2 The fast reactor

The previous paragraphs have illustrated the alteration to reactor parameters and the change in fuel cycle when plutonium is recycled in thermal reactors. Most of the information on plutonium recycle is for PWR and BWR, and reflects the importance which some countries that have or are developing reprocessing and mixed-oxide fabrication facilities are placing on plutonium recycle in thermal reactors. Other countries such as the UK, France, and Japan, which have little indigenous natural uranium, see the recycle of plutonium in the fast reactor as a better prospect for the reduction of their dependence on uranium imports.

As already seen in §2, fission cross-sections are relatively small and of the same order as capture cross-sections at high neutron energies. A high concentration of fissile material is therefore required in the case of a fast reactor for criticality to be achieved. Even an infinitely large core, i.e. one with no loss of neutrons by leakage, would need an enrichment of about 7 per cent ^{235}U or 5 per cent ^{239}Pu in ^{238}U to become critical. In practice, large amounts of fissile material are in short supply, and the amount required to operate a power reactor can be kept to an acceptable value by using enrichments in the range 15–30 per cent. Such a concentrated highly fissile fuel mixture, with no moderator separating the fuel elements, becomes critical at a much smaller size than that of a thermal reactor core. The power is normally extracted from a relatively small volume resulting in a specific power (MW/tonne of heavy atoms) an order of magnitude higher than in thermal reactors, such as AGRs and PWRs. High power density in fast reactors in turn leads to a finely divided core to provide a large heat-transfer area, and to the use of a non-moderating coolant with good heat-transfer properties such as a liquid metal. Even with fuel pins a few millimetres in diameter, fuel and core temperatures are high, so that a fissile fuel would tend to swell easily on the internal accumulation of fission product gases. This problem and the problem of achieving enhanced burn-up (the target being a peak burn-up of 10 per cent all heavy atoms) has led to the development of ceramic fuels such as mixed uranium–plutonium oxide canned in stainless steel, the latter being compatible with both fuel and coolant at high temperatures. Typical core parameters for an early 1250 MW(e) commercial reactor with oxide fuel and a possible later reactor with carbide fuel are indicated in Table 9.8.

Table 9.8 Fast reactor characteristics

	Units	CFR Oxide	CFR Carbide
Nominal output	MW(e)	1250	1250
Thermal efficiency	%	38.8	38.8
Steam cycle		indirect	indirect
Moderator		none	none
Fuel: type		oxide	carbide
geometry		pellet	pellet
pellet diameter	mm	6	10
cladding		SS	SS
peak core rating	W/g h.a.	229	170
mean core burn-up	% h.a.	6.9	5.2
initial weight core	t h.a./GW(e)	16.9	22.2
axial breeder	t h.a./GW(e)	14.3	18.8
radial breeder	t h.a./GW(e)	31.9	38.0
replacement weight core	t h.a. GW(e)$^{-1}$ y^{-1}	13.0	17.7
axial breeder	t h.a. GW(e)$^{-1}$ y^{-1}	11.0	15.0
radial breeder	t h.a. GW(e)$^{-1}$ y^{-1}	7.5	7.6
Enrichment ⎰ initial	% Pu(T)†	17.7	13.2
in core (average)⎱ and feed	% Pu(E)‡	14.2	10.6
Uranium utilization	%	50–80	50–80
Net plutonium production	t Pu(T) GW(e)$^{-1}$ y^{-1}	0.24	0.49
Breeding gain		0.21	0.44
Linear doubling time§ (75% load factor 0.5 % h.a. in residues)	years	28.6	14.5
Coolant		sodium	sodium
inlet temperature	°C	370	370
outlet temperature	°C	537	548
Pressure containment		not necessary	
Steam quality ⎰ temperature	°C	486	486
at TSV ⎱ pressure	MN/m²	16.8	16.8

† Pu(T) : total isotopes
‡ Pu(E) : ^{239}Pu equivalent
§ Linear doubling time is the time required for one reactor to produce enough plutonium in excess of its own requirements to provide a second reactor with its total (in-pile and out-of-pile) inventory.

Table 9.8 shows that fast reactors can be operated to produce more plutonium than they consume, up to a maximum of approximately 0.25 t Pu per GW(e) y for oxide-fuelled versions and up to 0.50 t GW(e)$^{-1}$ y^{-1} for carbide-fuelled versions. This net production is only of the order of 10 per cent (20 per cent in the case of carbide) of the annual plutonium feed, and less than half this as a percentage of the total plutonium inventory in the reactor and processing plants. Alternatively, if plutonium is no longer needed in the system, the reactors can be operated to consume more plutonium than they produce by removing the breeder elements. In'this way the amount of 'free' plutonium, i.e. material in excess of immediate needs, can be kept to a minimum.

As will be indicated later, when operated in the breeder mode, the 'quality' of the plutonium being recycled ultimately reaches an equilibrium as high-quality, low-irradiation plutonium continues to be produced in the breeder fuel to replace that consumed in the core. Furthermore, as the fast reactor requires only a small supply of depleted uranium during operation, the use of fast reactors in the place of thermal reactors reduces overall natural uranium requirements. Studies indicate that adopting such a fast reactor strategy over the world as a whole could possibly result in uranium requirements being supplied from low-cost high-grade sources without the need to exploit high-cost sources such as shales or sea-water. In this way fast reactors will assist in stabilizing the price of uranium by restricting demand and so assist in maintaining nuclear electricity generating costs at a steady value.

3.2.1 *Kinetics parameters of a fast reactor*
The possibility exists of some sodium being lost from the core or from a fuel sub-assembly due, for example, to boiling following a blockage, and consideration needs to be given to the kinetics parameters of a fast reactor to understand the safety implications this may have. Such a loss of sodium is termed sodium voiding.

In the analysis of the transient following sodium voiding, two parameters—the sodium-void coefficient and the Doppler coefficient—assume importance. Sodium voiding may be brought about either by a blockage of the flow, leading to localized boiling of the sodium, or by pin failure with the release of fission-product gas. Such voiding will cause an increase in fuel temperatures because of the decrease in cooling, and also because there might be an associated reactivity increase, which, in turn, will lead to a power increase. As the fuel temperatures rise, reactivity will tend to decrease again, thus limiting the power transient. The principal coefficient of reactivity associated with fuel temperature increases is, as already indicated in the case of thermal reactors, the Doppler coefficient.

3.2.2 *Sodium void effect*
The reactivity effect that occurs when sodium is voided from a reactor can be split into four components:
(1) a spectral component due to an increase in neutron flux in the MeV region, which in turn increases the fission rate in the isotopes ^{238}U, ^{240}Pu and ^{242}Pu; this component is always positive;
(2) a leakage component, which is caused by an increase in the number of neutrons escaping from the reactor and is always negative, thus leading to a decrease in reactivity;
(3) a capture component; and
(4) a fissile component.

Both (3) and (4) are due to the reduction of the neutron flux level at the resonance band of energies because of the general upward shift of all neutron energies. The capture component is positive (because there is a reduction in neutron capture) and leads to an increase in reactivity; it is larger than the fissile component, which is always negative (because there is a reduction in the number of fissions). The spectral component is highest where the neutron flux is highest, i.e. generally at the centre of the reactor. The leakage component is greatest at the core boundaries. The spectral and absorption components become increasingly positive as the fissile material concentration decreases. One consequence of this has been that for the earlier experimental fast reactors, which were small and fuelled with a high enrichment of ^{235}U, the leakage component dominated all others and the sodium-void effect was negative.

Commercial fast reactors will have large cores with mixed-oxide fuelling, and here by contrast the spectral component increases (because ^{240}Pu and ^{242}Pu have larger fission cross-sections than ^{238}U in the MeV region), and the leakage component decreases. The spectral component therefore dominates in the central regions of the reactor giving rise to a positive void effect. The void effect is still negative in the outside regions of the core and in the breeder regions because of the leakage effect.

3.2.3 *Doppler effect*

Just as in thermal reactors, the principal temperature coefficient of reactivity (the Doppler coefficient) arises from the Doppler broadening of fission and capture resonances. However the resonance cross-sections of importance are at higher energies (between 10 eV and 10 keV), whereas in thermal reactors the resonances of importance were in the near-thermal region of the neutron spectrum. Leakage in fast reactors at these energies is very small and the temperature coefficient is due almost entirely to the variation in fission and capture processes and hence in the neutrons produced per neutron absorbed. The heavy isotope which is present in the largest concentration, and which has the largest resonance absorption (mainly capture), is ^{238}U and hence the fuel temperature coefficient of reactivity is negative. The coefficient is less negative at higher enrichments because of the reduction in the amount of ^{238}U present.

3.2.4 *Actinide production in the fast reactor*

Most of the intense neutron activity of discharged fast reactor fuel comes from curium isotopes; the gamma activity of ^{241}Am is also important. The higher actinide production rates depend on the levels of ^{241}Pu and ^{242}Pu in the feed fuel of the reactor since the production rate of ^{244}Cm is largely determined by the amount of ^{242}Pu in the fuel, and that of ^{242}Cm depends on the level of ^{241}Am, which is itself formed by the decay of ^{241}Pu. These levels vary considerably according to the source of the plutonium. Table 9.9 repeats a range of plutonium compositions given in Table 9.6, which are typical of the fuel discharged from early Magnox, AGR, and PWR. If a four-year period between reprocessing of irradiated thermal reactor fuel and the irradiation of the contained plutonium in fast reactor fuel is assumed, the rate of higher actinide production would be as shown in Table 9.10.

Table 9.10 indicates that changing the feed from Magnox to AGR or PWR

Table 9.9 Composition in wt.% of various sources of plutonium (at discharge)

Plutonium source	^{238}Pu	^{239}Pu	^{240}Pu	^{241}Pu	^{242}Pu
Magnox	0.1	80.0	16.9	2.7	0.3
AGR	0.6	53.7	30.8	9.9	5.0
PWR	1.5	56.2	23.6	13.8	4.9

Table 9.10 Production of the higher actinides in a fast reactor (at one year after discharge)

Plutonium source	kg GW(e)$^{-1}$ y^{-1}			
	^{241}Am	^{243}Am	^{242}Cm	^{244}Cm
Magnox	11.0	0.71	0.23	0.12
AGR	39.5	9.7	0.94	1.88
PWR	46.7	9.2	1.18	1.8

plutonium increases the amounts of ^{241}Am and ^{242}Cm produced by a factor of 4, whilst the ^{243}Am and ^{244}Cm are increased by a factor of 15. Storing the plutonium between thermal fuel reprocessing and irradiation in fast reactors allows the ^{241}Pu contained in the fuel to decay to ^{241}Am. It would be expected therefore that varying amounts of ^{241}Am and hence ^{242}Cm would appear in discharged fuel from the fast reactor, depending upon the time fabricated plutonium-bearing fuel is stored prior to irradiation. This effect is illustrated in Table 9.11, which gives the higher actinide content of spent fuel, one year after discharge, in which the initial fissile loading was PWR plutonium.

The table shows that increasing the storage time of plutonium, prior to its irradiation, allows the ^{241}Am and ^{242}Cm levels to rise. In contrast the ^{243}Am and ^{244}Cm levels remain unchanged because they depend principally on the amount of ^{242}Pu in the fuel.

Comparisons of Tables 9.10 and 9.11 with Table 9.7 shows that in general the higher actinides produced in a PWR fuelled with enriched uranium are very much less than in a fast reactor. However, if plutonium-fuelling is undertaken in a PWR, then much higher levels of ^{243}Am and ^{244}Cm will result, and there will be a corresponding increase in neutron emission due to (alpha, neutron) reactions. As indicated in §2, the ratio of capture to fission

Table 9.11 Variation of the higher actinide production rates against plutonium storage time (at one year after discharge)

Storage time before irradiation (y)	kg GW(e)$^{-1}$ y^{-1}			
	^{241}Am	^{243}Am	^{242}Cm	^{244}Cm
1	27.4	9.2	0.50	1.8
4	46.7	9.2	1.18	1.8
10	77.0	9.3	2.28	1.9

cross-sections are higher for the plutonium isotopes in thermal reactors than in fast reactors; the plutonium produced in thermal reactors will, therefore, have much higher levels of ^{241}Pu and ^{242}Pu. The production rates of ^{242}Cm and ^{244}Cm will therefore be correspondingly higher. In the case of ^{242}Cm, this effect is counterbalanced because the much longer irradiations in thermal reactors allow most of this isotope to decay before the fuel is discharged. In addition, nearly all of the ^{242}Pu will be transmuted into ^{243}Am and ^{244}Cm in a thermal reactor, whereas some of it (i.e. the ^{242}Pu) will fission in a fast reactor.

3.2.5 *Continued recycle of fast reactor fuel*

During irradiation, current fast reactor technology allows a peak burn-up of 10 per cent with a mean burn-up around the 7 per cent mark. At this level of burn-up, approximately 25 per cent of the initial plutonium contained in the fuel will be fissioned and will produce thermal energy. This loss of plutonium will be more than offset by plutonium production in the axial and radial breeders and in the core itself. Fissile plutonium is also lost by the β^--decay of ^{241}Pu to form ^{241}Am and, as pointed out previously, this leads to additional handling problems along the fuel discharge route. When plutonium is repeatedly recycled, as will be the case with a large-scale fast reactor programme, its composition will eventually reach an equilibrium which is independent of the initial feed plutonium quality. For example, if a fast reactor is fuelled with Magnox plutonium which has a very low level of ^{241}Pu, and the fuel for each new irradiation is made up by mixing plutonium contained in discharged fuel sub-assemblies with the plutonium contained in discharged radial breeder sub-assemblies, there will be some deterioration in its quality as a breeding material mainly due to the growth in the amount of ^{240}Pu. But the level of ^{241}Pu will remain at the low level of Magnox plutonium because its very high fission cross-section ensures its rapid burn-up. On the other hand, if the initial feed plutonium is from discharged PWR fuel, plutonium recycling will bring about an improvement in the quality even though the level of ^{241}Pu drops. In both cases, after many recycles (about 20 years), an equilibrium is reached, where the composition has been estimated to be:

^{239}Pu—71.1 per cent,
^{240}Pu—24.1 per cent,
^{241}Pu— 3.3 per cent, and
^{242}Pu— 1.5 per cent.

4 Fabrication and reprocessing

Chapters 13 and 14 in this book deal respectively with fuel fabrication and fuel reprocessing, and it is the intention in the following section simply to comment on some additional issues associated with the handling and recycle of plutonium.

4.1 **Problems of fabricating and handling plutonium-enriched fuel for thermal reactors**

The manufacture of low-enrichment fuel for thermal reactors (e.g. up to 3.5 per cent ^{235}U) has been accomplished successfully on a commercial scale over the last 10 to 15 years in several industrialized countries. If part or all of the uranium enrichment were to be replaced by plutonium, a new set of engineering problems would arise. The methodology to overcome these has to some extent been worked out, but hitherto only on a small scale during the fabrication of fast reactor fuel for the prototype fast reactors, PFR and Phénix.

Recovered plutonium from thermal reactor fuels is a toxic material with a higher level of radioactivity than low-enriched uranium. When first recovered from thermal reactor fuels, it can be safely handled in sealed glove boxes operating at sub-atmospheric pressures. But, as seen in §3, if plutonium is continually recycled in thermal reactors, there is a gradual increase in the quantity of higher isotopes, which results in a corresponding increase in the more penetrating gamma radiation. Therefore, to eliminate increases in operator dose-rate it becomes necessary, in continuous recycling, to move to fabrication plants with heavier shielding, i.e. sealed concrete cells. This will necessitate plants being designed for remote operation and being as fully automated as practicable. This will lead to added complexity with the attendant problems of plant maintenance resulting from contamination of equipment with plutonium dusts.

As an example of what might be involved, a description of a plant for refabricating mixed uranium oxide—plutonium oxide LWR fuels follows. This is designed to cater for material arising from a reprocessing plant of the type shown in Fig. 9.4, typical of plants which have been operated successfully for many years.

4.2 **Mixed-oxide fuel fabrication plant**

Plutonium recycling in thermal reactors (LWRs) requires the fabrication of mixed uranium oxide–plutonium oxide fuels which consist of $(U–Pu)O_2$ pellets in Zircaloy cladding tubes. A typical scheme is given in Fig. 9.5. Plutonium oxide from the conversion plant has to be mixed with uranium oxide to form mixed oxide (MOX), which is then fabricated into fuel elements. The uranium oxide may be produced from either natural, recycled, or depleted uranium. The exact proportions of the plutonium and uranium oxides contained in the mixed-oxide fuel will depend on what blend is required for the reactor which will be using it, and on the isotopic proportions of the recycled plutonium. Typically the plutonium content of MOX could range from 3.5 per cent to 8 per cent total Pu (Pu(T)). It will, therefore, be necessary to vary the operation of the plant from one fabrication campaign to another to produce individual batches of fuel with various specifications.

Mixed-oxide fuels can be prepared in various ways, but the following

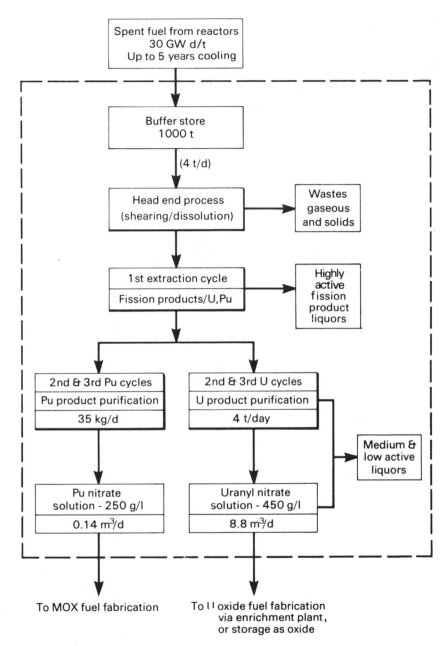

Fig. 9.4 Thermal oxide reprocessing plant—simplified flow diagram

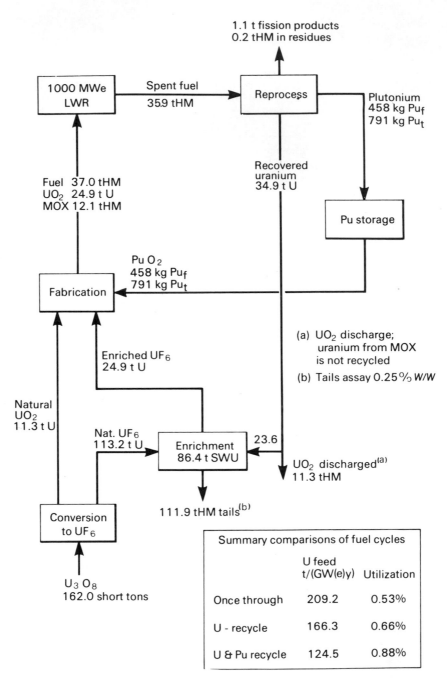

Fig. 9.5 Fuel cycle requirements per 1000 MW(e) year for LWR (PWR/BWR = 2/1) Equilibrium recycle; load factor 100%

Table 9.12 MOX fuel fabrication plant specification

Theoretical capacity†	300 t heavy metal/y
Effective capacity†	200 t heavy metal/y
(owing to the number of different fuels produced)	
Pellet density	10.1–10.6 g/cm³
Clean rejected oxide reprocessed	12–24 t/y
Dirty rejected oxide reprocessed	4–6 t/y
Pu-contaminated material (PCM) treated	1000 m³/y
PCM disposal	200–250 m³/y (at 100 g Pu/m³)
Pu-contaminated liquid (PCL) treated	300 m³/y
PCL disposal	20–25 m³/y (at 100 g Pu/m³)
Other active liquid wastes	800–1000 m³/y
Inactive liquid wastes	20 000 m³/y

† using six or seven fabrication lines and 12 t heavy metal per fabrication campaign

method can be described as the reference technology. Initially the UO_2 and PuO_2 powders are blended in the desired proportions to give a mixed-oxide powder, which is then precompacted and granulated into a free-flowing powder. This is formed into pellets, which are first sintered and then ground to produce a density and diameter within specified limits. The pellets are then heated to reduce their moisture and gas content. Finally they are loaded into Zircaloy tubes, which are sealed and assembled into fuel bundles. A diagrammatic representation of these processes is shown in Fig. 9.6 and details of the specification of the plant are given in Table 9.12.

Accurate quality assurance and control are an essential feature of the plant. All the rejected material is treated so that the out-of-specification oxides can be recycled and plutonium extracted from the waste streams.

4.3 Reprocessing plutonium-enriched thermal reactor fuel

The reprocessing of ^{235}U-enriched thermal reactor fuel has been carried out successfully for many years, although difficulties appear to have increased as fuel irradiation has gone up with a resultant increase in radioactivity. The presence of larger quantities of plutonium in plutonium-enriched thermal reactor fuel should present little extra difficulty as the principal hazard from radioactivity is due to the fission products. However, with the higher plutonium content, particular attention will have to be paid to criticality problems which may arise as plutonium is separated from the uranium and fission products. This might necessitate the use of plant made eversafe by limiting dimensions of components (pipes, vessels, etc.) to sizes in which a critical quantity of plutonium could not be accommodated.

4.4 Transport of plutonium

Plutonium-containing material will probably have to be transported from one location to another at different points in the fuel cycle. Such material

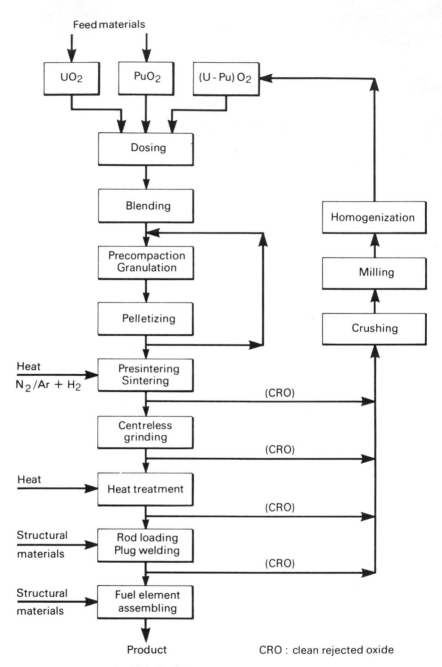

Fig. 9.6 Mixed oxide fuel fabrication process

may take different forms, such as new mixed-oxide fuel assemblies, spent fuel assemblies, plutonium-contaminated waste, or even as plutonium oxide or nitrate. The precise transport steps required will depend on whether any, or even all, of the associated plants are co-located. During the next twenty years or so of the UK nuclear programme, it is likely that most of the nitrate to oxide conversion plants (which link the processes shown in Fig. 9.4 and 9.6) will be co-located with the reprocessing plant, but that the MOX fuel fabrication plants and the reactors will be located at separate sites. It follows that most of the transport requirements will involve the transfer of plutonium oxide from a reprocessing plant to a fabrication plant and the delivery of fresh MOX fuel assemblies from a fuel fabrication plant to a reactor site. To complete the cycle, irradiated MOX fuel will have to be transported from the reactor sites to the reprocessing plants after an appropriate period of radioactive decay.

Plutonium may be transported by air, sea, rail, or road. For long journeys, the likely method will probably be by air (with road transport between airports and plants); for shorter journeys the likely method is by road. For road transport the number of plutonium oxide containers carried on one truck is limited by legal weight restrictions for normal vehicles and also by IAEA regulations. Typically, a standard transport package for plutonium oxide consists of a metal container sealed inside a plastic bag, which is in turn sealed within a metal can. Several metal cans are placed inside a wooden box lined with cadmium and externally covered in steel. The physical dimensions of each package are designed to ensure that a criticality incident cannot occur under any circumstances.

In transporting fresh reactor fuel enriched with plutonium, the number of fuel assemblies in close proximity must be limited to prevent criticality hazards in the event of a serious accident. Also, steps must be taken to prevent the misappropriation of MOX fuel.

To overcome such problems it may be thought that the irradiated fuel shipping containers could be used, since these are returning empty from reprocessing plant to reactor site. However, unless the containers could be thoroughly decontaminated before being used for the unirradiated fuel, such a use would give rise to fuel contamination problems. For this reason, special transport containers have been devised capable of accommodating two mixed-oxide fuel assemblies for an LWR. Only one such container, weighing around 3 tonnes, would be loaded on each road vehicle or rail wagon.

Similar considerations apply to the transport of unirradiated fast reactor fuel, but in this case the transport flasks used at present in the UK carry four fast reactor fuel assemblies and weigh around 4 tonnes.

The transport of irradiated MOX fuel will be carried out in massive flasks which will weigh up to a maximum of 100 tonnes. These flasks will meet the national and international design and test requirements with respect to criticality, shielding, heat dissipation, and containment, both under normal and accident conditions, as now laid down for thermal reactor fuel.

4.5 **Impact on fuel cost**

The incorporation of plutonium as a fissile material in thermal reactor fuel in place of enriched uranium brings with it all the special requirements of plutonium handling which have been discussed above. The economic effect of these special requirements is reflected in the fabrication cost of plutonium-bearing fuels, which can be a factor of up to four higher than the fabrication cost of enriched uranium fuel. The increased cost results from the following six main causes.

1. Plant capital costs are higher due to the provision of glove boxes or shielded caves with remote manipulators and possibly closed-circuit television viewing; even fully automated plant is a possibility. The cost of ventilation plant is also increased, since sub-atmospheric pressure must be maintained in the plant and absolute filters must be provided for the exhausted air to prevent any release of plutonium to the outside atmosphere. There must also be an appreciable increase in radiation-monitoring equipment.

2. Process operating costs are higher even in the case of fully automated plant, since all materials must be *posted* in and out of the plant, whilst the use of glove boxes or caves reduces throughput per operator. Supervision requirements will be increased as will the training necessary for new operators.

3. Process service costs are higher because of greatly increased radiation-monitoring requirements, and the need to service the absolute filters at regular intervals. The handling and analysis of samples is also more complicated and time-consuming when plutonium is present.

4. Maintenance costs are increased considerably due to the requirement to handle plutonium in fully isolated enclosures. Whereas, in the case of enriched uranium fuel, defective machinery can be roughly cleaned and then maintained *in situ* or removed readily to the workshops, in the case of plutonium-enriched fuel, the machinery either has to be repaired on the spot by men in unwieldy 'frog' suits with their own air lines or has to be completely decontaminated, often a difficult process, and then posted out to the workshops. In most cases, complete decontamination is impossible and replacements are necessary, the old equipment having to be removed to suitably shielded sealed permanent storage.

5. Most fuel fabrication processes produce residues of one sort of another and if these contain plutonium, their recovery, necessary for environmental as well as economic reasons, is a very costly process. The use of sealed plant produces an excessive amount of contaminated 'bits and pieces' such as swabs, worn gloves, polythene containers, and small tools such as brushes and mops. To recover plutonium from these and to reduce their volume, incineration in purpose-built incinerators is necessary, with acid extraction from the ash. Such processes carried out in sealed enclosures are expensive.

6. Care will be needed to ensure that the amount of plutonium in MOX fuel is appropriate for the isotopic composition of the plutonium being used, and some extra cost is anticipated.

4.6 Comparison with similar problems for plutonium-enriched fast reactor fuel

The main differences between plutonium-enriched thermal reactor fuel and plutonium fast reactor fuel lie in the concentration of plutonium in the fuel, i.e. the plutonium enrichment, and in the complexity of the fuel assemblies. In plutonium-enriched thermal reactor fuel, the concentration of total plutonium will be in the region of 3.5 per cent to 8 per cent, depending on the amount of recycling, although the concentration of fissile isotopes will only be in the range 2.5 per cent to 5 per cent. In the case of the fast reactor the average total plutonium concentration is in the range 15 per cent to 30 per cent, and this poses some problems of criticality control. This leads to the need to install completely separate plants for each type of fuel so that the criticality control is effected by the use of eversafe plant in each case.

Such control is assisted by the dimensions of the fuel pellets in the fuel assemblies, the diameters being around 10 mm in the case of the thermal reactors, and only 5 or 6 mm in the case of mixed-oxide-fuelled fast reactors. Typical thermal reactor fuel assemblies contain either 48–63 fuel pins (BWR) or 204–264 fuel pins (PWR) in relatively simple assemblies. The pins each contain 3.66 m of fuel pellets in a Zircaloy can. In the BWR the pins are loaded into grids in a square Zircaloy wrapper, whilst in the PWR the pins are located in spring-loaded spacer grids, which are themselves held in position by the tubes in which the cluster control rods operate. One tonne of MOX fuel would be contained in around five BWR fuel assemblies or two or so PWR fuel assemblies.

The fast reactor fuel assembly is more complex. The stainless-steel-clad fuel pin contains not only a 1 m length of core fuel pellets but also two 40 cm lengths of axial breeder pellets, one at each end of the core section. Either 325 or 271 such pins are contained in a hexagonal wrapper, which has a complicated sodium filter and gag at the lower end and a swirler and lifting device at the upper end. For the 271 pin design, 1 tonne of core fuel heavy atoms would be contained in 18 assemblies.

The more complex design and greater number of assemblies per tonne for the fast reactor leads to much higher fuel fabrication prices than those for the light water reactors. This is counterbalanced, however, by the much higher mean burn-up achieved in fast reactor fuel—approximately 70 000 MW d/t compared with 30 000–35 000 MW d/t in the light water reactors—which reduces the annual throughput per unit of electricity generated. Further factors that reduce fast reactor fuel costs per unit of electricity are that the overall efficiency of a fast reactor station will be 5–10 percentage points higher than an LWR, and the higher fuel rating (factor of around 5) leads to a much smaller initial inventory of core fuel.

5 Economics and the effect of plutonium recycle on a generating system

If irradiated thermal reactor fuel is not reprocessed, the residual uranium will be lost, and this will increase the requirement for mined uranium. It has already been indicated (in §3) that the recovery and recycle of uranium in PWR can save 20 per cent of this requirement. In the case of the AGR the net annual requirement of 162 t $GW(e)^{-1}$ y^{-1} of natural uranium quoted in Table 9.4 would be increased by over 30 per cent with no reprocessing. The result of not reprocessing will therefore be to increase the pressure of demand on the finite and indeed relatively small supply of natural uranium in the world, say not more than 10 Mt U in high-grade ores extractable at a cost of up to \$130/kg U. (However, as discussed in §7, actual known and identified real reserves in the World Outside Centrally Planned Economies Area (WOCA) that are economically extractable at the present time may be less than 1 Mt U.) Also failure to reprocess will deny the use of plutonium both in thermal reactors and for the initiation of fast reactor programmes.

5.1 Economics

The uranium recoverable by reprocessing irradiated thermal reactor fuel can be re-enriched and reused, and has therefore a value related closely to the price of natural uranium. Plutonium is more difficult to value. An assumption of a zero value for plutonium is convenient for the analysis of a nation's or group of nations' nuclear programme when treated as an independent closed system, because plutonium is created and used (or not used as the case might be) within the system. Consequently there is no need to ascribe a value to plutonium, since any debits or credits would be internal transfer payments which cancel when considering total system expenditure.

However, nations which are contemplating a thermal-reactor-only strategy may wish to introduce the concept of plutonium value to assess whether reprocessing is worthwhile to allow trading in plutonium. Similarly a nation considering the rapid introduction of fast reactors may wish to consider the value of plutonium in fast reactor inventories, in order to assess the possibility of buying-in plutonium.

For thermal reactors an economic balance can be struck between

    (1) the no-reprocessing case, involving the likely costs of encapsulating irradiated fuel and ensuring its safe storage and possible disposal, and

    (2) the reprocessing case, which will include the cost of reprocessing plus vitrification and storage of fission products, but reduced by the credit available for the recovered uranium and perhaps also for the recovered plutonium.

Neglecting the possible credit for plutonium, the break-even point between no-reprocessing and reprocessing will depend on the relative costs of each and on the cost of natural uranium. Recent studies by member countries of the International Nuclear Fuel Cycle Evaluation showed that reprocessing is more economic once the uranium price has risen above \$30–40/lb U_3O_8 even with no plutonium credit, whilst in the UK the break-even point was \$40/lb U_3O_8, a price already reached in sales of uranium on short-term contracts.

The economics of fast reactors are far more difficult to establish. To begin with, the relative capital costs of sodium-cooled fast reactors are likely to be higher than for thermal reactors because of their greater complexity (for example an additional sodium coolant circuit between the primary reactor coolant circuit and the steam-raising circuit). Also, as indicated in §4, the fuel-processing costs per tonne of fast reactor fuel will be higher than those for thermal reactors, but this is mitigated by the high rating and long burn-up to be achieved, so that the fuel-processing costs per unit of electricity generated could be lower for fast reactors. Finally, fast reactors fuelled with waste depleted uranium enriched with plutonium need no fresh natural uranium or separative work (isotope enrichment effort), and so are independent of the price of uranium.

To help decide whether fast reactors are economic relative to thermal reactors, it would be convenient if the target fast reactor capital cost to achieve break-even could be expressed as a function of uranium price. However, the other factor, the fuel cycle costs, is itself difficult to predict, particularly since the reprocessing is subject to further development.

To give an indication of the likely target, assume that the capital costs of fast reactors were not more than about 40 per cent above those of a light water reactor at that time, and if uranium prices were in due course to rise to about four times their present level, then fast reactors would appear to be competitive with LWRs. But because fast reactors need no fresh natural uranium, their introduction will limit the dependence of a nation's nuclear programme on obtaining fresh uranium supplies, and it may not be necessary for them to show a definite cost advantage at the time of commissioning. The introduction of fast reactors will depend, therefore, not only on how much weight is given to this feature of conservation of uranium supply, but also on the way in which uranium prices are expected to move (in real terms) in the 25 years or so of the reactor's life.

These aspects were also a subject of study by some of the member countries of the International Nuclear Fuel Cycle Evaluation, and the general conclusion that emerged was that fast reactors will become economic during the last decade of the present century if pressures on the available uranium supplies are such that U_3O_8 prices rise rapidly. In the event that prices do not rise rapidly, fast reactors will become economic during the first decade of the next century.

Some countries see this as a paradox; they argue that it is possible that uranium prices will tend to stabilize at a relatively low level if significant numbers of fast reactors are installed soon after the turn of the century, reducing demand on uranium, and thereby avoiding the necessity to mine low-grade uranium deposits. On the other hand, if fast reactors are not introduced early in the next century, demands on uranium will be high and its price will rise rapidly. It would then be too late for fast reactors to influence the uranium market to any marked degree. This implies that fast reactors need to be introduced and established before the turn of the century even though they are not demonstrably economic.

Finally, it must be stressed that the choice between one reactor strategy and another may not be entirely a matter of the economic saving that might be achieved, but also may reflect the desire to minimize future uranium imports, and hence the strategic dependence upon foreign uranium suppliers, by utilizing to the fullest extent the 'waste' uranium arising from early thermal reactor programmes.

5.2 **Effects of plutonium recycle on a generating system**

As indicated in the previous section, economics can justify reprocessing if uranium ore prices rise, even if no credit is taken for the value of plutonium. Whether this is done may well depend on non-economic aspects such as views on non-proliferation or ease of radioactive waste management. If reprocessing is undertaken, the resulting stock of plutonium which becomes available provides a valuable supply of fissile material that can be used as feed into thermal or fast reactors, with a resulting saving in ^{235}U and thus in uranium ore requirements.

With an all-uranium thermal reactor strategy, cumulative uranium requirements continue to rise as long as the installation programme proceeds. The gradual introduction of plutonium recycle in thermal reactors can reduce cumulative uranium requirements by around 20 per cent, in addition to the 20 per cent or so saving achieved by recycling recovered uranium.

Alternatively the plutonium recovered from the thermal spent fuel can be used to introduce a fast reactor component into the nuclear strategy. As already described, the fast reactor utilizes waste uranium for its operation and therefore makes no demand on uranium ore resources; its rapid development within a nuclear strategy is therefore of particular importance for reducing uranium consumption.

The rate at which fast reactors can be installed will depend on several factors: the amount of plutonium available from the thermal reactors already in operation, the amount of plutonium required for the first core of the fast reactor, the plutonium produced by the fast reactor itself, the plutonium held up out-of-pile, e.g. in cooling ponds, and the plutonium irrecoverably held in waste residues in the fuel-processing cycle.

Fast reactor designs may well aim for low inventory requirements, particularly in the early years to maximize the number of fast reactors that can be installed using existing plutonium stocks from thermal reactors, and for high rates of breeding, particularly once the early development problems have been overcome. The extent to which the fast reactor can dominate the nuclear sector, i.e. supplant the thermal sector, is governed by the relative doubling times of the fast reactor and the electricity generating system. If the fast reactor doubling time is less than that of the electricity generating system, fast reactors will eventually fully penetrate the system. It follows that the annual demand for uranium will run down to zero during the lifetime of the last thermal reactor to be commissioned, and the cumulative requirement will steady out at some level determined by the characteristics of the system. The uranium savings due to the use of fast reactors have been studied for closed systems for many years throughout the world and are illustrated in Fig. 9.7 for a nuclear installation programme that can be considered illustrative for the UK. Figure 9.7 considers three alternative thermal reactor (LWR) fuel cycles as follows:

(1) recycling uranium and stockpiling plutonium (for illustrative purposes only),
(2) recycling both the uranium and plutonium in the LWR, or
(3) using the plutonium in the fast reactor.

Figure 9.7 indicates that, as previously mentioned, recycling plutonium in the thermal reactor reduces total uranium requirements up to the year 2040 by about 20 per cent, from 550 to 440 kt U. Using the plutonium in the fast reactor, however, total uranium requirements for an on-going nuclear programme level out at around 250 kt as the fast reactor fully penetrates the nuclear system. The effect of delaying the introduction of the fast reactor is

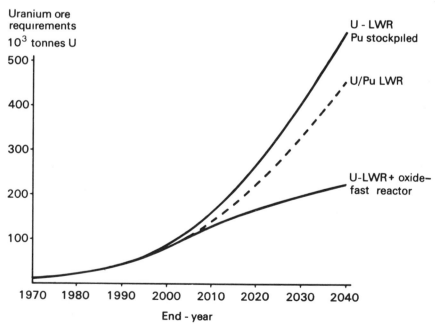

Fig. 9.7 Illustrative UK cumulative uranium requirements

of course to reduce the benefit obtained. With delays of 10 and 20 years, the cumulative uranium requirements rise to around 270 and 370 kt, assuming the same nuclear installation programme is adopted.

The introduction of more advanced designs of fast reactors at some later stage would effect some further reduction in uranium requirements. Thus changing from oxide to carbide fuel leads to increased plutonium production, shorter doubling times, and a more rapid penetration into the electricity generating system. However, delaying the start of a fast reactor programme markedly affects the ability of advanced fuels to reduce the demand for natural uranium. This is because the plutonium stockpile arising from the thermal reactors during the interim can become so large that, even with only an all-oxide strategy, relatively few further thermal stations are commissioned once fast reactors are established; there is thus less scope for a more advanced design of fast reactor to effect a further reduction in the number of thermal stations commissioned and therefore in the amount of uranium required.

The magnitude of the rate of increase in size of a nuclear electricity generating programme has a marked effect on natural uranium saving resulting from the use of advanced fast reactor fuel. This is because the superior market penetration capability of an advanced fuel becomes less important the smaller the rate of increase in electricity demand. Conversely the higher the rate of growth the more important the superior penetrating power, and hence the larger the relative reduction in uranium demand. If the fast reactor introduction is delayed, it is possible that in the interim period plutonium may be recycled in thermal reactors. Studies have shown that with interim thermal recycling of up to 20 years, little or no change in total uranium requirements results compared with merely storing the plutonium until needed in fast reactors. This arises because the uranium saved in thermal reactors by use of plutonium is balanced by the increase in demand later, owing to the need to install some thermal reactors in the place of fast reactors (there being no longer sufficient plutonium to install fast reactors alone).

Strategies which involve the recycling of plutonium lead to different distributions of plutonium in the nuclear system compared with a strategy in which the plutonium is stockpiled. An indication of the disposition of plutonium for the three strategies assumed in Fig. 9.7 is shown in Table 9.13.

Column (3) gives the quantities of plutonium contained in irradiated fuel awaiting reprocessing plus the small amount held up in the chemical reprocessing and fabrication plants for an assumed delay time of eight months (six months cooling, two months reprocessing and fabrication). Column (5) includes one month's buffer stock of new fuel assumed held at each station as well as the quantities of plutonium available for investment in new plutonium-fuelled reactors (after deduction of the inventory requirement of any reactors commissioned during the year).

The total quantities of plutonium in the system are similar for the thermal

Table 9.13 Illustrative distribution of plutonium in nuclear systems†

Strategy	Year	Tonnes plutonium (all isotopes)			
		Ponds and plants	Reactor	Stocks and new fuel	Total
(1)	(2)	(3)	(4)	(5)	(6)
Thermal +	1980	2	13	19	34
plutonium	1990	4	16	42	62
stockpile	2000	5	22	89	116
	2010	12	38	201	251
	2020	16	59	380	455
Thermal +	1980	2	13	19	34
plutonium	1990	4	16	42	62
recycle	2000	7	42	74	123
	2010	20	86	12	118
	2020	29	127	8	164
Thermal +	1980	2	13	19	34
fast reactor	1990	4	16	42	62
	2000	7	30	77	114
	2010	37	124	79	240
	2020	92	291	24	407

† In once-through mode plutonium in both ponds and stocks would be contained in stored unprocessed irradiated fuel

strategies with fast reactors and with plutonium stockpiled, because the net plutonium production of the thermal reactors and the mixed-oxide-fuelled fast reactors are about the same. The thermal reactor strategy with plutonium recycle has the minimum system inventory of plutonium because the thermal reactors (or parts of thermal reactors in SGR mode) on plutonium-enriched feed are net consumers of plutonium. The recycle of plutonium in fast and thermal reactors reduces the plutonium stocks to a minimum and effectively transfers them into the reactor cores.

6 Status of recycling programmes

6.1 Thermal reactors

6.1.1 *Status of refabrication technology*
The large-scale reprocessing of spent fuel to separate plutonium by solvent extraction (using the Purex process) has already been developed into a well-established industrial technology. Tens to hundreds of tonnes of LWR uranium oxide fuel has been reprocessed in each of the following countries: Belgium (Eurochemic-Mol), France (Cap de la Hague), the Federal Republic of Germany (WAK, Karlsruhe), Japan (Tokai Mura), the United Kingdom (Windscale), and the United States (NFS, West Valley). In addition several thousands of tonnes of irradiated Magnox natural uranium fuel have been reprocessed.

The design and specification of the reference MOX fuel fabrication plant outlined in §4 is based on experience gained over a number of years at smaller-scale plants in a number of countries, including Belgium, The Federal Republic of Germany, France, Italy, Japan, the United Kingdom, and the United States. Most of these plants have fabricated many tens of tonnes of plutonium-bearing (MOX) fuels.

Similarly, procedures and experience already exist for the safe transport and storage of plutonium, both as nitrate and as oxide, in tonne quantities. The transport of MOX fuel assemblies to LWRs might, however, need some modifications in the design of containers to allow for changes in toxicity, heat generation, and shielding needs.

Considerable numbers of MOX fuel pins and complete MOX fuel element assemblies have been irradiated to high burn-up in many light water reactors in the world, notably in Belgium, the Federal Republic of Germany, Italy, Japan, and the United States. Experience of the performance of mixed-oxide fuel assemblies is similar to that for uranium oxide fuels.

The experience which has already been gained with reprocessing, plutonium handling, and plutonium recycle in thermal reactors therefore gives confidence that scaling-up from present technology to an industrial scale, where this has not already been done, can be achieved satisfactorily.

UK experience in irradiating MOX fuel is comparable to that of the other countries listed above with respect to the high-temperature stainless-steel-clad gas-cooled reactor fuel, but experience with Zircaloy-clad fuel for LWRs is more limited. The interim conclusions which can be drawn from this UK programme on thermal reactor irradiations are given below.

6.1.2 *Irradiation experience of recycled fuel*
If recycled material is to be successfully used as a nuclear fuel, it must of course meet the same kind of standards of efficiency and freedom from adverse side effects as fresh fuel. As has already been explained in earlier sections, the theoretical aspects of the use of recycled material are well understood and adequate data are available to enable the neutronic performance in a given reactor to be predicted reliably. However, there are other factors which are not so easy to assess theoretically, such as the degree of irradiation swelling to be expected, and the rate of release of gaseous fission products which will pressurize the fuel can. Chemical effects in the complex conditions, such as the interactions between can and fuel or between the various impurities, are also uncertain. These issues lead to the necessity for demonstrations. Such demonstrations must of course be under representative conditions of irradiation temperature and time, and so far have been on a relatively small scale. In the UK the small Windscale AGR (30 MW(e)) and the larger Winfrith SGHWR (100 MW(e)) are used for experimental irradiation of gas-cooled and water-cooled reactor fuel respectively. These facilities have tended to be fully committed to the study of the performance of new uranium fuel in standard fuel elements, so that only a

few irradiations relevant to the recycling of plutonium in thermal reactors have been carried out.

6.1.3 *Recycling in UK gas-cooled thermal reactors*
The main experiments have involved the irradiation of stainless-steel-clad fuel pins in which the fissile component is plutonium taken from Magnox fuel reprocessing. These have all been carried out in the Windscale AGR, a total of about 350 fuel pins having been irradiated. Some of these experiments were not given a full post-irradiation examination; instead the fuel was reprocessed in the Dounreay Fast Reactor metallic fuel reprocessing plant to assist in identifying problem areas prior to its conversion to the Prototype Fast Reactor (PFR) MOX reprocessing plant. Others were involved in failures not connected with their non-standard composition. However, a significant number completed the planned irradation, and examination of these revealed no new problems. These post-irradiation examinations naturally concentrated upon features which were expected to give a different behaviour, such as the presence of a different spectrum of fission products, but the experiments do not seem to have resulted in a final performance either better or worse than that seen with uranium-enriched fuel. However, more work is needed in this field since the samples so far irradiated are too small for complete assurance.

6.1.4 *Recycling in UK water-cooled thermal reactors*
The main UK irradiation facility for water-cooled thermal reactor fuel elements, theWinfrith SGHWR (100 MW(e)), has been operating for a shorter time than the smaller-capacity Windscale AGR. Consequently, there have been fewer irradiation experiments relevant to recycling. The one experiment there has been with plutonium-enriched fuel did not reveal any physical or chemical effects produced by the plutonium which might adversely affect the performance of such fuel. This demonstration has, of course, not been adequate to give more than an initial assurance. The absence of adverse effects up to 12 000 MW d/t burn-up does however provide some confidence that, with regard to the use of plutonium enrichment at least, no major problems are likely, especially when this result is taken in conjunction with the large number of LWR MOX irradiations in other countries.

6.2 **Fast reactors**

6.2.1 *Fabrication of fuel*
The process for manufacturing fast reactor core fuel follows closely that for manufacturing plutonium-enriched thermal reactor fuel, the main difference being the plutonium concentration, which is in the range of 15 to 30 per cent rather than 3.5 to 8 per cent, leading to a corresponding increase in the americium-241 content. As waste or depleted uranium is used for breeder

fuel, it can be fabricated without difficulty in a normal thermal reactor fuel plant. However, as it is expected that fast reactor core fuel will be taken to a high mean irradiation, some 70 000 MW d/t compared with levels up to 30 000 MW d/t or so for thermal reactor fuels, the amount of fuel to be fabricated and reprocessed will be reduced proportionately.

6.2.2 *Fuel fabrication plant*

As noted previously, the plutonium content of thermal reactor fuel will depend on the 'quality' of the plutonium, i.e. on the number of recycles it has been through. In the case of the fast reactor the problem is compounded since the plutonium can arise from several sources: namely, from the re-processed core fuel, from reprocessed axial and radial blankets, and from irradiated thermal reactor fuels perhaps of several types, e.g. Magnox, AGR, PWR, each type having a specific fuel irradiation level and thus a particular blend of plutonium isotopes. But as explained in earlier sections all plutonium isotopes are fissionable in fast reactors, and the determination of the plutonium content of the fuel is assisted by ascribing to each plutonium isotope a ^{239}Pu equivalence and specifying a fixed ^{239}Pu equivalent content for the fresh fuel. As stated in §2 these equivalences are typically, ^{239}Pu $= 1.0$, ^{240}Pu and ^{242}Pu about 0.15, ^{241}Pu about 1.5.

In the UK, by Spring 1979, some 7 tonnes of uranium and plutonium in the form of mixed oxides had been successfully fabricated into 157 core fuel assemblies (51 000 pins) for the Prototype Fast Reactor at Dounreay. Irradia-tion of this core fuel has proceeded to the point where high fuel burn-up levels are being attained. Reference design pins operating at enhanced ratings have reached a peak burn-up of 10.5 per cent of heavy atoms (i.e. about 100 000 MW d/tonne of heavy atoms), whilst the main driver charge has attained 6.5 per cent burn-up (1 per cent burn-up of heavy atoms is approximately equal to 9350 MW d/t). No pin failures have occurred in this driver charge fuel. Some sub-assemblies in the core have been rotated through 180 °C, using machinery designed for the purpose, to counteract the bowing due to differential wrapper swelling, which increases with irradiation. No difficulties have been experienced in the removal and loading of fuel assemblies, the main fuel charge machine having carried out 275 single movements up to the end of 1979 without significant difficulty.

6.2.3 *Reprocessing of fuel*

Immediately on discharge from the reactor core, fuel elements have a high heat output owing to the radioactivity of the fission products. They are stored for some time in liquid sodium to allow the shorter-lived fission products to decay and the sub-assembly heat output to fall to about 10 kW. In the case of the PFR the irradiated fuel is being stored for reprocessing in the plant now commissioned on the Dounreay site. For commercial fast reactors it is likely that individual sub-assemblies will be sealed in sodium-filled stainless steel cans, which will then be placed in holes in massive solid steel flasks. These flasks will be designed to meet the latest IAEA regulations for the safe transport of radioactive materials, but with some consideration given to possible more onerous conditions in the future. On arrival at the reprocessing plant the residual sodium will be removed from the fuel assemblies by a mixture of inert gas and steam before breaking up the fuel pins and dissolving the fuel prior to reprocessing.

6.2.4 *Reprocessing plant*
The reprocessing to separate uranium, plutonium, and fission products will use the normal solvent extraction route for thermal reactor fuel, but the plant will be designed to cope with the special requirements of the high plutonium content. Geometrically limited equipment or fixed or soluble neutron absorbers as *poisons* will be included to prevent accidental criticality, and a process will be included to remove noble-metal fission-product alloys that are formed in high burn-up fuel, and which are unlikely to dissolve with the other products in the fuel. To facilitate active commissioning of the Dounreay reprocessing plant, long-cooled fuel is being reprocessed first, but as experience is gained it is anticipated that the necessary radioactive decay cooling will be reduced to a level where the amount of plutonium held outside the fast reactors is one-third to one-half of that held within the reactors, thus allowing the achievement of linear doubling times of less than 30 years.

7 Factors governing a decision for recycle

Before recycling fuel can be undertaken certain basic factors must be considered. These are:
    (1) the establishment of the basic feasibility;
    (2) the need for an industrial-scale demonstration programme;
    (3) economics;
    (4) proliferation aspects;
    (5) the availability and utilization of nuclear fuel.

7.1 Basic feasibility and industrial-scale demonstration programme

As far as recycling uranium fuel is concerned there can be no doubt that the procedure is established in practice, as irradiated thermal reactor fuel has been reprocessed to separate the uranium from the plutonium and the fission products. In the UK many thousand tonnes of recovered uranium product have been converted into a Hex feed to enrichment plants. Alternatively it could be converted to the oxide, blended with higher-enrichment material, and then refabricated into new thermal fuel elements.

Many plutonium-enriched thermal reactor fuel pins have been manufactured, assembled into fuel elements, and irradiated in various programmes throughout the world. Some of this fuel has been taken to burn-ups comparable to that achieved in uranium-235 enriched fuel with no undue difficulties emerging. The reprocessing of this fuel would be expected to be similar to the processing of irradiated uranium fuel, as the plutonium concentrations will be similar at a few per cent. It is only after repeated recycling of plutonium in thermal reactors that problems due to higher plutonium concentrations might arise. As far as the recycle of plutonium in thermal reactors is concerned therefore, the next stage would be the mounting of a large plutonium recycle programme in thermal reactors, which some countries favour because they argue that the experience would provide the build-up and continuity of a difficult technology that in the end will contribute to the solution of the even more difficult problems envisaged in the manufacture and reprocessing of plutonium-enriched fast reactor fuel.

With respect to the fast reactor, there are some 25 to 30 years' experience with experimental facilities, and several large prototype electricity-generating fast reactor stations are in operation or are being built in the world using fuel having plutonium contents of 20 per cent or so. It follows that experience in fabrication and irradiation of such fuels is being acquired rapidly. Reprocessing of this type of fuel still needs proving, and to this end the fast reactor plutonium-fuel reprocessing plant started up at Dounreay in 1980 should provide invaluable operating experience.

The next stage in development of recycling fuel in fast reactors is to build and operate large commercial reactors and their associated fuel-servicing plant. In France, Super Phénix is being built at Creys-Malville and is due for completion in 1984. In order to provide the necessary reprocessing facilities the COGEMA plant at Cap de la Hague is to be expanded to cope with fuel from at least three Super-Phénix-sized reactors. In the USSR a 600 MW(e) pool-type fast reactor (BN600 at Beloyarsk) started operation in 1980. In the UK, consideration is now being given to the next stage of fast reactor development, namely the construction of a commercial demonstration fast reactor (CDFR). Before this commences the Government is committed to a public inquiry and, if the go-ahead is sanctioned, construction of CDFR could commence in the mid-1980s for operation in the 1990s. The objective of such a project is to establish a capability to construct and prove the safe operation of a large fast reactor power station together with its supporting fuel cycle and waste-management plants, so that the option of commencing a programme of fast reactors after the turn of the century is available.

7.2 Economics

As already indicated in §5, the balance to be struck in deciding whether it is worthwhile to recycle fuel in thermal reactors is between (1) the cost of the once-through mode, involving large uranium and, where applicable, enrichment costs and the indefinite storage of spent irradiated fuel, and (2) the recycling mode, where the cost of reprocessing is offset by the reduced requirements for uranium and enrichment because of the value of the recovered and recycled uranium and plutonium. The break-even price of uranium for these modes is in the range \$30 to \$40/lb U_3O_8. It would therefore appear that thermal recycle of both uranium and plutonium should be economic by the time the industrial capability for reprocessing irradiated oxide fuel and manufacturing plutonium-enriched thermal fuel is established.

There is no doubt that the economics of fast reactors will be more difficult to establish, as the capital cost of early fast reactors will be substantially greater than for thermal reactors. Even in the long term it is possible that their capital cost will be some 25 to 40 per cent greater than for thermal reactors of the same electrical capacity. However, if future nuclear power programmes are based solely on thermal reactors, there will be increasing pressures on uranium resources and production, which in due course would be reflected in rising uranium prices. At some stage the prospective forward uranium price over the lifetime of a new power station will make the fast reactor a competitive prospect. Amongst those countries involved with a fast reactor programme, views vary as to when this stage will arrive, with an extreme range being from the early 1990s to around 2025, although the consensus of opinion amongst the majority of countries is that fast reactors will become economic around the turn of the century.

7.3 Proliferation and security aspects

Fissile material such as ^{235}U or ^{239}Pu could be considered as susceptible to diversion for use in weapons if it is contained in new unirradiated fuel elements. In the case of ^{235}U-enriched uranium, the ^{235}U could be obtained by further enrichment processes, and plutonium used as an enrichment in depleted uranium could be obtained by chemical separation. However, once the fuel is within a reactor being irradiated, or is newly discharged from a reactor when it is very highly radioactive, it can be considered inaccessible from the point of view of diversion or proliferation. Table 9.13 shows quite clearly that in strategies where plutonium is recycled, either in thermal or fast reactors, the amount of plutonium available in stocks (including stocks of fresh fuel) is extremely low compared to the first case given in the Table, where the plutonium is not utilized. Even in the once-through mode, it cannot be argued that the plutonium remains inaccessible due to high radiation hazards whilst it is contained in stored, unprocessed, irradiated fuel. After a comparatively few years the radioactivity of the irradiated fuel decreases to such an extent that misappropriation of the fuel cannot be ruled out, and so the plutonium could subsequently be recovered by reprocessing for use in weapons. Finally, it might be considered from a comparison of the second and third strategies in Table 9.13, that a nuclear programme of a given size utilizing thermal recycle of plutonium would be preferable to utilizing plutonium in fast reactors, as the total plutonium inventory is less and hence the potential for proliferation is smaller. However, as the utilization of uranium in thermal reactors is some 50 times less than that in fast reactors, such an argument would only stand up if there were sufficient supplies of uranium available to sustain the all-thermal programme. In addition, it must be remembered that the performance of a fast reactor can be finely tuned, so that only enough excess plutonium is produced to sustain a programme and no excess stocks need exist. Indeed, by removing the fertile blankets from around the core of a fast breeder reactor, it can be converted into a plutonium incinerator if the requirement to reduce the world's inventory of plutonium rapidly should arise.

7.4 Availability and utilization of nuclear fuel

Numerous studies in many countries in the world have shown that unless nuclear power plays an ever increasing part in electricity generation, there will be no escape from experiencing the disastrous economic effects that flow inevitably from an insufficient fuel base to power industry. But an increasing nuclear power programme for electricity generation assumes a continuing availability of nuclear fuel, namely uranium. For those countries without indigenous sources of uranium, a thermal reactor nuclear programme necessitates importing uranium, and hence they could be held to ransom by the uranium-supplying countries. As already indicated in earlier sections, the use of reprocessing and the recycle of recovered uranium and plutonium could reduce uranium requirements with respect to the once-through mode by 35–40 per cent or so. Again, as already indicated in previous sections, the introduction of fast reactors into nuclear programmes will ultimately allow a 50-fold or more increase in the utilization of uranium compared to that achievable with thermal reactors. But before a decision is taken to develop the technology required to recycle fuel in either thermal or fast reactors, it is first worthwhile considering the availability of uranium to investigate whether the option for recycle is one worth having.

Using the OECD-NEA/IAEA terminology, the 'reasonably assured resources' of uranium up to an extraction cost of \$50/lb U_3O_8 are assessed at 2.5 Mt U. A similar amount of uranium is included in the 'estimated additional resources'. The term reasonably assured resources indicates that there is a reasonable degree of assurance that the material is actually in place in the ore bodies, although the exact definition of the degree of assurance probably varies from country to country. The term, however, does not mean that the material can be mined at a profit or even that it can be mined. Assuming that the current price of long-contract uranium is \$25/lb U_3O_8, only a fraction of the 2.5 Mt reasonably assured resources could be extracted by mining companies economically, i.e. at an acceptable profit; this fraction may be less than 50 per cent. That is to say the true economic reserve of uranium at the present time is, say, around 1 Mt U, whereas the identified but subeconomic resource is also around 1 Mt, but steps would not be taken by mining companies to extract this latter material until the price of uranium moves upwards. The term estimated additional resources applies to material as yet undiscovered, material which may or may not exist, and which may not be found even if it does exist. It is important to emphasize therefore the profound difference between the uranium resource which is tangible, always available for use at some cost of extraction and at some (presumably higher) price, and the large mass of material that may or may not be real, and which may not be found and brought to use in any reasonable time frame even if it is real. These undiscovered 'sources' are therefore nothing but a number no matter how carefully contrived or how complicated and abstruse the method of achieving it. It is ridiculous to attribute to these numbers so real a status as to conclude that the need for reprocessing and the need for the introduction of fast breeder reactors can be ignored.

The picture that emerges therefore is that in a world with increasing thermal reactor programmes, demands for uranium will soon outstrip supply unless vast, new reserves are discovered and exploited. The lead times for the establishment of new mining capacity are long, and for countries which rely solely on importing uranium the risk of continuing with large thermal reactor programmes, with or without recycling, seems large. The conclusion is drawn therefore that such countries should develop the option to utilize to the full their stocks of plutonium extracted from their thermal reactor programmes, together with their stocks of depleted and waste uranium in fast reactors. The full deployment of the fast reactor option will take some considerable time because of the learning processes and the large investment involved in both reactor and fuel-cycle plants. At the time it is exercised, the pressure on other fuel resources may well be great, and if the fast reactor is to be most effective in meeting energy demands it is necessary to complete the demonstration and development phases in good time. It is therefore imperative that the current development programmes, including the construction of a commercial demonstration fast reactor together with its associated fuel servicing plants, should proceed as early as practicable, so that the option of recycling waste uranium and plutonium in fast reactors is available as an established technology for the generation of electricity when the effects of the depletion of the gaseous and liquid hydrocarbon fuels begin to bite.

Appendix   Equivalent plutonium-239 worths

In an unreflected nuclear assembly of constant composition, the multiplication factor, k, that is to say, the probability of a fission event causing a subsequent fission event, can be expressed by the relation:

$$k = \frac{\nu \Sigma_f}{\Sigma_a + L},\tag{1}$$

where

ν is the number of neutrons released per fission event,

Σ_f is the total (or macroscopic) fission cross-section of the fuel atoms in a unit volume,

Σ_a is the total absorption cross section of all the atoms in a unit volume,

and

L is the neutrons lost by leakage from a unit volume.

For the assembly to be just critical k must be equal to 1, giving

$$\nu \Sigma_f - \Sigma_a = L.\tag{2}$$

Now, $\Sigma_a = \Sigma_{a \text{ fuel}} + \Sigma_{a \text{ parasitic}}$, where $\Sigma_{a \text{ fuel}}$ equals the total absorption cross-section of all the heavy atoms (fuel), and $\Sigma_{a \text{ parasitic}}$, or Σ_{a_p}, equals the total neutron absorption cross-section of all other material atoms in the assembly, such as fuel cladding, coolant, and construction materials. Hence eqn (2) may be written:

$$\nu \Sigma_f - \Sigma_{a \text{ fuel}} = L + \Sigma_{a_p}.\tag{3}$$

For a reactor containing a total number of heavy atoms (i.e. fuel atoms) equal to N_h per unit volume and comprising uranium enriched with plutonium, the following expression is obtained:

$$\begin{aligned}(\nu \Sigma_f - \Sigma_a)_{\text{fuel}} = N_h[&\varepsilon\{\lambda_9(\nu\sigma_f - \sigma_a)_9 + \lambda_0(\nu\sigma_f - \sigma_a)_0 + \lambda_1(\nu\sigma_f - \sigma_a)_1 \\ &+ \lambda_2(\nu\sigma_f - \sigma_a)_2\} + (1 - \varepsilon)\{\lambda_5(\nu\sigma_f - \sigma_a)_5 \\ &+ \lambda_8(\nu\sigma_f - \sigma_a)_8\}],\end{aligned}$$

where

σ_f is the microscopic fission cross-section of an atom,

σ_a is the neutron absorption cross-section of an atom,

ε is the proportion of N_h comprising plutonium,

λ_i is the fraction of isotope i in plutonium or uranium,

and the suffixes $9, 0, 1, 2, 5$ and 8 refer to ^{239}Pu, ^{240}Pu, ^{241}Pu, ^{242}Pu, ^{235}U, and ^{238}U, respectively. By putting $(\nu\sigma_f - \sigma_a) = y_i$, eqn (3) can be written:

$$N_h[\varepsilon\{\lambda_9 y_9 + \lambda_0 y_0 + \lambda_1 y_1 + \lambda_2 y_2\} + (1-\varepsilon)\{\lambda_5 y_5 + \lambda_8 y_8\}]$$
$$= L + \Sigma_{a_p}. \qquad (4)$$

Subtracting $N_h y_8$ from each side, then

$$N_h[\varepsilon\{\lambda_9(y_9 - y_8) + \lambda_0(y_0 - y_8) + \lambda_1(y_1 - y_8) + \lambda_2(y_2 - y_8)\}$$
$$+ (1-\varepsilon)\lambda_5(y_5 - y_8)] = L + \Sigma_{a_p} - N_h y_8. \qquad (5)$$

Dividing eqn (5) by $y_9 - y_8$ and putting $\dfrac{y_i - y_8}{y_9 - y_8} = w_i$ (so $w_9 = 1.0$ and $w_8 = 0$), the following expression is obtained:

$$N_h[\varepsilon(\lambda_9 w_9 + \lambda_0 w_0 + \lambda_1 w_1 + \lambda_2 w_2) + (1-\varepsilon)\lambda_5 w_5] = \frac{L + \Sigma_{a_p} - N_h y_8}{y_9 - y_8} \qquad (6)$$

= constant, very nearly, since the sum of the second and third terms in the numerator are slowly varying for any given fast reactor design with equal core size. Hence given the critical size of a fast reactor with fuel of a certain isotopic composition, the plutonium enrichment of alternative compositions can be determined by using the left-hand side of eqn (6). The quantity w can therefore be defined as the reactivity worth of any isotope (or mixture) in forming a critical mass and as derived above

$$w_9 = 1 \text{ for } {}^{239}\text{Pu},$$

and all the other w_i can be defined as the 'equivalent-^{239}Pu worth' for a unit mass of isotope i. Typical values for current designs of oxide fast reactors are:

$w_5 = 0.7$
$w_9 = 1.0$
$w_0 = 0.15$
$w_1 = 1.5$
$w_2 = 0.15.$

Acknowledgements

The author has been considerably assisted by his colleagues in writing this article. In particular he wishes to acknowledge the help of Mr. H. Barton. Dr. R. Burstall, Mr. A.N. Knowles, Mr. B.L. Richardson, Mr. C. Scrogg and Mr. J.E.R. Shallcross.

Suggestions for further reading

Section 2

1. HUGHES, D.J. *Neutron cross-sections. BNL* 325 1st, 2nd, and 3rd edns Pergamon Press, London, (1957).
2. REICH, C.W. Status of beta- and gamma-decay and spontaneous-fission data from transactinium isotopes. In *Proc. of an IAEA Advisory Group Meeting on Transactinium Isotope Nuclear Data, Karlsruhe, 1975*, vol. 3, paper B7. International Atomic Energy Agency, Vienna (1975).
3. LEDERER, C.M. *Table of isotopes* (6th edn). John Wiley, New York (1967).
4. INTERNATIONAL ATOMIC ENERGY AGENCY. *Index to the literature on microscopic neutron data. CINDA-75*, vol. 2. IAEA, Vienna (1975).
5. GREEN, A. Recycle of Plutonium in UK thermal reactors. In *Proc. of a Panel on Plutonium Utilisation in Thermal Power Reactors*, Karlsruhe (1974), p. 313. IAEA, Vienna (1974).
6. CHAMBERLAIN, A. and MELCHES, A. Prospects for the establishment of plutonium recycle in thermal reactors in the FORATOM countries: status and assessment. In *Proc. of IAEA Con. on Nuclear Power and its Fuel Cycle, Salzburg, (1977)*, vol. 3, p. 271. IAEA, Vienna (1977).
7. MARSHAM, T.N. BAINBRIDGE, G.R., FELL, J., ILIFFE, C.E., JOHNSON, A. The technical problems and economic prospects arising from the alternative methods of using plutonium in thermal and fast breeder reactor programs, In *Proc. of the 4th Int. Con. on the Peaceful Uses of Atomic Energy, Geneva, 1971*, vol. 9, p. 119. IAEA, Vienna (1972).
8. STORY, J.S. JAMES, M.F., KERR, W.M.M., PARKER, K., PULL, I.C., SCHOFIELD, P. Evaluation, storage and processing of nuclear data for reactor calculations. In *Proc. of the 3rd Int. Con. on the Peaceful Uses of Atomic Energy, Geneva, 1964*, vol. 2, p. 168. United Nations, New York, (1965).
9. BESANT, C.B. CHALLEN, P.J., MCTAGGART, M.H., TAVOULARIDIS, P., WILLIAMS, J.G. Absolute yields and group contents of delayed neutrons in the fast fission of ^{235}U, ^{238}U and ^{239}Pu. *J. Br. Nucl. Energy Soc.*, **16**, no.2 (1977).
10. BAKER, A.R. and ROSS, R.W. Comparison of the value of plutonium and uranium isotopes in fast reactors, In *Proc. of the Con. on Breeding, Economics and Safety in Large Fast Power Reactors, October 1963. ANL* 6792. National Technical Information Service, Springfield Va., (1963).

Sections 3 and 6

1. UNITED STATES ATOMIC ENERGY COMMISSION. *Generic enviornmental statement mixed oxide fuel (recycle plutonium in light water-cooled reactors)*, vols. 3 and 4. Washington-1327. United States Atomic Energy Commission (1974).

Section 4

1. INTERNATIONAL NUCLEAR FUEL CYCLE EVALUATION. *Final report of Working Group 4—Reprocessing, plutonium handling, recycle*. Vienna, INFCE (1980).

Section 5

1. VAUGHAN, R.D. and FARMER, A.A. The fast breeder reactor—energy without depletion of natural resources. In *Proc. Instn. Mech. Engrs. (Nucl. Energy Group)*, **19030/76** vol. 190, no. 30, pp 163-167 (1976).

2. MARSHAM, T.N. The fast reactor and the plutonium fuel cycle. *Atom*, no. 253 (1977).

3. MARSHALL, W. Nuclear power and the proliferation issue. Graham Young Memorial Lecture, University of Glasgow (February 1978).

4. FARMER, A.A. AND HUNT, H. Important factors in fast reactors economics. *Nucl. Engng. Int.*, *23* (273) July 1978. pp. 43-45.

5. NICHOLSON, R.L.R. and FARMER, A.A. The introduction of fast breeder reactors for energy supply. In *Proc. of Uranium Institute 4th Annual Symposium, London, September 1979*. Uranium Institute, London (1979).

Section 7

1. ORGANIZATION FOR ECONOMIC CO-OPERATION AND DEVELOPMENT (NUCLEAR ENERGY AGENCY) IAEA. *Uranium resources production and demand. OECD, Paris (1979)*.

2. DAVID, S., ROBERTSON, D.S.‡ *The interpretation of uranium resource data*. Uranium Institute, London (1977).

3. MARSHALL, W. The Use of Plutonium.† *Atom* no. 282 p. 88-103 inc. (1980).

‡ Robertson and Associates Ltd., Toronto, Canada
† Lecture given to the Royal Institution, February 1980

10

Uranium and thorium raw materials

S.H.U. BOWIE

Uranium is widely distributed in the Earth's crust. The average crustal abundance is however only some 2–3 p.p.m, and only highly localized deposits with a concentration of 500 p.p.m. or more can be mined economically. This chapter reviews the main types of uranium deposit, the technologies for the recovery of uranium, and the prospects for its future production.

The main source of low-cost uranium reserves is found in sandstones, notably in the Colorado–Wyoming uranium province of the USA, though vein-type deposits have assumed an increasing importance with the development of the large Alligator Rivers deposits in Australia. Other main categories of deposit also reviewed include quartz-pebble conglomerates, quartz alkali–feldspar pegmatites, and nepheline syenites.

The technology for the discovery of uranium still mainly rests, as in the past, on detecting the gamma radiation associated with the ^{238}U decay series. The use of Geiger–Müller counters for this purpose has however been effectively superseded by the use of more sophisticated scintillation counters and gamma spectrometers. Additionally, increasingly sensitive geochemical methods have been developed, which show considerable future potential. Other promising future techniques include the use of thermal sensors, to measure radiogenic heat, and of enhanced Landsat imagery.

Identified uranium resources are quite substantial compared with present production rates, but it is noted that there may nonetheless be significant constraints on the rate at which production can be expanded to meet future demands unless there continue to be substantial new discoveries of low-cost resources, and much may depend on the development of new prospecting methods. Finally, a brief look is taken at thorium production, noting that, while identified reserves are limited, there is no doubt that they could be greatly increased in response to an increased demand.

Contents

1 Uranium and its occurrence in deposits

Uranium is widely distributed in the Earth's crust and is always associated with oxygen. It is more abundant in rocks composed of light-coloured minerals, such as granites, in which the average content is 4 to 5 p.p.m. Rocks composed of dark-coloured minerals, such as basalts, average 1 p.p.m. U as compared with the crustal average of 2 to 3 p.p.m.

Uranium occurs in two isotopic forms, ^{238}U (99.3 per cent) and ^{235}U (0.7 per cent), and is accompanied by the decay products of both parents. Those of ^{238}U are shown in Table 10.1.

In primary ores occurring at depth uranium and its daughters are usually in equilibrium, but in an oxidizing environment, as at the Earth's surface, disequilibrium is common. This results mainly from the relative solubility of uranium under such conditions compared with daughters such as ^{230}Th or ^{210}Pb. It is the radioactivity of the daughters rather than the parent isotopes that is used in the detection of uranium deposits. For example, gamma spectrometers normally measure the concetration of ^{214}Bi with its important gamma emission at 1.76 MeV. Knowledge of the state of equilibrium is therefore important when quantitative measurements of uranium are being made. Ideally, a chemical determination of uranium should always be made as a check.

Uranium is an essential constituent of about 100 minerals of varying degrees of complexity, but the bulk of uranium mined has been from only twelve species. Seven are primary and five secondary species formed by the

1.	Beaverlodge	13.	Poços de Caldas	25.	Shinkolobwe
2.	Clufflake	14.	Figueira	26.	Rössing
3.	Wollaston belt	15.	Salta	27.	Witwatersrand
4.	Elliot Lake/Agnew Lake	16.	Malargue	28.	Singhbhum
5.	Bancroft	17.	Ranstad	29.	Alligator Rivers
6.	Makkovik	18.	Massif Central	30.	Westmoreland
7.	Ilímaussaq	19.	Salamanca	31.	Mary Kathleen
8.	Spokane	20.	Urgeiriça	32.	Yeelirrie
9.	Wyoming basins	21.	Hoggar	33.	Frome/Yarramba
10.	Uravan	22.	Agadès	34.	Itataia
11.	Grants	23.	Bakouma		
12.	Texas	24.	Mounana		

Fig. 10.1 Geographical distribution of uranium deposits

Table 10.1 U-238 decay series

Nuclides in order of decay sequence	Principal mode of decay and energy in MeV			Half-life
	α	β	γ	
^{238}U	4.19		0.048	4.51×10^9 y
^{234}Th		0.19	0.030	24.1 d
^{234}Pa		0.14	0.043	6.75 h
^{234}U	4.77		0.053	2.47×10^5 y
^{230}Th	4.69		0.068	8.0×10^4 y
^{226}Ra	4.78		0.186	1602 y
^{222}Rn	5.49			3.82 d
^{218}Po	6.00			3.05 min
^{214}Pb		0.67	0.053	26.8 min
^{214}Bi		1.51	0.61, 1.76	19.7 min
^{214}Po	7.69			1.64×10^{-4} s
^{210}Pb		0.015	0.046	22 y
^{210}Bi		1.16		5.01 d
^{210}Po	5.31			138.4 d
^{206}Pb		Stable		—

solution and re-precipitation of primary minerals. The primary phases are: uraninite (UO_2) or the less-well-crystallized variant, pitchblende; coffinite $(USiO_4)_{1-x}(OH)_x$; brannerite $(U,Y,Ca,Fe,Th)_3 Ti_5O_{16}$; betafite $(U,Ca)(Nb,Ta,Ti)_3O_9 \cdot nH_2O$; davidite $(Fe,Ce,U)(Ti,Fe)_3 (O,OH)_7$; uranothorite $(Th,U)SiO_4$; and uranothorianite $(Th,U)O_2$. Secondary phases are carnotite $K_2(UO_2)_2 (VO_4)_2 \cdot nH_2O$; tyuyamunite $Ca(UO_2)_2 (VO_4)_2 \cdot 9H_2O$; torbernite $Cu(UO_2)_2(PO_4)_2 \cdot 12H_2O$; uranophane $Ca(UO_2)_2Si_2O_7 \cdot 6H_2O$; and autunite $Ca(UO_2)_2(PO_4) \cdot 12H_2O$.

Uranium is concentrated well above the normal levels in some granites (0.001–0.002 per cent U), nepheline syenites (0.025–0.03 per cent U), marine shales (0.002–0.03 per cent U), phosphate rocks (0.01–0.02 per cent U), and lignites (0.05–0.1 per cent U). However, such concentrations are inadequate, except in special circumstances, to permit the contained uranium to be recovered economically except as a by-product or co-product of some other element. There is no cut-off value that can be universally applied, but as a general rule uranium has to be concentrated to 0.05 per cent to be recovered as the sole product. If adequate tonnage is available, such concentrations make up the deposits of the world as opposed to the occurrences. The geographical distribution of uranium deposits is shown in Fig. 10.1.

2 Types of uranium deposit

It is convenient to classify uranium deposits into four main types: vein-type deposits, uranium in sandstones, uranium in conglomerates, and other uranium deposits. More than 90 per cent of low-cost (< $80/kgU) uranium occurs in deposits in Precambrian (> 600 million years (My)) rocks or in younger rocks immediately overlying the crystalline basement. Dating by U, Th, Pb methods has established that most deposits formed over a long period of time. The earliest formative period was around 1800 My to 1700 My ago, when there was also widespread granite formation and metamorphism. For example, ages of deposits in Northern Australia commence around 1800 My; in the Beaverlodge district of Northern Saskatchewan at 1780 My; in Sweden at 1740 My; and in the Franceville basin Gabon, at 1740 My. Successive ages at the same locality span more than 1000 My, but it is uncertain as yet whether there was repeat mineralization or remobilization of an earlier concentration. The ages of deposits tend to be coincident with orogenic events in several continents, and in Europe there are significant developments of uranium associated with the Hercynian and Alpine mountain-building events at around 300 My and 50 My, respectively.

2.1 Vein-type deposits

The first vein deposits to be worked for uranium were those of Bohemia, Saxony, and Cornwall. Tonnages were relatively small, but grades were at

0.2 per cent U_3O_8 or above. The 200 or so tonnes of uranium produced in Cornwall were used mainly as a source of the daughter product ^{226}Ra, while the uranium was utilized as a glass pigment and as a photographic image intensifier. By contrast, the Ranger vein deposits in the Alligator Rivers region of Northern Territory, Australia, contain well over 100 000 t U at a grade of between 0.21 and 0.38 per cent U_3O_8. The main uranium minerals of vein-type deposits are usually pitchblende and coffinite.

The major vein deposits of the world occur in the Shield areas of Canada and Australia. The main Canadian deposits are in northern Saskatchewan at Beaverlodge, Rabbit Lake, Cluff Lake, and Key Lake and range in age from 1780 My to 100 My. The three last mentioned are districts with deposits associated with the pre-Athabaska unconformity and are sometimes called unconformity-type vein deposits.

The deposits of Northern Territory, Australia, also tend to be associated with an inter-Precambrian unconformity which was a controlling factor in the localization of the ore deposits. The main localizing feature of all such deposits, however, are major geofractures in the host rock through which the uranium was introduced. Important deposits occur at Ranger, Jabiluka, Koongarra, and Nabarlek in the Alligator Rivers region and contain over 80 per cent of Australia's reserves (see §5) of 290 000 t U.

Uranium in vein deposits totals 21 per cent of low-cost reserves.

2.2 Uranium in sandstones

Deposits in sandstones were for many years regarded as the main source of the element and they still contain 41 per cent of low-cost reserves. However, the recent discovery of vein-type deposits in Canada and Australia has resulted in such deposits assuming a new significance.

The main deposits in sandstones are in the Colorado–Wyoming uranium province of the U.S.A. in sediments ranging from Triassic to Eocene in age. The main uranium minerals are pitchblende and coffinite, though several secondary species make a contribution to the uranium recovered. In the early years of mining most deposits were worked by open-cast methods (see Fig. 10.2), but, as concentrations less than 200 m deep are depleted, mining will move underground to depths of 700 m or more.

The deposits generally conform to the bedding of the host sandstone and are best developed where there is an abundance of carbonaceous material or sulphide minerals which act as reducing agents allowing uranium to precipitate from solution. They also tend to be associated with unconformities, troughs or ancient stream channels as well as interfaces between shale and sandstone horizons. Three types of deposit are recognized: (a) roll deposits formed at oxidation-reduction boundaries, (b) tabular deposits associated with permeability changes, (c) lensoid pockets in depressions.

The uranium content of deposits a decade or more ago averaged around 0.2 per cent U_3O_8 but has since declined to about 0.1 per cent U_3O_8.

Fig. 10.2 Lucky Mac mine, Gas Hills, Wyoming, USA

Other deposits in sandstone are those of Niger which occur in Carboniferous and younger sediments of the Agadès basin on the western margin of the Precambrian massif of Aïr. The main deposits at Arlit, Akouta, and Imouraren have many similarities with those of the Colorado–Wyoming province. Smaller concentrations in sandstones occur at Frome Lake, South Australia; Figueira, Brazil; Sierra Pintada, Argentina; Nagalia Basin, Australia; Ningyo-Toge and Kurayoshi, Japan; Val Rendena, Italy; and Zirovski-Vhr, Yugoslavia.

2.3 Uranium in conglomerates

Uranium occurs in Precambrian conglomerates in only two major provinces, namely, Witwatersrand, South Africa and Elliot Lake–Blind River, Canada. Other deposits are known in Cambrian conglomerates in the USSR and Algeria, but their uranium contents are not known with any certainty.

The uranium grade in the Elliot Lake–Blind River region is about 0.1 per cent U_3O_8, but in the Witwatersrand it averages about 0.02 per cent U_3O_8. This means that generally uranium is recoverable only as a by-product of gold.

Deposits in quartz-pebble conglomerates in the Witwatersrand (Fig. 10.3) occur in the Dominion Reef and at five main horizons in the overlying sedi-

Fig. 10.3 Quartz-pebble conglomerate, Witwatersrand, South Africa

ments of the Witwatersrand System. The main ore mineral in the Dominion Reef is a thorium-rich uraninite which has been dated at 3050 My, making it the oldest known uranium mineral to form a deposit. The main mineralization of the Witwatersrand, however, is of pitchblende dated at 2040 My and generally associated with carbonaceous material and various sulphides, particularly pyrite.

The Elliot Lake-Blind River deposits are concentrated in a quartz-pebble conglomerate horizon close to the crystalline basement and are very similar in physical form to the tabular deposits in the sandstones of the Colorado Plateau. However, their mineralogical composition is very different. The main uranium species are uraninite and brannerite, but at some localities monazite (Ce,La,Nd) (PO$_4$) and uranothorite also occur. The presence of these minerals changes the U:Th ratio from 4:1 to 1:4 and makes the ore more refractory—hence more costly—to work. Between 65 and 70 per cent of Canada's reserves of 215 000 t U is in quartz-pebble conglomerates.

Uranium in both provinces constitutes 21 per cent of low-cost reserves.

2.4 Other uranium deposits

The most important deposits in this category are the quartz, alkali-feldspar pegmatities of Rössing, South West Africa, which occur as dyke-like structures in Precambrian schists an other metamorphosed sediments. The main ore minerals are uraninite with subsidiary betafite which give the ore an average grade of 0.035 per cent U_3O_8. Because the deposits are near surface and can be worked open cast on a large scale, they are economically viable.

Rarely are concentrations of uranium formed by magmatic differentiation rich enough to be considered economic. However, at Kvanfield in the northern part of the nepheline syenite intrusion of Ilímaussaq, South Greenland, both uranium and thorium are enriched in this way. The main radioactive mineral is Steenstrupine—a complex silicate and phosphate of rare earths, sodium, niobium, tantalum, thorium, and uranium. The occurrence has reasonably assured resources, recently estimated to be 27 000 t U with 16 000 t U as 'estimated additional resources'. (See §5.) The future of this occurrence depends mainly on whether or not the ore can be upgraded economically. Somewhat similar occurrences are associated with nepheline syenite rocks at Poços de Caldas, Brazil.

A unique uranium deposit is that of Mary Kathleen, Queensland, which is of high-temperature replacement type, in which uraninite is associated with rare-earth silicates. The grade is about 0.12 per cent U_3O_8 and the reserves some 7000 t U.

The important secondary deposits of uranium, such as the carnotite concentrations of Colorado, are not normally considered separately from the associated primary deposits whence they were derived. An exception to this is the carnotite found dispersed through calcrete at Yeelirrie, Western Australia. This deposit, which was discovered in 1972, has an average grade of 0.15 per cent U_3O_8 and reserves of 40 700 t U.

Uranium normally occurs in sedimentary rocks in the range of 1–4 p.p.m. In some sediments, however, it is enriched to as much as 0.25 per cent U_3O_8. The best known large-tonnage example is the Upper Cambrian alum shales of Southern Sweden. In the Västergötland and Närke districts resources total a million tonnes of uranium at a grade of 0.03 per cent U_3O_8; but, because of mining constraints, only about 300 000 t U could be recovered.

Most phosphate rock deposits of the world contain appreciable amounts of uranium because of the chemical affinity between uranium and phosphate complexes. In the Phosphoria and Bone Valley formations of Idaho, Montana, Utah, and Wyoming uranium contents range from 0.001 to 0.0065 per cent U_3O_8 and in some horizons reaches 0.01 to 0.02 per cent. Similar phosphate deposits occur in a broad belt stretching from Morocco through Algeria, Tunisia, Egypt, Israel, and Syria at an average grade of around 0.015 per cent U_3O_8. Even richer phosphatic sediments occur in Cabinda, Angola, and Central African Republic, where uranium contents are between 0.05 and 0.25 per cent U_3O_8. At present there is no economic method of recovering uranium from such rocks except as a by-product of the manufacture of triple superphosphate.

In favourable environments, lignites are also enriched in uranium. Those of North and South Dakota, for example, average 0.2 per cent U_3O_8 over widths of 0.5 m. In Europe the Permo-Carboniferous coals of Bohemia contain 0.2 per cent U_3O_8 or more; and the Tertiary lignites of the Ebro Valley, Spain, contain large tonnages of uranium at grades between 0.05 and 0.25 per cent U_3O_8. Recovery of uranium is not economic except as a by-product by utilizing lignite as a fuel.

3 Prospecting methods and their application

3.1 Scintillation counters

Since 1946 most uranium deposits have been found by the recognition of geologically favourable regions and use of Geiger–Müller or scintillation counters (Figs. 10.4 and 10.5) These have either been carried in aircraft or motor vehicles, or by prospectors or geologists on foot. The scintillation counter had virtually replaced the G–M counter by 1960, and some five years ago a trend began towards sophisticated gamma spectrometers, many with spectral-stripping facilities, which enabled estimates of equivalent-uranium, thorium or potassium to be made at surface or from the air. Such instruments, with sodium-iodide detectors up to 50 litres in volume are used for large-area reconnaissance, flying typically at a ground clearance of 125 m, a speed of 200 km/h, and a line spacing of 5 km. Instruments with smaller detectors are used from fixed-wing aircraft or helicopters mainly in follow-up surveying.

Carborne techniques, using a gamma spectrometer with recorder, are valuable where there is a good network of roads. Uranium:thorium ratios can be established more accurately than from the air and samples can be taken to establish equilibrium conditions. Some instruments are equipped so as to allow samples to be crushed and analysed on the spot.

On-foot surveys are made less frequently than hitherto. Nevertheless, sensitive light-weight scintillation counters are invaluable in the hands of a geologist during follow-up investigations. Total-count instruments are available with either a ratemeter or digital read-out display and an alarm device that can be preset at a given level above the regional background. Portable gamma spectrometers are still rather heavy for field use, but with electronic developments the weight is being reduced, and at the same time reliability and mechanical robustness are being improved.

3.2 Radon measurement

[222]Ra was first used to detect buried ore bodies in the late 1950s, but a decade passed with only a few pioneers persisting in improving the technique. Today radon detection is universally applied. Four main types of detector are in use: (1) alpha-sensitive probe (Fig. 10.6); (2) pump monitor; (3) integrating alpha meter; and (4) cellulose-nitrate film. All four methods measure radon in ground air.

Measurement is facilitated by making a shallow hole, usually with an auger

Fig. 10.4 Portable γ-spectrometer with digital display and probe extension

Fig. 10.5 γ-ray detectors assessing uranium content of trucks

Fig. 10.6 Radon probe monitor inserted in shallow hole

or iron bar, to a depth of between 0.5 and 2 m. The alpha detector is then inserted into the hole and the radon concentrations measured directly by a zinc-sulphide phosphor. Alternatively, air is pumped from the hole and radon concentrations measured in either an ionization chamber or one internally coated with zinc sulphide. In the case of the integrating meter, counts are allowed to accumulate for a day or more.

The simplest method of measuring the alpha emission of radon is by a cellulose-nitrate film detector. In this technique a small piece of film is placed in an inverted cup near the top of the hole and left for two or three weeks. The film is then recovered and etched with a dilute caustic soda solution to show up the damage caused by the radon alpha-particle tracks. The tracks are counted under a microscope to obtain a measure of the average alpha flux. If radon measurement were quantitative—which it is not—this method would be the most suitable one for eliminating diurnal effects caused by variations in humidity, barometric pressure, and water content of the overburden material.

Radon can also be measured in stream or lake waters by a suitably designed pump monitor.

3.3 Geochemical methods

Geochemical methods of detecting uranium deposits have not been employed as widely as they might, mainly because of the tardiness in introducing analytical techniques of sufficient sensitivity and accuracy to measure small amounts of uranium. This changed gradually with the introduction of the delayed-neutron method which has a sensitivity down to $0.03\,\mu\mathrm{gU}$ and a high degree of precision. The uranium contents of water, stream-sediment, soil, and rock samples can all be measured by this method. The analysis of relatively few run-off water or stream-sediment samples is both rapid and effective at indicating whether a catchment area is likely to be part of a uranium province. Follow-up by detailed soil or rock samples can be used to locate mineralization.

Recently a laser fluorimeter has been developed with a similar sensitivity and the capability of being used in a field laboratory. With such improvements in analysis there is little doubt that geochemical search methods will gradually supplant the scintillation counter. Already, uranium deposits have been found by geochemistry that were not detected by detailed air-borne surveys.

As multi-element analytical methods develop (e.g. plasma-source spectrometer), elemental associations are beginning to be recognized that could be indicative of uranium mineralization too deep below surface to be detected by gamma radiation. Well-known associations are of uranium with vanadium, selenium, molybdenum, fluorine, titanium, and iron.

3.4 New methods

Preliminary work on isotope ratios in common rock-forming minerals such as Feldspars or sulphide minerals such as Galena have shown that in a uranium province they carry abnormal amounts of radiogenic $^{206}\mathrm{Pb}$ and $^{207}\mathrm{Pb}$. If this can be shown to be universally applicable, then it opens up a new approach in the recognition of a uranium province in which deposits can be expected to occur.

The measurement of the isotope $^4\mathrm{He}$ may also become an important technique in the discovery of uranium provinces. $^4\mathrm{He}$ is an end product of uranium decay, which, because it is inert and highly mobile, may give indications of uranium at depths more effectively than by measuring radon. Some total helium measurements have met with success, but the best prospects are from measurements of $^3\mathrm{He}{:}^4\mathrm{He}$ ratios. The combined use of helium isotope ratios and of $^{222}\mathrm{Rn}{:}^4\mathrm{He}$ ratios might well enable the depth of an ore body to be predicted.

Another method which would seem to have promise is the measurement of heat flow resulting from radiogenic heat in a uranium province. Measurements could either be made on the ground by thermal sensors or by high-resolution thermal infra-red scanners in an aircraft.

It is well known that radon escapes to atmosphere and decays to ^{214}Bi, which has to be corrected for in high-sensitivity air-borne spectrometric surveying. What has not yet been tested is whether ^{214}Bi or other of the longer-lived decay products, such as ^{210}Pb, ^{210}Bi, or ^{210}Po, can be collected and measured. Such information could provide a history of atmospheric particulate matter and could enable uranium provinces to be located with greater precision than hitherto.

Finally, a method which is non-specific, but which could have important applications in locating favourable structures for uranium mineralization, is enhanced Landsat imagery.

4 Uranium mining

Uranium mining methods differ little from those conventionally used in the winning of copper, lead, zinc, gold, or any other metal. Mining is either by the open-cast method as in many of the Wyoming deposits or in Niger, or underground as at Beaverlodge or in the Elliot Lake–Blind River district. Open-cast methods are normally applicable to flat-lying beds near surface. In such cases, overburden can be a problem, because if it is more than 100 m thick underground methods may have to be utilized. Open-cast operations do not usually exceed 200 to 300 m in depth. Underground mining is normally undertaken if the ore body is between 300 m and 3000 m in depth and can be by any of the recognized methods such as stoping, room-and-pillar, longwall retreat, or panel mining.

The main hazard in mining uranium as compared with any other metal is the presence of radon gas. This is usually not important in open-cast workings but can be very troublesome in underground mining. Radon, in itself, is not a serious health hazard, but it decays to various daughter products (see Table 10.1), which are solids. Of these the two alpha emitters ^{214}Po and ^{210}Po are the main sources of excessive radiation dosage to the lungs. The method of overcoming this risk is air filtration together with adequate ventilation, which not only reduces radon to below tolerance levels, but also minimizes the dust hazard and the effects of diesel exhaust fumes where such equipment is used. The internationally accepted tolerance level for radon in equilibrium with its daughters is 100 pCi/litre of air. Radon could collect in open-cast workings especially under atmospheric-inversion conditions, and where rich ore is being mined some form of turbulence might have to be induced.

A relatively new form of mining is known as solution mining. In this, lixiviant solutions are injected into the ore body through drill holes. The solutions percolate through the ore and are withdrawn through suitably placed production holes. In solution mining, a relatively impermeable layer at the base of the ore horizon is necessary if the leach liquor is not to be lost.

Even so, there are many difficulties due to jointing and fracturing, permitting the leach solutions to get into ground water. One of the main advantages of solution mining is that because less than one per cent of the ore body is leached, there are no problems of subsidence, and, if the volume of rock leached is flushed with water before the area is abandoned, there is less hazard as a result of mining than existed before operations commenced.

One of the most important aspects of uranium mining is the safe disposal of tailings after the uranium has been extracted. ^{226}Ra is only partly leached from the ore, where it remains relatively harmless in the solid tailings. However, some radium is present in the leach residue, and this waste liquor must be treated with barite or barium chloride to immobilize the radium. The solid wastes are normally neutralized with lime. Organic solvents used in the extraction of uranium must also be removed or absorbed on activated charcoal. Even when all precautions are taken to make a tailings dam safe, steps are subsequently taken to reduce the seepage of radionuclides into ground water. This involves periodic monitoring of the tailings area, and the eventual covering of the tailings with soil to eliminate gamma radiation and to reduce radon escape to a minimum. Stabilization of the whole area can then be effected by the use of suitable vegetation.

5 Reserves, resources, potential, and production

5.1 Reserves

The term 'reserves' relates to material that is known with a high degree of geological certainty and on the basis of measurements of the size and grade of the ore bodies (Table 10.2). The term also has an economic connotation, presently restricting reserves to uranium that can be produced at a cost of < \$80/kg U (\$30/lb U_3O_8). Uranium reserves are in the ground, except for a relatively small quantity in stockpiles, and may not be available when needed by the industry.

Approximately 90 per cent of reserves in the World outside Communist Areas (WOCA) are in seven countries, namely, Australia, Brazil, Canada, Namibia, Niger, South Africa, and the USA and this could restrict uranium availability. Where the ore grade is marginal, constraints on availability might result from the rate of production being affected by physical difficulties, production itself only being possible as a by-product or co-product of another element, the country's own requirements, or political factors. Availability of uranium is much more difficult to assess than reserves. The simple calcuation using requirements as the divisor and reserves as the dividend does not give a realistic indication of how long uranium will be available. To offset this, additions to reserves are being made on a continuing basis.

Table 10.2 Uranium reserves at production costs < $80/kg U(NEA/IAEA)[1]

Country	Reserves (kt U)
Australia	290
Brazil	74
Canada†	215
France	40
Gabon	37
India	30
Niger	160
South Africa (including Namibia)	364
USA	531
Others	109
Total	1850

† mineable at prices up to $CAN 125/kg U

5.2 Resources

These are divided into two categories: 'reasonably assured resources' and 'estimated additional resources'. The former equates with reserves, except in that the cost of production is from $80–130/kg U ($30–50/lb U_3O_8). The latter is estimated mainly on the basis of geological evidence. (Further information is given in Chapter 11.) Few, if any measurements have been made, nor has the extent of the ore bodies been defined. Included in estimated additional resources are predicated extensions to known deposits, partly explored occurrences, and deposits thought to exist in a uranium province from comparison with known deposits.

Similar reliance can be placed on reasonably assured resources as on reserves, but estimated additional resources have their main value in giving a guide as to how much uranium could be produced within the cost categories of < $80/kg U and $80–$130/kg U at present-day currency values.

Reasonably assured resources at a cost of $80–$130/kg U are given in Table 10.4 estimated additional resources at < $80/kg U in Table 10.3, and estimated additional resources at $80–$130/kg U in Table 10.4.

5.3 Potential uranium

Potential uranium is even more speculative than estimated additional resources. Figures are of little value therefore, except in that they indicate that all of the uranium in the accessible parts of the Earth's crust is not included in reserves and resources. The availability of potential uranium is extremely nebulous, as no account is taken of how or when potential ore will be discovered—if at all.

The term 'speculative resources' is synonymous with potential when the cost limit of $130 kg U is applied. A category of potential uranium not included in speculative resources is material that could be made available say at $130–400 kg U. This includes uranium, the whereabouts of which is

Table 10.3 Estimated additional resources at production costs
< $80/kg U(NEA/IAEA)[1]

Country	Reserves (kt U)
Australia	47
Brazil	90
Canada†	370
France	26
Niger	53
South Africa (including Namibia)	84
USA	773
Others	37
Total	1480

† mineable at prices up to $CAN 125/kg U

Table 10.4 Reasonably assured resources and estimated additional resources,
tonnes U × 10^3 (NEA/IAEA)[1]

Country	RAR $80–130/kg U	EAR $80–130 kg/U
Canada†	20	358
France	16	20
India	—	23
South Africa (including Namibia)	160	108
Sweden‡	300	3
USA	177	385
Others	67	73
Total	740	970

† mineable at prices within the range $CAN 125–175/kg U
‡ not likely to be available at more than 1000 t U/y

well known, in marine shale, phosphate rock, and lignite. Uranium in the ground totals more than 20 Mt U, but, because of the low grade, output from all sources is unlikely to exceed 25 000 t U/y.

5.4 Production and availability

A peak year of uranium production was 1959 when 34 265 t U were mined. From that time it gradually declined to 17 600 t U a decade later and is now running at just over the 1959 level.

Although reserves and estimated additional resources at less than $80/kg U total 3.3 Mt U, output from known uranium provinces is unlikely to exceed 100 000 t U/y by the end of the century. The reasons for this are the many constraints imposing limits on output. For example, the reserves and estimated additional resources at < $80/kg U in South Africa total 301 000 t U, yet production is unlikely to exceed 11 000 t U/y because uranium yield is essentially geared to the main product, gold. In Canada, over 50 per cent of reserves and estimated additional resources at < $80/kg U are in the Elliot

Table 10.5 Probable uranium availability, t U × 10³/year

	1979†	1985	1990	2000
Australia	0.6	10	15	20
Brazil	0.1	0.5	1	1
Canada	6.9	12	13	12
France	2.2	4	4	4
Gabon	1.0	1	1	1
Niger	3.3	5	10	10
South Africa (including Namibia)	8.9	11	14	15
USA	14.8	20	30	25
Others	0.6	5	6	14
Totals (rounded)	38.4	70	95	100

† planned production

Lake–Blind River conglomerates. Moreover, half of this tonnage is in one mine property and the nature of the mining imposes physical restrictions on annual production. In the major province of Colorado–Wyoming, USA, the average grade has declined from 0.2 per cent U_3O_8 in 1969 to 0.1 per cent in 1979. Reserves and reasonably assured resources available at a cost of less than \$130/kg U total 708 000 t U, but the average grade of this material is only 0.06 per cent U_3O_8. The lowering of grade, together with increasing mining costs mainly resulting from the necessity to mine deeper deposits, will create availability problems in the future.

An extreme example of a resource that is likely to produce only a relatively small tonnage of uranium before the end of the century is that in southern Sweden. Total uranium in the Ostergötland, Västerogötland, and Närke districts is some 1 Mt. However, because of mining constraints, recoverable uranium is only about 300 000 t U. More importantly, because of the low grade—0.025 to 0.032 per cent U_3O_8–huge tonnages of shale have to be moved. This, together with the necessity of removing a thick overburden, imposes real environmental problems to the extent that the annnual output is unlikely to exceed 1000 t U.

Other limitations on availability can arise from geographical location and from environmental difficulties related to tailings disposal. All important restrictions, other than political climate and some environmental that are difficult to appreciate, are taken into account in estimating probable uranium output to the end of the century (Table 10.5). This is compared in Fig. 10.7 with the less realistic maximum attainable figures of NEA/IAEA and with predicted demand.

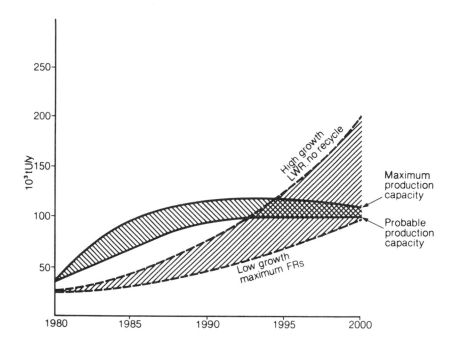

Fig. 10.7 Comparison of uranium supply and demand

6 Thorium: its occurrence, reserves and resources

Thorium is approximately four times more abundant than uranium in crustal rocks. However, it is an essential constituent of only a few minerals, of which two, namely, Thorite ($ThSiO_4$) and Thorianite (ThO_2) are important. In other ore or potential ore minerals thorium is not an essential constituent. These are: Monazite (Ce,La,Nd,Th) PO_4, Brannerite (U,Y,Ca,Fe,Th)$_3Ti_5O_{16}$ Uranothorite (Th,U)SiO_4, and Uranothorianite (Th,U)O_2.

Production of thorium in the past has been mainly from placer (detrital) deposits with vein deposits making a small contribution. The beach sand deposits of India, Australia, Brazil, Egypt, and Liberia have important reserves mainly in the form of monazite. Vein deposits of thorite and monazite occur in USA, Turkey, and South Africa. By-product thorium is available from the Elliot Lake-Blind River uranium deposits, where reserves are of the order of 100 000 t U.

Reserves and resources of thorium have not been assessed with the same precision as those of uranium. Figures are tentative and appreciably less than could be expected if there were active prospecting programmes for the element.

Reserves and estimated additional resources are given in Table 10.6.

Table 10.6 Thorium reserves and estimated additional resources, tonnes Th $\times$ 10^3 (NEA/IAEA)[1]

Country	Reserves <$75/kg Th	Estimated additional resources <$75/kg Th
Australia	21	—
Brazil	68	1200
Canada	—	295
Egypt	15	280
India	320	—
Norway	132	132
South Africa	10	—
USA	108	261
Totals (rounded)	675	2165

7. Conclusion

Uranium and thorium are both widely distributed in the Earth's crust in the parts per million range. Concentrations to economic grade, however, are relatively rare. Uranium is much more mobile than thorium, occurs as an essential constituent in many more minerals, and is concentrated in a greater variety of ways. Uranium deposits range in size from small pockets to massive developments of the primary minerals pitchblende and coffinite each with reserves of 200 000 t U. Most thorium deposits are of placer origin, with monazite the main ore mineral.

The most effective way of prospecting for radioactive minerals has been by G–M counters and more recently by scintillation counters or gamma spectrometers. The latter instruments have their most important application in regions where mixed uranium–thorium ores occur. Deposits that are buried too deep below overburden to give a gamma signal at surface can be detected by the measurement of ^{222}Rn in ground air.

In future, geochemical methods of exploration are likely to replace many of the classical physical methods, especially when used in conjunction with enhanced Landsat imagery to give structural information.

The main problems in mining uranium are associated with the decay products of radon and with dust. Both can be kept to below tolerance limits by increased ventilation, air filtration, and water sprays. Solution-mining offers hopes for the recovery of uranium from deposits not amenable to conventional mining methods, but prudence is essential in its application. The disposal of the waste products of uranium mining is not universally satisfactory. Periodic inspection of tailings is necessary to ensure that there is no significant uptake of radionuclides by plants or animals. Stabilization of the tailings area at mines is also of great importance.

Reserves and resources of uranium would seem to be adequate to meet demands to the end of the century, provided there are no undue political constraints on availability. However, if requirements grow to over 100 000 t U/y then substantial new discoveries will have to be made on a continuing basis. The historic rate of discoveries of reserves of 65 150 t U/y since 1945 will have to be more than trebled if predicted demand into the first quarter of the next century is to be met. The annual availability of uranium concentrate is much more important than reserves in the ground. Sound geological knowledge, supported by one or more of the various prospecting techniques, is required in the discovery of more near-surface deposits. The reserves required for the future, however, will depend much more on the development of new techniques of detecting hidden ore bodies. Research into methods of identifying uranium provinces in which ore bodies are likely to occur is also very important.

Thorium will continue to be produced mainly from placer deposits or as a by-product of uranium. Should the demand for thorium increase because of its use in thorium-based reactors, there is no doubt that reserves could be substantially increased.

Suggestions for further reading

1. INTERNATIONAL ATOMIC ENERGY AUTHORITY. *Proc. Panel on Uranium Exploration Geology*, IAEA, Vienna, 1970.

2. BOWIE, S.H.U., JONES, M., and OSTLE, D. (eds.) *Uranium prospecting handbook*. Institution of Mining and Metallurgy, London (1972).

3. INTERNATIONAL ATOMIC ENERGY AUTHORITY. *Proc. Symp. on Formation of Uranium Deposits, Athens, 1974*. IAEA, Vienna (1974).

4. *Proc. Panel on Nuclear Techniques in Geochemistry and Geophysics, Vienna, 1974*. IAEA, Vienna (1975).

5. JONES M.J. (ed.). *Geology, mining and extractive processing of uranium*. Institution of Mining and Metallurgy, London (1977).

6. IAEA, *Proc. Tech. Committee on Recognition and Evaluation of Uraniferous Areas*, IAEA, Vienna (1977).

7. KIMBERLEY, M.M. (ed.) *Uranium deposits, their mineralogy and origin*, University of Toronto Press, Toronto (1978).

8. THE ROYAL SOCIETY. *Theoretical and practical aspects of uranium geology*. The Royal Society, London (1979).

9. NUCLEAR ENERGY AGENCY/INTERNATIONAL ATOMIC ENERGY AGENCY. *Uranium resources, production and demand*. OECD, Paris (1979).

10. OECD NUCLEAR ENERGY AGENCY. *World uranium geology and resource potential*. Miller Freeman Publications Inc./OECD Nuclear Energy Agency (1980).

11

Uranium supply

A. W. HILLS

This chapter considers first the features which determine the special nature of the uranium market, discussing briefly how these have interacted in the past. It highlights the importance that Government initiatives can have in altering the natural interplay between demand and supply. The factors influencing demand and supply are scrutinized, and it is noted that a number of these will operate only over long periods because of the lead-times involved on both sides. At the same time the chapter identifies some short-term flexibilities, before drawing on recent major studies to discuss possible features of the evolution and pattern of uranium supplies in the future. A number of uncertainties make precise analysis impossible, but it is shown that in the longer term a crucial role could be played by uranium-conserving technologies, particularly by the fast reactor. The evolution of a political framework within which international trade in uranium can develop in a stable manner will also be of prime importance.

Contents

1   Introduction

Uranium was first identified as a separate element by Klaproth in 1789, and, apart from very small amounts of plutonium, is the heaviest of the known naturally occurring elements on Earth. It is widely distributed throughout the world, but most of the large deposits contain less than 0.1 per cent of uranium. It is never found in its elemental state but always in a chemical combination with other elements, with which it forms over 100 known minerals. Uranium can therefore be produced as a by-product of the extraction of other minerals with which it is associated. In the United States, for example, significant quantities are obtainable in copper-mining operations. Again uranium can be obtained as a by-product from wet process phosphoric acid, used in the production of phosphates. In South Africa, most uranium has been extracted as a residual by-product of gold-mining operations, where the two elements are found in the same geological formations. The greater part of uranium production has, however, stemmed from purpose-built mines, exploiting different types of deposit. After the existence of an ore-body of suitable size and grade has been established, the uranium is mined by open pit or underground methods similar to techniques used in coal-mining.

Before the Second World War, uranium was of relatively little commercial interest, and as late as the 1946 edition of the *Encyclopaedia Britannica*, the principal use quoted for uranium was as a colouring agent in ceramic and glass manufacture. But the years immediately following the war saw a major programme of research and development in many of the advanced nations of the world, who had perceived the potential use of uranium to provide energy. The last twenty years have seen the growing utilization of nuclear power as a component of national energy strategy. The nuclear reactors that have been constructed and continue to be built depend upon a supply of uranium to fuel them. The development of peaceful nuclear power has therefore been paralleled by the growth of a uranium production industry, and the creation of a market through which uranium can be bought and sold.

The attraction of natural uranium as a fuel for generating electricity lies in its very high specific energy content. This is about 3 million times that of coal or 1.7 million times that of fuel oil. This means that a small quantity of uranium can produce a very large quantity of energy. The technology exists for this vast energy potential to be tapped in such a way as to relieve for the foreseeable future the dangers of a major shortfall in world energy supplies. The extent to which countries call upon nuclear power as a major contributor to energy supply will depend upon the evolution of world economic patterns and energy demand, but the prospect of a major increase in nuclear power, and thus in the demand for uranium, is one of the characteristics of the uranium market. Those responsible for the discovery and exploitation of uranium resources have to carry out their work according to their best judgements of the rate of expansion of nuclear power and thus the rate of expansion of demand for their product. The history of the industry in the last 20 years has demonstrated the difficulty of making judgements on such matters.

2 The uranium market

There are a number of features of the uranium market which determine its character.[2] Among these are:
  (1) the limited range of applications for uranium,
  (2) the difficulty of substituting other materials for uranium in these applications,
  (3) the fact that these applications lie in the production of energy,
  (4) the nature of the techniques required to make uranium available,
  (5) the strategic value of the uranium as an energy raw material and its use in the production of nuclear weapons, and
  (6) the geographical distribution of uranium supply and demand.

Uranium has a markedly limited range of applications. Leaving aside its potential use for defence purposes, the uses for uranium other than as a raw material for the production of nuclear power are few and small. This means that plans for the production of uranium will be geared directly to the perceived prospects for nuclear power. Misjudgements will be especially painful, since over-production cannot be diverted into other markets. Equally, under-production cannot be readily made good by the use of other materials, because of the difficulty of substituting any other material in the short term. With the present designs of thermal reactor which form the basis of the world's installation programmes, a failure to procure uranium would be followed inexorably by the close-down of power stations. If this happened, the ability to replace nuclear electricity with electricity derived from other fuels would probably be limited, as changes in the power station mix in a system take time.

In due course this problem could be eased by the adoption of nuclear fuel cycles involving plutonium (or to a lesser extent, thorium), but such a transition would have to be carefully planned and would take many years. In practice there are a number of palliative measures which can be and are taken to reduce the risk of uranium shortages occurring, such as stockpile policies and the diversification of sources of supply, but these could not avert the consequences of a major prolonged failure to produce the required quantities of uranium.

Particular importance is attached nowadays to uranium because of its usefulness as a fuel. This has several consequences. Energy is likely to be one of the most critical areas for political decision and economic progress for the next few decades. This has always been true to some extent, but in the past large and cheaper sources of energy have become available as the need arose, and this availability has been an important precondition for economic development. Nuclear power using uranium offers the best hope at present perceived for continuing this process, and the importance of its raw material reflects this potential.

Uranium also has some special features as an energy-producing raw material. The concentration of potential energy in uranium relative to that in other raw materials has already been mentioned. In addition, unlike other fuels, only a relatively small part of the price of fuel to the consumer arises from the extraction and refining of the raw material. The greater part of the costs of the fuel cycle relate to the costs of special processes before or after its use. Naturally occurring uranium contains only a small proportion of the energy-producing fissile ^{235}U isotope, and in most reactor systems it is necessary to increase the percentage content of this isotope by a physical process known as enrichment, which is relatively expensive. After use, spent fuel can be reprocessed, not only to recover the unused uranium and the plutonium produced during the nuclear reaction, but also to convert the waste fission products into a more manageable form. Therefore, in addition to the basic cost of the uranium, the cost of these processes will influence the cost of the material as a nuclear fuel.

Moreover, the economics of nuclear power are such that the key element determining total generating cost is the capital cost of the power station. The cost of nuclear power is relatively insensitive to the cost of the fuel cycle, which typically contributes only some 35 per cent of the total cost of the electricity. At current prices, the uranium cost itself represents less than half of this. The fuel costs for a coal-fired station, on the other hand, may well account for 75 per cent of the cost of the electricity produced. The fluctuation in uranium prices will have much less effect on the eventual cost of the electricity than fluctuations in coal prices. Conversely, however, the capital commitment involved in a programme of nuclear power is such that the investment will not be made in the first place without reasonable assurance that uranium will be available, and at prices that will make the investment pay off.

Uranium is a mineral found in a variety of ore bodies, and its availability on the market will depend upon the continued development of suitable technologies to locate and extract it. It is a finite resource, and as mining takes place, the resource in the ground is irrevocably depleted. The extraction of uranium is subject to the normal lead-times applicable to all mining activities for prospecting and developing a resource, although some features of the uranium-mining business, such as the imposition of environmental requirements and the need to make proper provision to cope with the radioactive materials encountered during mining and present in the spoil, can lengthen these lead-times. It can now take 15 years from first exploring for uranium to putting a mine into operation.[3] Suitable techniques for extracting and processing the ore are required, which vary from mine to mine, and here again, special characteristics can apply to uranium mining because of the radioactive aspects of its production.

Uranium's initial use on a major scale was for defence purposes, and it remains a material of considerable strategic importance. While uranium in its natural form cannot be used as a weapons material, it is the ultimate raw material in weapons manufacture. In addition, any country which has opted for nuclear power as a major source of its energy supply will attach considerable importance to the assurance of a reliable and adequate provision of uranium. These features make uranium a mineral of great interest to governments, and can lead to policies that will influence the market, directly or indirectly, especially on the part of producing countries.

A further feature of the uranium market is the concentration of currently identified resources in a relatively small number of supplying countries. This geographical distribution is, of course, not uncommon for mineral deposits, but has an important influence on the character of the market. The relative youth of the uranium production industry contributes to the geographical concentration in a few areas of the currently identified major deposits. This point is of particular importance because any rapid market development must depend on activity in a few countries. This is discussed in greater detail in Chapter 10, and Fig. 10.1 shows the major deposits in non-Communist areas. There is, moreover, only limited correlation between the global distribution of uranium resources and the requirements for them. Some countries, such as Australia and Canada, have resources considerably in excess of their foreseeable needs for many years to come, while others, for example the United Kingdom and Japan, have totally inadequate indigenous resources to meet their present and forecast requirements.

The existence of all of these features differentiates the uranium market from that of every other commodity. Some of these features may apply to other commodities, but their combination does not. The interdependence of the producers and consumers has encouraged them to evolve a market capable of overcoming the problems these special features create, but not surprisingly this process has not been without perturbations, and continuing mutual efforts will be necessary in the future in order to smooth out the instabilities in the market.

3 History of the uranium market

At this point it is probably helpful to review the history of the development of the market. It is a relatively young market. Until the late 1950s the market for uranium was almost exclusively based on military needs. The two principal buyers of uranium in the western world were the governments of the United States and the United Kingdom, who co-ordinated their activities through a Combined Development Agency. The CDA sponsored geological surveys for uranium which covered the five continents and the Antarctic. For a period of some 20 years, until the termination of the CDA in 1963, the Agency bought most of the world's uranium from many different producers (but from a few countries) at many different prices. There then followed a process of negotiation to determine the allocation of the supplies between the two participants. As part of the process of stimulating exploration, guaranteed prices were offered for long periods during the 1950s.

By the early 1960s the military requirements had been met, and the civil requirements, though building-up, were still small. The 1960s saw therefore a recession in the uranium industry. In the United States there was an attempt to protect the indigenous uranium industry from the effects of this recession by barring the American utilities from importing uranium (1968–1977). While this offered an element of protection to the US uranium industry, it accentuated the effects of the recession for other producing countries. Exploration was at a low level, mining operations were shut down or survived only through subsidies or special arrangements. There were attempts in the early 1970s by the producers to bring about a more satisfactory price for their product, but the significance of these moves was overtaken in 1973/4 by the repercussions of the first major OPEC oil price increase. The announcement of a number of ambitious nuclear programmes at that time promised the rapid growth of demand for uranium. The lack of exploration in the preceding years and the long timescales required before new major discoveries could be tapped combined to suggest that some of this demand might be hard to meet. The result was a marked increase in price; an indication of this is given by the price of uranium on the 'spot market', which involves a small percentage of the total but gives a good guide to underlying trends in the whole. The price moved from about $21/kg U in early 1973 to over $105/kg U in 1976. ($U_3O_8$ is the *yellowcake* form in which uranium is usually sold and shipped.) These higher price levels in turn stimulated more exploration.

Since 1974, it has become apparent that the economic recession which had marked the 1970s, and was in itself partly triggered by the oil price increases, has retarded the growth of demand for energy, thereby reducing the impetus behind the nuclear programmes announced earlier. The progressive lowering of forward demand estimates, coinciding with some large new discoveries, has caused the market price for uranium to stabilize and indeed fall in real terms. Some of the implications of this will be touched on later in the paper in the discussion of the outlook for future supplies.

This history has illustrated the impact of many of the factors indicated earlier. In particular it can be seen that government actions have had a direct and marked effect on the market. It was the policies of the British and American governments which determined the initial development of the industry, and it was the decision of the United States Government in 1968 to impose an embargo on uranium imports that had the most damaging effects on the fortunes of the uranium industry elsewhere. The characteristics of uranium as a mineral, the lack of alternative uses, and the lack of suitable substitutes, all combine to make the fluctuations more extreme. Throughout, the most important characteristic of the market was the fact that it was demand-led: i.e. prices were determined by the changing perceptions of demand rather than by a fluctuating level of supply. Unless the consuming countries perceived a requirement for uranium, there was no way that the producers could stimulate a need. Their only option would have been to impose a negative influence on the customer's plans, by denying him the material when he required it. It has yet to be demonstrated conclusively that in practice even this could be achieved, given the degree of competition that has existed between the various potential suppliers.

4 Factors affecting demand

It is worth examining briefly the main factors that can determine the demand for uranium.[4] Of these the most important is the long-term prospect for the growth of nuclear capacity. This is complicated to assess. The energy requirements of a country will be geared to its economic growth, and a supply of energy at the right price will be a pre-condition for continued growth. The way in which the energy requirement can be met will vary from country to country depending upon the purpose for which the energy is required, upon the range of possible energy sources available to each country, and upon the costs of those sources. The share of energy demand that may be met by electricity has to be predicted, and then that contribution has to be analysed to see how it will be generated. With all these factors impacting on the requirements for nuclear power, it is hardly surprising that forecasts of future nuclear capacity have varied, as economic and political circumstances change and relative costs interact.

Once an estimate of the total nuclear capacity has been made, there can be significant variations in the demand for uranium depending upon which reactor type, or mix of types, is adopted. Figure 11.1 shows the requirements of various reactor types for 1 GW(e)y of operation, and it can be seen that the way the choice of reactors is made around the world can lead to variations in the total amount of uranium required. Research and development work on nuclear power continues in many countries, and the figure includes two types of reactor which are still under development: the 'improved' pressurized-water reactor and the fast reactor. The widespread adoption of either of these, particularly the fast reactor, would have a pronounced effect on uranium demand.

Another technical and economic factor is the choice made of tails assay in enrichment. Enrichment is a process required for most reactor systems before the uranium is suitable for use as fuel, and involves the treatment of the uranium to increase the proportion of its fissile isotope uranium-235. In its natural state uranium only contains 0.7 per cent uranium-235, with virtually all the remainder consisting of the non-fissile isotope uranium-238. When the uranium is enriched, a portion of the uranium is given a higher proportion of ^{235}U, while the rest, known as depleted uranium, has the proportion of ^{235}U reduced. In producing uranium of a particular enrichment, more or less of the ^{235}U can be extracted from the depleted uranium before it is released from the process, the proportion of ^{235}U left in the depleted uranium being known as the *tails assay*.

The choice of the level of tails assay is influenced by the trade-off between,

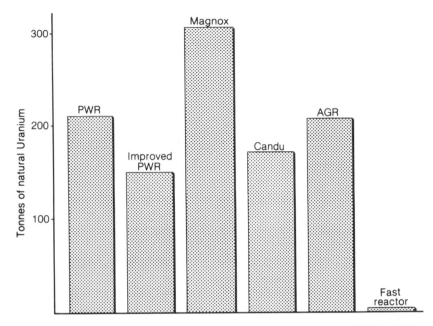

Fig. 11.1 Uranium requirements for various reactors for 1 GWy, 100% load factor, 0.2% tails)

on the one hand, the cost of operating the enrichment plant and, on the other, the cost of finding the uranium to use as the feed. The higher the tails assay (i.e. the less of a given quantity of uranium is subject to the enrichment process), the higher will be the total quantity of uranium feed required to produce a given quantity of enriched uranium. This is illustrated in Table 11.1.

The tails assay can be, and has been, varied within relatively short periods and may also be reduced if the enrichment capacity outstrips the requirement at the original assay, because of the desirability of operating these highly capital-intensive plant at full output. At times the requirements for uranium, as determined by the contracted supply for enrichment purposes, may differ

Table 11.1[2] Enrichment plant variables (showing possible variations for the production of 3.25% enriched uranium)

Tails assay (%)	Natural U requirement[†]
0.3	110.3
0.25	100.0
0.2	91.7
0.15	84.9
0[‡]	70.3

† Requirement at 0.25% tails = 100
‡ for illustrative purposes, but not at present technically feasible

from the requirements that would be derived simply from projected reactor programmes, as seems likely, for example, to be the case in the early 1980s.

An important influence on long-term uranium demand will be the extent to which recycle is adopted. Figure 11.2 shows the importance of decisions on whether to recycle the uranium and plutonium contained in spent fuel after its first use in a reactor. Recycle entails the treatment of fuel, after its use in a reactor, in a reprocessing plant in order to extract the unused uranium and the plutonium created in the reactor, which can be incorporated in new fuel. The operational characteristics and scale of recycling activities after reprocessing is thus one important feature of the nuclear fuel cycle for determining uranium requirements.

There are a number of other features that can affect the short-term demand for uranium. One of these is the policy adopted by consumers towards maintaining a stockpile of uranium to provide additional assurance of supply. With such stockpiles typically amounting to two or more times the consumer's annual requirement, the amount of uranium involved is appreciable. Stockpiles can act to moderate the effects of excess demand if a temporary pressure on the market interrupts supplies or pushes the price up in the short-term, and, equally, an attempt to build up stockpiles may moderate falls in demand which might otherwise depress prices more seriously.

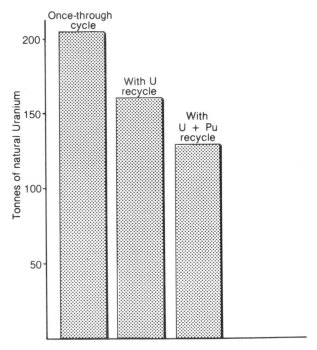

Fig. 11.2 Effects of alternative recycle regimes (for current LWR at 100% load factor and 0.2% tails)

Although important, stockpiling is more of an additional factor to be borne in mind than one that has a decisive effect on market conditions.

Finally demand will be influenced at any given time, though not in the long-term, by its balance in the market-place with supply as reflected by price. In practice most requirements for uranium are covered by long-term contracts, but a proportion is still the subject of short-term arrangements based on the spot market. More importantly, a prudent customer may be influenced in the timing of his attempt to secure long-term supplies by his view of the prospect of obtaining a better deal within the time-scale imposed by prudence. If he considers that an imbalance leading to a rise in price may be about to occur, he may seek to anticipate the market by the early purchase of the uranium he requires. This interplay is a normal feature of any commodity market.

5 Factors influencing supply

Just as many factors help to determine the future demand for uranium, there are a similar number of factors that will influence the supply of uranium to meet this demand. Ultimately the most important will be the discovery at the right time of sufficient amounts of uranium which can be exploited at prices the market will bear. This process must involve a number of steps. First the likely existence of sufficient uranium resources has to be estimated, inferring possible areas of potential from general geological knowledge. The most promising areas then have to be explored, and only then will it be known whether a suitable uranium deposit exists. Ultimately, there comes a time when resources available at lowest cost become scarce, thereby putting an upward pressure on price and making attractive those fuel cycles which minimize the use of uranium, but have higher initial costs than those currently in use.

When a uranium reserve has been delineated by test drillings there then needs to be further action to produce uranium for commercial use. Suitable techniques have to be chosen for mining and processing (*milling*) the ore to extract the uranium, and there has to be adequate skilled expertise and investment capital available. The need to observe environmental restrictions and the nature of those restrictions will also help determine whether a deposit is exploitable. For example the substantial uranium resources in the Ranstad shales in Sweden may not be exploited because of the impact this would have on the environment. Many uranium deposits have been located in inhospitable and remote areas, such as those in Canada near the Arctic Circle and those in the Sahara desert, and these require the provision of new transport facilities and heavy overhead costs to provide vital requirements, such as water and food, for acceptable working conditions. All of these factors means that a decision to develop a deposit will not be made unless there is a reasonable expectation that the uranium produced can be sold at a price sufficient to cover costs and give a reasonable profit.

Even when a mine and the associated milling capacity have been created, there remains some ability to use short-term flexibilities in the production process to vary the supply. Lower-grade ores in the deposit can be extracted than were originally planned, thereby increasing the supply, or conversely more lower-grade ore can be left in the ground or discarded. Such adjustments in the short-term will of course have repercussions later. The mill can be operated for varying periods and aim at higher recovery, although increases in production secured in this way will also involve increases in costs. In this connection, it is worth noting that typical recovery rates for milling have been falling from past levels of around 95 per cent to the current average of 92 per cent, and have been predicted to fall further. Nevertheless, the United States Department of Energy believes that increased milling losses will in practice have no significant effect on resource utilization and depletion profiles because of the lack of accuracy of the resource estimates.[5] Where uranium is extracted as a by-product from another activity, as with the South African Witwatersrand gold operations, there is some scope for raising production by reprocessing previously mined and treated wastes providing prices are sufficiently attractive. Moreover, there is scope for greater extraction of uranium as a by-product from processes now largely untapped, such as phosphoric-acid production and copper production, although the scope for this has been variously estimated and will again depend on economic factors.[6]

6   Adjusting the balance of supply and demand

There is wide scope for variations in supply and demand, and these assume great importance because of the lead-times involved, which have lengthened in recent years, heightening the importance of uncertainties as their impact increases with time. The lead-times for a reactor from planning to full power can be eight to ten years, and the timescale for a mine, from confirmatory exploration to delivery of the product in a suitable state for use in a reactor, is of the same order. A need for initial exploration could add a further five years. There would therefore be considerable uncertainties in evaluating the adequacy of future uranium supply even without considering uranium's special characteristics. The existence of these, however, means that the market will also reflect government policies. Suppliers will also want to ensure that adequate provision is being made for their own domestic needs, that the supply of uranium does not entail unacceptable environmental penalties, and that local enterprise is given a reasonable stake in the market. For example, the governments of Australia and Canada have held public inquiries to ensure these aspects have been adequately covered before they have allowed particular projects to proceed, and they have also issued a number of policy guidelines. Governments of consumer countries will wish to check that supplies of uranium will be forthcoming at the times they are needed and at a fair price. They may also want to encourage their own mining industries to become involved in uranium extraction, and more recently this concern has extended to the participation by electricity utilities in joint ventures to explore for and develop uranium deposits. If a nation has other significant nuclear fuel business interests, such as in enrichment or reprocessing, the government might also want to consider ways of influencing the market to the benefit of those activities.

But the most important interest of governments in recent years in the uranium market has stemmed from concern over the non-proliferation of nuclear weapons. The supplying nations in particular have wished to guarantee that the uranium they supplied would not be used in any way to contribute to the spread of nuclear weapons. This has led to the imposition of conditions on the terms on which uranium could be exported and, if agreement was not secured to those conditions, the consequence has been an interruption of supply. This can clearly upset the normal operation of the market. The legitimacy of the concern to safeguard all the operations of nuclear power has long been recognized, and was a dominant motive in the creation of the International Atomic Energy Agency (IAEA). The desire to limit the spread of weapons technology also led to the Non-Proliferation Treaty in 1968. Recent years, however, have brought pressure both to strengthen the IAEA and to supplement it with specific international understandings between supplying and consuming countries. Most recently the whole inter-relationship between proliferation and nuclear power has formed a central part of the International Nuclear Fuel Cycle Evaluation (INFCE), which has included an examination of uranium demand and supply.[7]

7 Future patterns

Having identified a number of factors bearing on the uranium market, it is time to examine the prospects for uranium supply in the future. This will be done by summarizing an illustrative view of uranium requirements and supplies, and then indicating some of the unknowns and differing views which indicate alternative ways for the supply pattern to develop in the future.

The most widely used estimates of uranium resources are those published jointly every two years by the Organization for Economic Co-operation and Development (Nuclear Energy Agency), abbreviated OECD(NEA), and the International Atomic Energy Agency (IAEA). Typical (1979) figures, given in Tables 11.2 and 11.3,[3] show about 2.59 million tonnes in the 'reasonably assured resources' category (of which 1.8 million tonnes are 'reserves') and an additional 2.45 million tonnes as 'estimated additional resources', each recoverable at costs less than $130 kg U ($50/lb U_3O_8). These cost bases include not only the direct costs of mining and processing, but also the cost of capital spent in providing and maintaining the production unit. Past and current explorations costs are not included, nor is any allowance made for commercial return to the mining company.

Table 11.2[3] Reasonably assured resources per 1000 tonnes U from data available at 1 January 1979

Cost range	Reserves at c. $80/kg U	$80–130/kg U	Total at c. $130/kg U
Algeria	28	0	28
Argentine	23	5.1	28.1
Australia	290	9	299
Austria[2]	1.8	0	1.8
Bolivia	*	0	0
Botswana	0	0.4	0.4
Brazil	74.2	0	74.2
Canada[1]	215	20	235
Central African Republic	18	0	18
Chile	0	0	0
Denmark	0	27	27
Egypt	0	0	0
Finland	0	2.7	2.7
France	39.6	15.7	55.3
Gabon[2]	37	0	37
Germany, Federal Republic of	4	0.5	4.5
India	29.8	0	29.8
Italy	0	1.2	1.2
Japan	7.7	0	7.7
Korea, Republic of[4]	0	4.4	4.4
Madagascar[2]	0	0	0
Mexico[3]	6	0	6
Namibia	117	16	133
Niger[2]	160	0	160
Philippines[2]	0.3	0	0.3
Portugal	6.7	1.5	8.2
Somalia[3]	0	6.6	6.6
South Africa	247	144	391
Spain	9.8	0	9.8
Sweden[5]	0	301	301
Turkey	2.4	1.5	3.9
United Kingdom	0	0	0
United States of America	531	177	708
Yugoslavia	4.5	2	6.5
Zaire[2]	1.8	0	1.8
Total (rounded)	1850	740	2590

* Less than 100 tonnes U

Notes
1. The material reported as reserves is mineable at prices up to $CAN 125/kg U and other reasonably assured resources are mineable at prices between $CAN 125 and $CAN 175/kg U
2. Source of data: *Uranium resources, production and demand*. OECD/IAEA, Paris (1977)
3. Data refer to resources *in-situ*, rather than recoverable
4. Reported as 13 000 000 tonnes of ore with an average grade of 0.04% U_3O_8.
5. No uranium production allowed in a deposit of 300 000 tonnes U because of a veto by the local authorities for environmental reasons

Table 11.3[3] Estimated additional resources per 1000 tonnes U from data available at 1 January 1979

Cost range	$80/kg U	$80–130/kg U	Total at $130/kg U
Algeria	0	5.5	5.5
Argentine	3.8	5.3	9.1
Australia	47	6	53
Austria[2]	0	0	0
Bolivia[2]	0	0.5	0.5
Botswana	0	0	0
Brazil	90.1	0	90.1
Canada[1]	370	358	728
Central African Republic	0	0	0
Chile	5.1	0	5.1
Dénmark	0	16	16
Egypt	0	5	5
Finland	0	0.5	0.5
France	26.2	20	46.2
Gabon[2]	0	0	0
Germany, Federal Republic of	7	0.5	7.5
India	0.9	22.8	23.7
Italy	0	2	2
Japan	0	0	0
Korea, Republic of	0	0	0
Madagascar[2]	0	2	2
Mexico[3]	2.4	0	2.4
Namibia	30	23	53
Niger[2]	53	0	53
Philippines[2]	0	0	0
Portugal	2.5	0	2.5
Somalia[3]	0	3.4	3.4
South Africa	54	85	139
Spain	8.5	0	8.5
Sweden	0	3	3
Turkey	0	0	0
United Kingdom	0	7.4	7.4
United States of America	773	385	1158
Yugoslavia	5	15.5	20.5
Zaire[2]	1.7	0	1.7
Total (rounded)	1480	970	2450

See table 11.2 for *notes*.

Reasonably Assured Resources[3] refers to uranium that

occurs in known mineral deposits of such size, grade and configuration, that it could be recovered within the given cost ranges with currently proven mining and processing technology. Estimates of tonnage and grade are based on specific sample data and measurement of the deposits and on knowledge of deposit characteristics. Reasonably Assured Resources have a high assurance of existence and in the cost category below $80/kg U ($30/lb U_3O_8) are considered as reserves.

Estimated Additional Resources refers to uranium

in addition to Reasonably Assured Resources that is expected to occur, mostly on the basis of direct geological evidence, in

extensions of well-explored deposits;
little explored deposits;
undiscovered deposits believed to exist along well-defined geological trend
with known deposits.

Such deposits can be identified, delineated and the uranium subsequently recovered, all within the given cost ranges. Estimates of tonnage and grade are based
primarily on knowledge of the deposit characteristics as determined in its best-known
parts or in similar deposits. Less reliance can be placed on the estimates in this
category than for Reasonably Assured Resources.

It is important to note the predominance in these figures of a few countries
as identified sources of uranium. Seven countries (Australia, Canada,
France, Namibia, Niger, South Africa and the United States) are estimated
to possess over three quarters of the reasonably assured resources and more
than ninety per cent of the estimated additional resources. France and
probably the United States will require their resources primarily for internal
use, leaving four major areas of uranium resource at present believed with
confidence to exist with major quantities available for export. This is illustrated in Fig. 11.3; the large resource shown for Sweden consists predominantly of the Ranstad shales, which, as noted already, are unlikely to
be exploited in the foreseeable future.

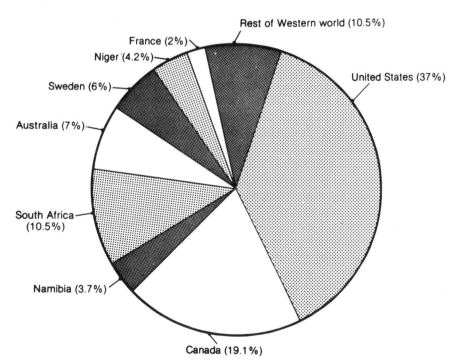

Fig. 11.3 Regional distribution of Western World uranium resources (RAR − EAR
below $130 kg U)

There may well in time prove to be additional uranium-exporting areas. A major international project is under way to attempt to establish more clearly the potential resources of uranium. The report of Phase I of this exercise, known as the International Uranium Resources Evaluation Project (IUREP), has been published.[8] In it are presented the results of a paper review of geological and geographical data, which resulted from an examination of 185 countries. The next stage of the project is an intermediate phase of investigation into the resources of the most promising countries in order to prepare the data required for any more detailed field investigation. Such field investigation, involving ground and aerial surveys, would form Phase II of the project, and it would be some time before this could be completed.

In the meantime, Phase I also involved an attempt to estimate the possible range of resources exploitable at costs of up to \$130/kg U that might exist. These were termed 'speculative resources', and in putting them forward the group responsible for the report stressed a number of caveats on the figures, as follows:

In presenting this general impression of Speculative Resources that were thought to exist in the world, a number of factors must be emphasized in addition to the highly subjective nature of the exercise. The totals are not meant to indicate ultimate resources of uranium, since our perspective is restricted by current knowledge, which in the case of many areas of the world is severely limited. For this reason our judgements of Speculative Resources may change as cumulative geological knowledge increases. It is important also to emphasize that even if these Speculative Resources exist, there is no guarantee the resources will be discovered, or if discovered, that they can be made available. In other words, these estimates imply nothing about *discoverability* or *availability*. Serious constraints may well arise on many fronts, such as the restriction of access to the favourable areas for the purpose of exploration, inadequacies in uranium exploration techniques, the rate of discovery once exploration has commenced, conflicts in land use where discoveries are made, difficulties of capital formation and allocation, and the availability of trained manpower. Also, the rate of discovery will be controlled primarily by the incentive to explore. The incentive to explore for an export market will be strongly influenced by the forecasts of market prices for uranium and also by the efforts of some countries to establish uranium reserves within their own borders. Given these constraints and in view of the long lead-times for exploration, mine development, and production it is likely that a major part of these Speculative Resources will not be discovered and brought into production during the first quarter of the 21st century.

In view of the qualifications noted above it is essential to emphasize that the tonnages of Speculative Resources should not, under any circumstances, be used for nuclear power programme planning purposes. Rather, they should be viewed as a qualitative measure of the present state of geological knowledge with all the inherent uncertainties and looked upon as a guide for establishing priorities for future evaluation efforts.

With these important reservations, the speculative resources identified by continent are shown in Table 11.4.

Table 11.4[8] Speculative resources listed by continent

Continent	Number of countries	Speculative resources (million tonnes U)
Africa	51	1.3– 4.0
America, North	3	2.1– 3.6
America, South and Central	41	0.7– 1.9
Asia and Far East†	41	0.2– 1.0
Australia and Oceania	18	2.0– 3.0
Western Europe	22	0.3– 1.3
Total	176	6.6–14.8
Eastern Europe, USSR Peoples Republic of China	9	3.3–7.3‡

† Excluding Peoples Republic of China and the eastern part of USSR
‡ The potential shown here, is 'estimated total potential' and includes an element for reasonably assured resources and estimated additional resources, although those data were not available to the steering group.

Notes

1. Speculative resources refers to uranium in addition to estimated additional resources, that is thought to exist, mostly on the basis of indirect indications and geological extrapolations, in deposits discoverable with existing exploration techniques. The location of deposits envisaged in this category could generally be specified only as being somewhere within a given region or geological trend. As the term implies, the existence and size of such resources are highly speculative.
2. The distinction drawn between reasonably assured resources, estimated additional resources, and speculative resources based on differing degrees of geological evidence, make it essential that each category be regarded as a discrete entity. Therefore, great care should be taken in the use of resources estimates (e.g. taking the sum of the estimates of each of the categories to obtain 'total resources'). The resource level of estimated additional resources should be considered as having a potential for later conversion to reasonably assured resources as the possible result of further exploration effort.

In line with the recommendations of the Report, these estimates will not be involved in the illustrative comparisons of demand and supply that follows, but their existence will be of relevance to the conclusions.

The OECD/IAEA biennial reports derive estimates of forward requirements for uranium from projections of nuclear reactor capacity. In their 1979 report[3] they were able to draw on data assembled for the International Fuel Cycle Evaluation Exercise (INFCE), and the resulting capacity projections are summarized in Table 11.5 for the World Outside Centrally-Planned Areas (WOCA). They are derived from questionnaires circulated during INFCE which asked individual nations to provide low and high projections for nuclear capacity up to the year 2000. When figures beyond that date were required, these were derived by applying growth-rates to the questionnaire response from the year 2000. The effect of this extrapolation is shown in Figure 11.4. Any projections must involve uncertainty, which must be of a high degree when they extend half a century into the future. These projections should be regarded as illustrative; all long-term nuclear growth projections are subject to substantial changes, and the actual amount

Table 11.5³ INFCE estimates of nuclear power growth in the world† (installed capacities of plants in operation at end of year indicated in GW(e))

Region or country	1980		1985		1990		1995		2000	
	L	H	L	H	L	H	L	H	L	H
Australia	—	—	—	—	—	—	—	—	—	—
Austria	0.7	0.7	0.7	0.7	0.7	0.7	0.7	0.7	0.7	0.7
Belgium	2.6	3.5	5.5	5.5	6.8	9.4	8.1	12.0	9.4	14.6
Canada	6.1	6.1	12.0	12.0	20.0	22.0	33.0	39.0	52.0	67.0
Denmark	—	—	—	—	0	2.6	0	5.2	0	6.5
Finland	2.2	2.2	2.2	2.2	2.2	3.2	3.2	5.2	4.2	7.2
France	17.5	17.5	39.0	39.0	59.0	67.0	73.0	84.0	86.0	106.0
Germany, FR	11.2	13.6	20.2	23.8	35.9	40.7	49.1	55.8	53.8	72.8
Greece	—	—	—	—	1.2	1.2	2.2	2.2	3.2	3.2
Iceland	—	—	—	—	—	—	—	—	—	—
Ireland	—	—	—	—	—	—	—	—	—	—
Italy	1.4	1.4	5.4	7.4	25.9	33.4	32.0	52.0	43.0	81.0
Japan	7.0	17.0	26.0	33.0	45.0	60.0	70.0	100.0	100.0	150.0
Luxembourg	—	—	—	—	—	—	—	—	—	—
Netherlands	0.5	0.5	0.5	0.5	0.5	2.5	0.5	3.5	0.5	4.5
New Zealand	—	—	—	—	—	—	—	—	0	0.5
Norway	—	—	—	—	—	—	—	—	—	—
Portugal‡	—	—	—	—	1.8	1.8	3.6	3.6	5.4	5.4
Spain	4.9	4.9	9.8	11.8	15.8	20.5	28.0	33.0	38.0	47.0
Sweden	3.75	7.4	3.75	9.4	0	10.4	0	10.4	0	10.4
Switzerland	1.9	1.9	2.8	2.8	2.8	2.8	2.8	2.8	2.8	6.7
Turkey	—	—	0.6	0.6	0.6	0.6	0.6	0.6	0.6	0.6
United Kingdom	7.4	7.4	9.4	9.4	12.3	12.3	19.7	19.7	27.6	40.2
United States	62.3	66.3	100.3	122.3	157.3	192.3	200.3	275.3	255.3	395.3
OECD Total	139.4	150.4	238.1	280.4	387.8	483.4	526.8	705.0	682.5	1019.6
Rest of world total†	5.0	9.0	19.0	23.0	45.0	50.0	90.0	100.0	150.0	186.0
World total†	144.4	159.4	257.1	303.4	432.8	533.4	616.6	805.0	832.5	1205.6
INFCE adjusted world total†	144.0	159.0	243.6	272.0	374.0	460.0	551.6	771.6	833.6	1206.6

† data not included for Eastern Europe, USSR nor China
‡ data received since the INFCE projections were made, indicate for Portugal: 0 GW(e) before 1995, 1.8 GW(e) in 1995, and 4.7 GW(e) in 2000

Note
The individual country data given in this table are official estimates made during the first half of 1978 in replies to the INFCE questionnaire; they have not been revised (upward or downward) in light of more recent information. The omission of a figure indicates that there are no official plans for nuclear power plants. The 'rest of world' total is based on the replies from 14 countries.

of installed nuclear capacity in future years could be outside the range indicated in the projections.

The projections are then used to derive estimates for the requirements for uranium, the values obtained varying depending on the demand of reactor types that are assumed to be installed. Figure 11.5 compares the cumulative uranium requirement for two scenarios, both based on the low and high projections of capacity together with the identified uranium resources, and

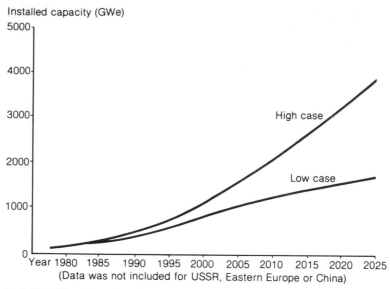

Fig. 11.4 INFCE adjusted nuclear power growth in the world

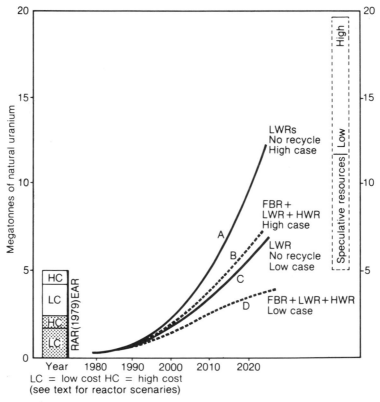

Fig. 11.5 WOCA cumulative requirements (1975–2025) with no forward commitments

also showing, for information, the possible impact of the speculative resources, which have not as yet been proved to exist. The two scenarios, based on rather extreme assumptions to demonstrate the possible impact on uranium resources, were derived from INFCE projections. The reactor scenarios were defined in the following ways.

1. *The light water reactor (LWR) scenarios.* All reactors built in the period 2001–2025 are current-technology LWRs using the once-through mode of operation; construction is constrained only by the limit of the nuclear power estimate.

2. *The LWR, heavy water reactor (HWR), and fast reactor (FR) scenarios.* Liquid-metal-cooled fast reactor (LMFR) capacity increases at a constant rate to 150 MW(e) in 2005 for the high nuclear growth projection, with the same rate being applied to the low nuclear growth forecast. Beginning in 2006, the FR market share is limited only by plutonium availability and the growth in nuclear power demand. The pre-2000 mix of LWRs (current technology) and HWRs (natural uranium) is extrapolated, with each type maintaining its proportion of the non-FR additions; LWRs and HWRs use the once-through mode of operation. FRs are oxide fuelled, using slow recycle (out-of-pile time for hold-up of plutonium is two years).

More relevant scenarios are likely to fall in between the curves.

Figure 11.5 demonstrates the importance of the choice of reactor system, and the reduction in requirement that could be made with the use of the FR. It also shows the pressure that is likely to arise on the currently identified resources, and this is particularly significant when it is noted that some of the estimated additional resources have yet to be discovered.

When comparing the resources against the requirement, it is also important to bear in mind that the uranium production from the resource has to be adequate on a year-by-year basis to meet the purchasing needs of the nuclear power plants. The OECD/IAEA also consider this aspect, and have warned that a major and sustained effort will be necessary to boost exploration and develop reserves. While they have estimated a growth in production capacity which should be sufficient until around 1990, further increases in maximum production capacity thereafter are likely to become increasingly difficult as ore-bodies become less easy and more costly to work because the higher-grade ores will become depleted and deposits mined out. This underlines the need for timely and large-scale exploration for new reserves to sustain any further expanded production.

The figures show the significance of the adoption of fuel recycle from 1985. The possible role of alternative fuel cycles, and in particular, the significance of the fast reactor for long-term uranium requirements, is considered in more detail at the end of this chapter.

Before turning to these long term issues, it is worth examining in more detail the nearer term, as demonstrating the types of issue that could affect uranium supply. The OECD/IAEA conclusion that, assuming no perturbations in the market, uranium supply should be adequate until 1990 to meet the demand, was broadly endorsed by a study by the Uranium Institute.[4]

Although events after the production of that study, and after the preparation of data for INFCE, are likely to bring about revisions in forecasts, particularly in the medium term, the general points made by these studies remain valid.

The Uranium Institute is an organization whose members include uranium producers, representatives of the electrical generating industry, specialists in uranium conversion and nuclear fuel reprocessing, and others concerned with the use of uranium in nuclear power. It is thus able to represent the viewpoints of many interests actively involved in the uranium market. In its study the Institute derived its own projection for uranium requirements by two methods. The first was comparable to that used by OECD/IAEA, working from forecast capabilities, although the actual figures were different. The second was to work from enrichment feed requirements, which suggested this would bring a higher demand than the figure set by the first method up to 1985, the relationship between the two between 1985 and 1990 depending on the operating regime adopted for the enrichment facilities and particularly the level of tails assay. The relationship between the estimates produced by the two methods, and the significance of the choice of tails assay, is shown cumulatively until 1990 in Fig. 11.6.

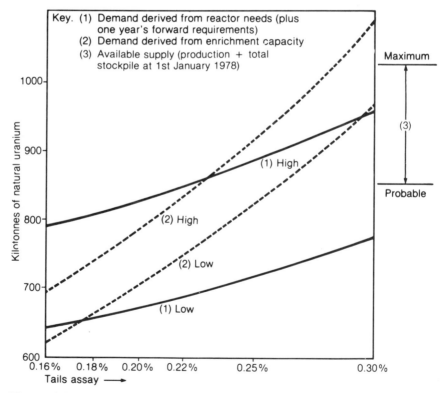

Fig. 11.6 Uranium supply and demand balance (1990): cumulative effect of tails assay variation (no recycling), assuming constant tails essay of 0.2% up to 1981

Figure 11.6 reflects the broad conclusions of the study that supply should be adequate to meet demand, and that, for much of the period, flexibility available in terms of tails-assay adjustment is greater than the range of uncertainty in the level of installed nuclear capacity. But the Institute emphasizes that this balance will depend on satisfactory market conditions. Producers will need to have the confidence to proceed with their development plans, and suppliers will wish to provide against short-term imbalances by prudent stockpiling policies. A large number of uncertainties could threaten this balance.

The most difficult area of uncertainties concerns the involvement of governments in the uranium trade. The relative concentration of uranium resources has already been noted. As already mentioned, the countries requiring uranium do not correlate to those in a position to supply it and Fig. 11.7 shows some regional balances between supply and demand.[9] It shows clearly that supplies from North America and, particularly, Australia are more than adequate to meet those countries' requirements, but that supplies for Western Europe and Japan will have very largely to be found from outside those areas. This will depend on international trade, in which governments can become involved to distort normal commercial market arrangements.

The intervention of governments in the past has had a marked effect, both

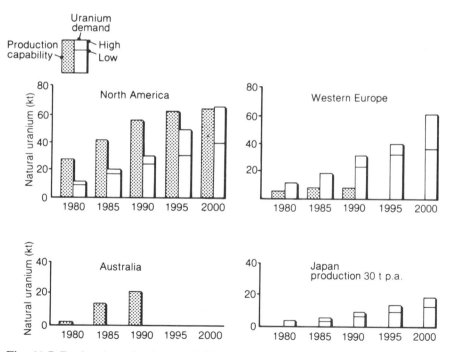

Fig. 11.7 Regional production capability and uranium demand in a once-through fuel cycle

on the actual operations and on the general level of confidence among consumers concerning their assurance of supply. Unilateral embargoes have interrupted supplies from Canada and the United States, and Australia has sought to impose a stringent set of conditions in the export of its uranium. These interventions have in the main been made as part of broader non-proliferation policies, although the link between uranium supply and weapons proliferation is indirect. The United States has taken a particularly strong line in seeking to convince other nations of the force of its non-proliferation policy, and took a leading part in initiating INFCE. It is hoped that, following the conclusion of the INFCE study, the international community will evolve a set of agreed principles to act as a framework for the future evolution of nuclear power, including the uranium market. These should provide for stability by centring on open and established rules governing export of, or trade in, uranium, made after discussion with all the parties concerned. These rules will need to be compatible with the existence of long-term contracts between suppliers and consumers if the necessary mutual confidence is to exist. It would also be helpful if embargoes were seen as the last resort, and only to be adopted after full consideration and exploration of alternative means of achieving the intervening government's objectives.

The achievement of an international consensus on these types of issues would be a major step in ensuring the stable development of a balance between uranium demand and supply, but there are, of course, many other uncertainties. With production concentrated in such a few countries, the loss of one supplier, because of other political or economic reasons, would gravely prejudice the stability of the market. Figure 11.8 shows for illustrative purposes a comparison for 1990 of two ranges of demand for uranium with the estimated production broken down by source.[10] It can be seen that the loss of one of these sources could have a major disruptive effect.

It is because of these primarily political considerations that the next decade is likely to bring an intensification of a trend already apparent, for purchasers of uranium to seek to diversify their sources of supply in order to minimize their vulnerability to sudden changes of policy by a single producer. One element of this could well be a significant entry on to the world market by electricity utilities in the United States. In the past these utilities have looked to indigenous production to meet their requirements, but foreign suppliers are appearing increasingly attractive and the US already imports a significant proportion of its needs.

In this context, it is useful to note the conclusions of a major study of the prospects for uranium supply to the United States from other countries, carried out for the US Electric Power Research Institute.[11] This concluded that there were considerable potential advantages for the utilities in diversifying their supplies to include foreign sources, while noting the element of extra risk because of the potential involvement of two (or more) governments. It indicated the prospect of possible production capacity available to

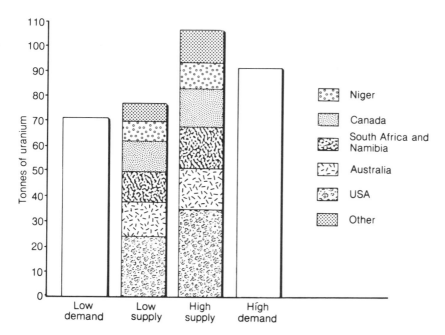

Fig. 11.8 A comparison of two ranges of demand for uranium, with the estimated range of production broken down by source, for 1990

meet US needs, particularly but not exclusively in Australia, although this capacity was not immediately available and would depend on the progress made in developing production capacities over the next few years. Their incentive to do this would be much increased if the United States were to become a major net importer of uranium.

The outcome of political consideration and the relative demands on the world market of the United States and of the uranium-import-dependent nations are two important areas of uncertainty which could affect the balance of supply and demand over the next decade.[12] Of importance also will be all the special features of the uranium market already described, which will determine the scope and timescale for adjustments. One important unknown is the impact of environmental considerations both on the numbers of nuclear reactors constructed and put into operation and on the mining industry's ability to work economically identified deposits.

If an imbalance does develop in the period up to 1990, it could be in the form of an excess of supply because of continuing cutbacks in programmes of nuclear power. However, a consequence of this could be the diversion of mining expertise, particularly that concerned with exploration, away from uranium into other minerals. A decline in exploration might result in the retardation of the expansion of production capability that would become necessary should demand subsequently pick up and nuclear programmes as a whole return to higher levels.

These considerations are of growing importance as the perspective shifts beyond 1990. An expanding contribution from nuclear power to world energy supplies is likely to make increasing demands on uranium production, and it eventually becomes essential to take some steps to modify this demand in order to allow nuclear power to continue to develop. This can be done through the use of more efficient utilization of uranium by improved thermal reactors and fuel cycles, and above all by the use of fast reactors.

Figure 11.5 showed the effect of introducing alternative reactors and fuel cycles on the demand for uranium. This becomes more marked if the lifetime requirements of the reactors in operation are calculated, as shown in Fig. 11.9 using the same reactor scenarios. It can be seen that in time the fast reactor, if adopted on a suitable scale, would reduce markedly the requirements for uranium. Indeed in due course, a progressive introduction of fast reactors until they make up the whole system could reduce the requirement for new supplies of uranium from that time onwards to negligible levels. Existing stocks of depleted uranium would fuel the world reactors for centuries. The point at which it will be seen as necessary to introduce fast

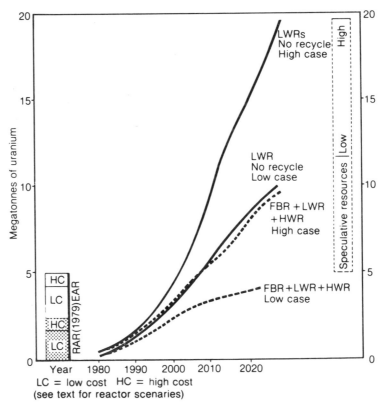

Fig. 11.9 WOCA lifetime uranium requirements of LWRs in operation at date indicated (1979–2025)

reactors will vary from country to country, although sufficient impact on world demand will need to have been made before scarcity of suitable uranium acts as an overriding constraint on the installation of nuclear plants. The case for their introduction will involve consideration of the relative economics of thermal and fast reactors, especially the development of the total demand for uranium and the available supply, as reflected in the price. But it will also involve considerations of the advantages of maximizing the energy extracted from the uranium and, for uranium-importing countries, the benefits of reducing dependence on imported energy. The development of the fast reactor, and of thermal reactors with improved uranium utilization, is not yet complete, and it will be some years before the timing and scale of their contribution can be predicted with confidence. Progress with their development will, however, be crucial to the long-term development of the uranium market.

Another important factor will be the role of the 'Centrally-Planned-Economy' countries which may become either net importers or exporters of uranium. There is virtually no information on their long-term supply or demand prospects.

The extent to which the speculative resources actually exist and can be brought into production will be crucial. A failure to find and develop at least a significant part of these resources would lead inexorably to rising uranium prices and pressure on supplies early in the next century or sooner, unless fast reactor development and installation had proceeded far enough to modify significantly the total uranium requirements. Even if a considerable part of the speculative resources is made available, this would mean the postponement of pressure on uranium supplies rather than its removal. This is indicated in Fig. 11.9, which shows the inroads into these speculative resources made by continuing development of nuclear power if fast reactors are not available.

8 Conclusions

This chapter has suggested that uranium is a special commodity, with many features that can cause uncertainties in forecasting supply and demand patterns. The uranium industry faces particular problems in setting its investment right because of the lead-times involved in mine development and the absence of alternative markets and alternative uses. The great political importance attached to the activities for which uranium forms the essential raw material serves to add an extra dimension of uncertainty and to create a major risk of short-term market discontinuities. These factors have in the past caused the uranium market to be subject to wide fluctuations, although ultimately adequate supplies to meet requirements have been forthcoming.

For the next decade a reasonable balance between supply and demand should be achievable, but it will be influenced by the ability of the world to evolve a framework of political understandings which will provide a stable and dependable basis for the evolution of the market. The numerous uncertainties could cause imbalances, and a production surplus for some of the period may occur unless nuclear power programmes recover a steady rate of growth. Even with restricted programmes, however, there will be a significant increase in requirements, posing a challenge to mining expertise and resources.

In the longer term mining facilities will be extended still further as nuclear power programmes continue to expand. Massive exploration and development programmes will be needed, as the more accessible and higher-grade deposits are used up, and a considerable expansion of the resource base beyond that currently identified will be needed. In time the availability of nuclear power as a source of energy available at a suitable price to foster economic growth will depend on the successful development of alternative fuel cycles and reactors to those now in use and, most critically, on the timely installation of adequate numbers of fast reactors.

References

1. Bowie, S.H.U. Chapter 10 *Nuclear power technology*. Oxford University Press (1983).
2. Langlois, P. The uranium market and its characteristics. In *Uranium Institute. Proc. Int. Symp. on Uranium Supply and Demand, London*, pp. 18–44. Mining Journal Books Ltd./Uranium Institute, London (1978).
3. Organization for Economic Co-operation and Development/International Atomic Energy Agency. *Uranium resources, production, and demand*. OECD, Paris (1979).
4. Uranium Institute. *The balance of supply and demand 1978–1990*. Mining Journal Books Ltd., London (1979).
5. US General Accounting Office/US Department of Energy. Correspondence in *Nuclear Fuel*, April 30, 1979, pp. 2, 3.
6. James, H.E. and Simonsen, H.A. *Ore-processing technology and the uranium supply outlook*. Mining Journal Books Ltd. London (1978).
7. (a) International Nuclear Fuel Cycle Evaluation. Reports of INFCE working groups 1 to 8. International Atomic Energy Agency, Vienna (1980).
    (b) Uranium Institute. Government influence on international trade in Uranium. Uranium Institute, London (1978).
    (c) Uranium Institute. The international fuel cycle evaluation report: initial comments by the Uranium Institute. Uranium Institute, London (1980).
8. Nuclear Energy Agency/International Atomic Energy Agency. *World uranium potential, an international evaluation*. Nuclear Energy Agency, Paris (1978).
9. International Nuclear Fuel Cycle Evaluation. *Report of Working Group 4: Reprocessing, plutonium handling, recycle*. IAEA. Vienna (1980).
11. McLeod, N.B. and Steyn, J.J. *Foreign Uranium Supply—EPRI EA-725*. Electric Power Research Institute, Palo Alto (1978).
12. Warnecke, S. *Uranium, non-proliferation and energy security* Atlantic Paper 37, Atlantic Institute for International Affairs, Paris (1979).

12

Uranium enrichment

J. H. TAIT

This chapter discusses the development of uranium enrichment processes. In the introduction there is a brief history of uranium enrichment, followed by a summary of the criteria used for the assessment of an isotope separation process, e.g. the separation factor, separative power, and the power consumption of a separating element.

This is followed by a discussion of the two main processes used, i.e. gaseous diffusion and centrifugation. The reason for the change from diffusion to centrifugation in the UK, mainly on power costs, is discussed. The development potential of centrifuges is also assessed.

Other processes which have been developed up to pilot stage are described, e.g. the Becker jet nozzle and the South African process. This is followed by a description of some plasma-based methods.

The next topic is concerned with chemical exchange methods and an attempt is made to assess their potential in the enrichment scene from published information.

This chapter concludes with a discussion of the advanced laser isotope-separation methods. The two approaches, i.e. the atomic and the molecular routes are discussed again using published information. This information is insufficient to give a complete assessment of the methods, especially the molecular route, but is enough to give indications of their potential.

Contents

Introduction

Uranium enrichment involves the partial separation of natural uranium into its two isotopes ^{235}U and ^{238}U, yielding the *enriched* fraction containing more than the naturally occurring 0.711 per cent of ^{235}U and a *waste* fraction of tails containing less than the natural concentration of ^{235}U. This enriched uranium is introduced into nuclear power stations to reduce capital costs by using different combinations of coolants, moderators, and structural materials.

The early history of uranium enrichment is, of course, associated with the production of nearly pure ^{235}U for the atomic bomb. Many methods were considered early in the war. These included centrifuges, gaseous diffusion, chemical, and photochemical methods, all of which are described in detail later in the chapter in addition to a number of other techniques, such as electromagnetic separators and distillation. These methods for general isotope separation had all been identified by Aston and Lindemann, in 1919.[1] Of all of these processes the most feasible technically seemed to be electromagnetic separation, gaseous diffusion, and gas centrifugation. The electromagnetic method was deployed during the early period of fissile material development for the US weapons programme, but since that time has not been used extensively for uranium enrichment, its use currently being confined to the supply of other stable isotopes. The method is illustrated in Fig. 12.1, and relies on the use of a magnetic field, ions in the field following different trajectories depending on their charge to mass ratio. The main difficulty of this technique for commercial production of enriched uranium lies in the small throughputs that can be achieved with it.

The main route for uranium enrichment after the war was through gaseous diffusion, and although there was early interest, going back to the beginning of the 1940s in the use of centrifuges, the technical and engineering problems encountered during the early years led to the temporary curtailment of this route during the immediate post-war period. In the UK likewise, the main route of commercial development, following an initial period of interest in centrifuges, was towards gaseous diffusion techniques. Development work on diffusion in the UK began in the early 1940s, though it was halted for a period in favour of the diffusion plant in the US. UK work restarted in 1947 and a plant for the production of highly enriched uranium was built at Capenhurst between 1950 and 1956. The rundown from military production took place in 1960 and an improvement for civil use was completed in 1967.

An aerial view of the Capenhurst plant is shown in Fig. 12.2. Similar programmes seem to have been carried out in the USSR. A diffusion plant was set up at Pierrelatte in France for military production, and this has been followed by a large commercial plant at Tricastin. A diffusion plant has been built also in China.

The use of centrifuges for commercial production came into its own only in relatively recent years with the development of new materials and bearing

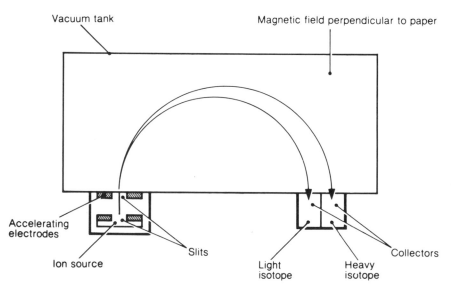

Fig. 12.1 The electromagnetic process

Fig. 12.2 Capenhurst plant showing centrifuge plant extension

systems. In Holland, centrifuge development was started in the mid-1960s. Work was carried out in Germany during and after the war. In the UK, work was restarted in about 1960 and intensified in 1968, when diffusion plant development was abandoned in favour of centrifuges. In 1970, the Treaty of Almelo was signed covering both technical exchange and joint marketing of centrifuges between the UK, the Netherlands, and Germany. There was also a major programme of centrifuge development in the US, and in April 1977, President Carter announced that the next US uranium enrichment plant would be a gas centrifuge plant. Centrifuge development is also being carried out in other countries, particularly Australia, Japan, and France.

The above is a brief outline of uranium enrichment development. It is of interest to examine these processes in more detail and attempt some comparison with those under current development, such as laser processes.

In order to make this comparison, the criteria used for judging an isotope enrichment process must be discussed, such as the separation factor and value function.

1.1 Cascade

In most separation processes, there is a small physical difference between the isotopes to be separated. The amount of separation which can be achieved in one stage is small, and to obtain the necessary separation the stages must be linked in a cascade. For example, to produce uranium enriched to 4 per cent ^{235}U and a waste depleted to 0.25 per cent would require about 1200 stages in a gaseous diffusion plant.

1.2 Separation factor

In all separation methods there is an elementary effect characterized by a separation factor α, which indicates a difference in the mole fraction N of the desired isotope (and, naturally, that of $1 - N$ of the other isotope) in different parts of the system. Instead of dealing with the mole fraction, it is convenient to use the abundance ratio,

$$\frac{N}{1-N}. \tag{1}$$

The separation factor α is defined as the ratio of this quantity after a stream has passed through a separating element to the initial value, i.e.

$$\alpha = \left(\frac{N_P}{1-N_P}\right)\bigg/\left(\frac{N_F}{1-N_F}\right) \tag{2}$$

where N_P and N_F are the concentrations in the product and feed streams. For example, in a process such as gas diffusion α depends upon the ratio of the flow velocities of the two species.

The factor, α, is independent of composition but may depend upon the throughput L (*feed*) and the *cut*, θ, the latter being the fraction of the feed into a separation stage which emerges as product as shown in Fig. 12.3. Similarly, a factor β can be defined which determines the concentration of the waste stream.

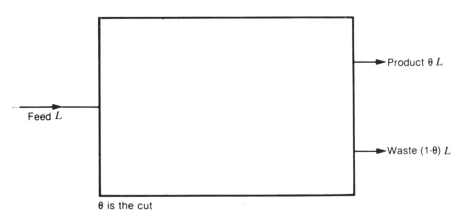

Fig. 12.3 Schematic of separating element showing the division of the flow into the product and waste streams

$$\beta = \left(\frac{N_F}{1-N_F}\right) \Big/ \left(\frac{N_W}{1-N_W}\right) \tag{3}$$

where N_W is the mole fraction in the waste stream. The single separation factor which determines the concentration change across each stage in then $\alpha\beta$. Sometimes the notation β_1, β_2 is used in place of α and β.

1.3 Basic cascade theory

1.3.1 *Separative power*
The concept of *separation factor*, while providing information about the efficiency of a separating element, contains no information about the amount of material that it is capable of enriching. In order to include this latter factor, which is clearly essential in characterizing a separative device, we need the further concept of *separative power*.

Separation processes may be considered as removing the disorder or entropy caused by the mixing of isotopes.

It seems, then, that the rate of change of the mixing entropy of the isotopes could be used as a measure of the performance of a separating device. However, this would depend on the composition of the mixture and would vary according to its stage position in a cascade. For example, if there were no light isotope there would be no change in entropy.

P.A.M. Dirac introduced a quantity related to the mixing entropy which can be used to measure the performance of a separating element. In the usual methods of separation of ^{235}U and ^{238}U, the work done by a device is approximately independent of the nature of the two molecules. For a mixture of two isotopes with mole fractions N and $1 - N$ of the light and heavy fractions respectively, the entropy change is proportional to $N(1 - N)$. To understand why this is so, one can imagine the molecules in the mixture being conceptually paired off, and in this case it is clear that only a fraction $2N(1 - N)$ will consist of unlike pairs, which are capable of being separated. A quantity

$$\frac{\Delta S}{N(1-N)} \tag{4}$$

where ΔS is the entropy change per second, is a measure of the performance of the machine. For a machine handling L mol/s, the quantity

$$\frac{L\,\Delta S}{RN(1-N)} \tag{5}$$

is defined as the separative power of the machine, R being the gas constant.

The entropy per mole of a mixture is greater than the constituents by an amount

$$S(N) = R\{N \ln N + (1-N)\ln(1-N)\}. \tag{6}$$

It can be shown that for a separation device which processes L gram-mol/s at fraction N and produces a fraction θL of product and a fraction $(1 - \theta)L$ of waste, the separative power of the device is given by

$$\delta U = \frac{\theta}{1-\theta} L \frac{\epsilon^2}{2} \text{ mol/s}, \tag{7}$$

where ϵ is equal to $\alpha - 1$. It can be seen that δU independent of N. The derivation is given in Appendix A.

1.3.2 *Value function*

The separative power of a device has been defined in the previous section. Also of interest is the total amount of separation that an enrichment plant can achieve. This can be done by introducing the quantity separation potential or value function. The material in a plant at concentration N is then said to have a value $V(N)$. To calculate the form of V consider the separative work produced by an elementary separating device where a feed of L mol/s and concentration N produces a product of θL mol/s at fraction $N + \Delta N_1$ and a waste of $(1 - \theta)L$ mol/s at a fraction $N - \Delta N_2$. Then V satisfies the equation

$$\delta U = \theta L V(N + \Delta N_1) + (1 - \theta)L V(N - \Delta N_2) - L V(N).$$

Again, by simple algebraic manipulation it can be shown that if ΔN_1 and ΔN_2 are small this is given by

$$\delta U = \frac{L}{2} \frac{\theta}{1-\theta} \, \epsilon^2 N^2 (1-N)^2 V''(N), \tag{8}$$

where V'' is the second derivative with respect to N.

Then this agrees with (7) if

$$N^2(1-N)^2 V''(N) = 1 \tag{9}$$

On integration it is found that

$$V(N) = (2N-1) \ln \left(\frac{N}{1-N} \right) + bN + a. \tag{10}$$

One way of choosing the constants is to assume that $V(N_0) = 0$, when N_0 is the concentration of the natural mixture. Moreover, V should be positive for material of increased content of either light or heavy isotope and this requires $V'(N_0) = 0$.

$V(N)$ is then given by the expression

$$V(N) = (2N-1) \ln \left(\frac{N}{1-N} \cdot \frac{1-N_0}{N_0} \right) + \frac{(1-2N_0)}{N_0(1-N_0)} (N-N_0). \tag{11}$$

Alternatively it may be assumed that $V(0.5) = V'(0.5) = 0$, and in this case $V(N)$ reduces to the simple formula

$$V(N) = (2N-1) \ln \left(\frac{N}{1-N} \right). \tag{12}$$

If a plant produces a mass P of product at a concentration N_P and a mass W of waste at concentration N_W from a mass feed F of concentration N_F, then the separative work is defined as

$$PV(N_P) + WV(N_W) - FV(N_F). \tag{13}$$

The unit used for measuring the above is termed *separative work unit* (SWU). It has dimensions of weight as V is dimensionless, and kg SWU is a usual unit, or 1000 kg SWU = tonne SWU. It is *not* the weight of enriched material drawn from a plant. As an example, a plant of 1 tonne SWU/y using a feed of 0.7 per cent isotopic enrichment and producing waste of 0.25 per cent enrichment will produce 0.27 tonnes/y of product of 3 per cent enrichment. The separative capacity or power of a plant is measured in kg SWU/y. The energy consumption is measured in terms of kW h/kg SWU.

1.3.3 *Ideal Cascades*

The theory of cascades has been developed by several authors, London,[2] Cohen,[3] Benedict and Pigford,[4], Pratt,[5] and Villani,[6] and recently the results have been conveniently summarized in an article by B. Brigoli.[7] In isotope separation plants a countercurrent scheme cascade is generally adopted, in which the tail fraction of each stage is subjected to further fractionation in lower stages; this results in a much higher yield. A countercurrent cascade is termed symmetric when the product stream of a stage goes to feed the next upper stage, while its tail stream is recycled at the inlet of the next lower stage. If the product stream of stage s goes to feedstage $s + n$ and the tail stream is recycled into stage $s - m$, the countercurrent cascade is non symmetric. The latter cascade is sometimes referred to as a *jumped cascade*. When the separation factor does not depend on the cut of the separating element, the most convenient scheme of connecting the stages corresponds to the symmetric cascade. However, certain aerodynamic separation processes, e.g. the separation nozzle, or the South African stationary wall centrifuge show a significant increase in the separation factor for small values of the cut. Then the optimum connection of stages is not the symmetric one. For example, in the South African process, the product from stage s goes to stage $s + 1$ and the waste goes to stage $s - 4$.[8]

A large-scale isotope separation plant consists of a large number of separating units arranged in a cascade. The problem is to choose the flows and cuts in such a way that one has the most efficient cascade. This is achieved when the separating units are arranged in the form of an ideal (or no-mixing-of-materials-of-different-mole-fractions) cascade. The ideal cascade is as shown in Fig. 12.4. Cohen's book gives the derivation of various properties of this ideal cascade, two of which are included here.

1. If the separation factor is independent of composition and close to unity then the total number of stages in the cascade is

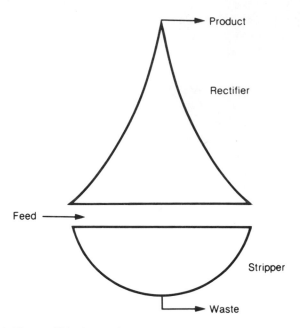

Fig. 12.4 Shape of ideal cascade

$$\frac{2}{\epsilon} \ln \left[\frac{N_P/(1-N_P)}{N_W/(1-N_W)} \right], \tag{14}$$

remembering that each stage is being run for maximum separative power and not maximum enrichment.

2. The total inter stage flow is given by

$$\frac{2}{\epsilon^2} P \cdot V(N_P, N_0). \tag{15}$$

1.3.4 *Material hold-up and equilibrium time*
In most cascades the total amount of material in process in a section of a cascade, the *hold-up*, is proportional to the flow in the stage. If the hold-up per unit flow is denoted by h then the hold-up of the desired isotope is

$$\int_{N_F}^{N_P} h L_s N_s \, ds, \tag{16}$$

where L_s is the interstage flow and h is the average process time per stage. It is also possible to define the net desired material hold-up, which is the amount of desired isotope in a plant over and above that which would be contained in a like quantity of material at normal abundance.

The equilibrium time can be defined as the number of days' production that is lost between starting-up time and steady-state production. This is equal to the material hold-up divided by a suitable average transport of desired isotope, which for the rectifier could be taken as $P(N_P - N_F)$.

1.3.5 *Squared-off cascades*

The important feature to note is that all of these quantities are inversely proportional to ϵ^2. The preferred cascade minimizes the number of separating elements. This is evident from eqn (7) where it can be seen that for a given separative power the flow L is inversely proportional to ϵ^2.

An ideal cascade requires that each stage should be of a different size. For this reason practical cascades deviate somewhat from ideal cascades.

The outline of a practical cascade follows that of an ideal cascade but without a smoothly varying flow function, which is virtually impossible to realize in practice. A practical cascade is therefore one in which each section is itself a *square cascade*.

This is one in which the flow turnover is the same in all stages, as distinct from the ideal cascade, where this is graded continuously from stage to stage. The advantage is evident in processes such as distillation and chemical exchange. There is less incentive to operate at maximum separating power because the power consumption within the column is usually small, compared with that occurring at the ends, and any change in flow requires a separate boiler, condenser or chemical converter.

The determination of optimum conditions for a squared-off cascade is a very complex problem.

1.4 **The power consumption of a separating element**

This is discussed in London's book,[2] where he distinguishes between reversible and irreversible processes. In the former, e.g. in a centrifuge, the separation of isotopes is achieved before any product and waste flow is established in the system, i.e. in the spinning of the rotor. In an irreversible process, e.g. in a diffusion plant, flow must be established before there is any separation of the isotopes (through the membranes). The former type of process obviously uses less power than the latter.

1.4.1 *Reversible processes*

From the second law of thermodynamics a machine which decreases the entropy of mixing by ΔS without producing any other change in the physical state of the substance, and works at a temperature T must consume at least the energy

$$Q = T\Delta S. \tag{17}$$

From the definition of separative work given previously,

$$U = \frac{\Delta S}{RN(1-N)} \qquad (18)$$

Then,

$$Q = T\Delta S = \delta U\, RTN(1-N). \qquad (19)$$

where R is the gas constant, T the absolute temperature. Being a purely thermodynamical quantity, the free energy of mixing is independent of the particular separation process and the separation factor α does not enter. The free energy change depends on N.

The above quantity is small, and therefore in reversible processes the actual power dissipated in a stage depends on technical conditions, e.g. in a centrifuge on friction losses in running the machine.

1.4.2 *Irreversible processes*
London[2] gives the formula

$$Q = (4/\epsilon^2)RT\,\delta U \qquad (20)$$

as the lower limit to the power consumed in a irreversible process; Q varies as in the interstage flow, i.e. as ϵ^{-2}, and is also proportional to $T\,\Delta S$. London then suggested a factor four in the formula, which fitted the value for the power consumption in ionic migration. It is only approximate. For a diffusion plant the ideal power consumption is

$$Q = \frac{5.05}{\epsilon^2}\, RT\,\delta U, \qquad (21)$$

where ϵ_0 is equal to

$$\sqrt{\frac{M_1}{M_2}} - 1, \qquad (22)$$

M_1 and M_2 being the masses of the heavy and lighter isotopes. This formula is derived in §2 dealing with the diffusion plant. In some aerodynamic processes Q can be a large quantity.

2 Gaseous diffusion

2.1 **Principle and history of the process**

If a vessel contains a mixture of gases in thermal equilibrium with its surroundings, the average kinetic energy of all molecules in the vessel is the same, and the lighter ones travel faster than the heavier ones. The lighter molecules strike the walls more frequently than the heavier molecules, and if the walls of the vessel contain fine holes, such that individual molecules can escape but bulk flow is prevented, the escaping gas will be enriched in the lighter component. This process, known as molecular diffusion, was first demonstrated by Graham[9] in 1846. The practical separation processes based on this principle are known as gaseous diffusion processes.

The process was used in practice by Aston in 1920 to enrich neon in the ^{22}Ne isotope, and subsequently by other workers to enrich hydrogen in deuterium and carbon in ^{13}C.[10] The use of a 24-stage cascade by Hertz[11] helped to improve the neon-22 enrichment.

During World War II, the gaseous diffusion process was developed for large-scale production of uranium highly enriched in its isotope ^{235}U as part of the US Manhattan Project[12] and the British Tube Alloys Project.[13] The first gaseous diffusion plant was built at Oak Ridge in 1945. Since then plants have been built in the USA at Paducah and Portsmouth. A plant was built at Capenhurst between 1950 and 1956. In France, there is a diffusion plant at Pierrelatte and one at the adjacent site, Tricastin. There are also plants in Russia and China. The American diffusion plants dominate the enrichment scene. The capacity is about 17 000 tonnes of separative work per year with a total power consumption at full power of about 6000 MW. The technology of these plants is being improved and their capacity increased to 27 000 tonnes of separative work per annum with a power consumption of about 7000 MW.

The membrane barrier is of course the key to the diffusion process. Its method of manufacture and performance characteristics remain highly classified information. However, it is worthwhile to review the process in general to determine what further improvements are possible and to compare it later with other processes such as centrifugation.

In Appendix B some results of the theory of flow through a membrane are discussed. It is shown how the membrane efficiency S, defined as

$$S = \frac{\Delta N}{\Delta N_{\text{ideal}}},$$

can be calcuated, where ΔN is the actual concentration change across the membrane and the ideal concentration change is defined as

$$\Delta N_{\text{ideal}} = \sqrt{\frac{M_H}{M_L}} - 1, \tag{23}$$

where M_H and M_L are the molecular weights of the heavy and light molecular species. In the case of uranium hexafluoride (Hex) the value of $\Delta N_{\text{ideal}} = 0.00429$.

The purpose of Appendix B is to give the derivation of some simple formulae which are used later in the determination of the lower bound to the energy consumption in a diffusion plant. The model used in these calculations assumes that the flow through a porous membrane is similar in its essential features to flow through long capillaries rather than flow through orifices in a thin sheet. The flow is assumed to be the sum of two terms, namely a Knudsen term which is separative and a Poiseuille term which allows for the collisions between the species in the pores of the barrier, (see Figs. 12.5-12.7)

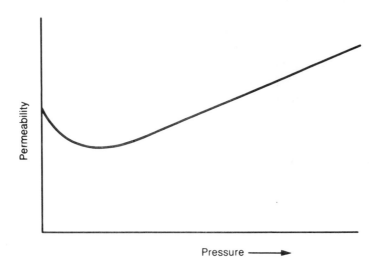

Fig. 12.5 Flow in a long capillary

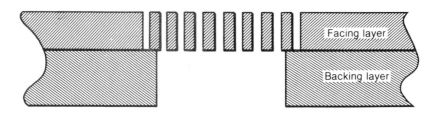

Fig. 12.6 Composite two-layer barrier

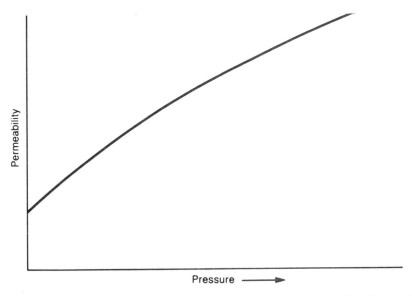

Fig. 12.7 Permeability of composite layer sometimes shows downward bond depending on parameters of layers

The lighter molecules flow slightly faster than heavier ones through the walls of a porous tube. According to Knudsen's law the rate of diffusion of each isotopic species through the barrier is inversely proportional to the square root of the molecular mass. This is true provided the molecules only collide with the walls of the pore. If they collide together in the gas in the pore then separation is reduced.

The only suitable uranium compound which is gaseous at room temperature around 60 °C is uranium hexafluoride (Hex). It has the further advantage that fluorine, which is its second constituent, is mono-isotopic.

The mean free path calculated from viscosity data is about 20 nm at 1 bar at 20 °C. This small value gives an idea of the diameter of the pores which must be obtained in the membranes to enable a plant to operate at a reasonable pressure with the membranes at a high efficiency—in practice with pore-diameters of the order of 10 nm.[16]

The diffusion process is dominated in its design and operational aspects by the chemistry of uranium hexafluoride. Hex is a strong fluorinating agent: it reacts with metals to form metal fluorides, with organic compounds to form fluorinated organic compounds, and with water to form highly corrosive hydrogen fluoride. The diffusion plant has to be designed to a high degree of vacuum tightness in order to prevent decomposition of the uranium hexafluoride, which could result in the blocking of membrane pores.

Massignon[14] lists in his comprehensive article many physical effects which have to be considered in a more exact treatment than those of Bosanquet and Present and de Bethune. These are the effects of pore geometry, the effect of the law of molecule–wall collisions, and the effect of adsorption and surface flow. The porous structure is extremely complex, see for example Fig. 12.6 and this has resulted in a multiplicity of models. Alderton[16] has pointed out that capital costs in a diffusion plant might be reduced by increasing the uranium hexafluoride pressures and so reducing the compressor size. However, this would require even smaller holes in the porous membrane, and a limit is set to this by migration in the adsorbed layer on the internal surfaces of the pores, which detracts from the degree of separation. A study of the surface effects is therefore of considerable importance in assessing the potential of the diffusion process.

2.2 *Adsorption and surface flow*

The residence time of a molecule on a surface is given by the formula

$$\tau = \tau_0 \exp\ (Q/RT) \tag{24}$$

where τ_0 is the vibration time of the molecule on the surface, $\approx 10^{-13}$ seconds, and Q is the energy of adsorption. Chemisorption involves the formation of a chemical compound between the Hex gas and the surface material. The value of Q can be very large, i.e. greater than 170 kJ/mol, and the residence time is very long.[17]

Physical adsorption involves a much lower energy, Q, the molecules being held by long range van der Waals forces between the molecules. Adsorption multilayers can be formed, and these reduce the pore radius and can result in pore plugging by capillary condensation. A pressure gradient along the pore can cause surface diffusion by either a hopping mechanism or by the motion of a two-dimensional gas, depending on whether or not the energy barrier for surface motion is greater than or less than kT.

When there is a high surface coverage, the adsorbed molecules behave like a sliding film upon the surface under a spreading bidimensional pressure gradient related to the pressure gradient through the barrier. This flow is non-separative. At low surface-coverage a hopping mechanism may be operative and result in separation of the isotopes comparable to the separation by molecular flow. This has been discussed by Higashi.[18] Experiments confirm Higashi's theory.

2.3 **Separation stage**

To use the separation process in a stage of a cascade, the process gas must be made to flow across the surface of a membrane so that a part of it is drawn through the membrane and enriched in the light component, while the remainder is depleted in the light component. The simplest geometrical case to consider is that of two parallel sheets separated by a small gap forming a membrane pair. As the diffused gas is drawn through the membrane it is enriched in the light component, and the gas film immediately above the membrane surface is depleted in this component. A concentration gradient

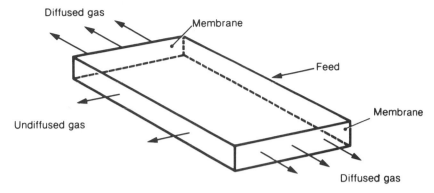

Fig. 12.8 Simple membrane pair

is therefore set up across the gas flow in the barrier tube, and the light component at the membrane surface will be replenished by the resulting diffusion from the bulk of the gas as shown qualitatively in Fig. 12.8.

At the membrane itself, there will be a jump in the concentration due to its separation. The change $N_1 - N$ between the diffused gas concentration N_1 and the average concentration N of the gas stream will be less than the change $N_1 - N_m$ across the membrane itself, where N_m is the concentration at the barrier face. The ratio of these changes in concentration is termed the point mixing efficiency E, i.e.

$$E = \frac{N_1 - \bar{N}}{N_1 - N_m},$$

where E may be calculated from the physical conditions in the channel. If the channel flow is laminar, the mixing will be entirely due to molecular diffusion. If the channel is turbulent, the mixing will be due both to eddy diffusion and molecular diffusion.

Figure 12.9 depicts the components of a large stage of an American plant in operation, including the tubular membrane pack.

The discussion so far has been qualitative. However, it is possible to be more specific in the estimation of power requirement, especially the calculation of the lower bound to the energy consumption.

2.4 Lower bound to the power consumption

In an ideal cascade with a cut of 0.5 the diffused gas will be one half of the total flow say J'. Let, with the previous notation,

$$r = p_B/p_F. \tag{26}$$

Then if the heat of compression is isothermal, the power will be

$$W = RTJ' \ln \frac{1}{r}. \tag{27}$$

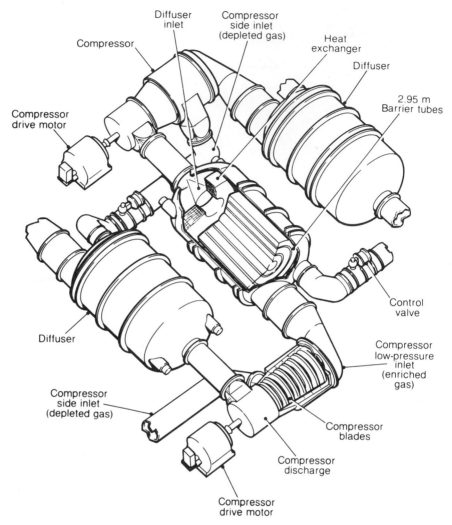

Fig. 12.9 Components of a larger stage in an American plant in operation

Half the total flow is given by

$$J' = \frac{4\phi}{\psi_S^2},$$

where ψ_S is the point separation factor and ϕ is the separative power (derived from eqn (15)).

Using the formula

$$\psi = \frac{\epsilon(1-r)}{1+(1-r)(p_F + p_B)/p_C} \tag{29}$$

(derived from eqn (72) in Appendix B), it is found that the value of r which makes the ratio ψ_s/ϕ a minimum is 0.285, with p_F and p_B zero, i.e. with ideal separation through the membrane.

The lower bound to the energy consumption is then

$$\frac{(5.05)RT\phi}{\epsilon^2}, \tag{30}$$

where

$$\epsilon = 0.004289.$$

Taking $T = 338.16$ K (Massignon's case), an energy consumption of

$$897.0 \text{ kWh/kg SWU} \tag{31}$$

is obtained. Massignon obtains 913 kWh/kg SWU as he has used a slightly different formula for ψ_s'. For adiabatic compression the energy consumption is slightly higher, but if a heat exchanger is placed between the enriched outlet of the diffusion cell and the compressor inlet the energy consumption is marginally lower.

In practice, the power consumption will be greater due to compressor inefficiencies, heavy fraction channel and pipe pressure losses, etc. Benedict *et al*[19] give 2331 kW h/kg SWU as a practical figure for the specific energy consumption for an advanced diffusion plant. There is a discrepancy between their quoted theoretical minimum and ref.(20). Some narrowing of the gap between practical and ideal conditions may be achievable. However, it is obvious that the potential for further energy savings is limited. In the UK the rising price of electricity supplied from the National Grid owing to fossil fuel price increases, caused a substantial increase in the cost of separative work from the diffusion route. In addition, the UK demand is insufficient for an optimum (i.e. large) diffusion plant. On the other hand, a centrifuge plant has a small optimum size and modular expansion possibilities.

In Fig. 12.10, there is a schematic diagram of the 1200 gaseous-diffusion stages in a plant with a capacity of 8.75 million SWU/y designed to produce uranium enriched to 4 per cent ^{235}U with tails at 0.25 per cent concentration.

Some information has been released of the French plant at Tricastin. A picture of their axial compressor alongside an integrated stage comprising a compressor, a diffuser containing the barrier, and a heat exchanger is shown in Fig. 12.11. This unit is large, as can be seen from the figure.

3 Centrifuges

The use of the gas centrifuge to separate uranium isotopes has been discussed extensively in the literature.[21,22,23] Historically, work started on liquid centrifuges about 100 years ago with the invention of the first cream separator by the Swedish engineer de Laval. One of the most important features of his invention was his idea of self-balancing using flexible shafts. In his work, top circumferential speeds of 400 m s^{-1} were achieved. Among other work on

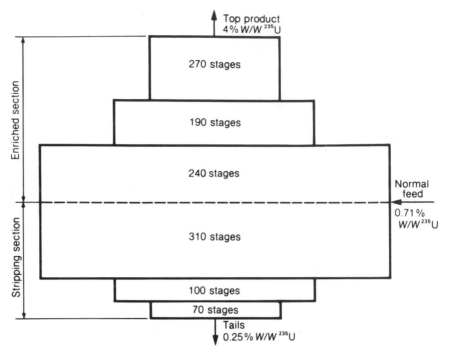

Fig. 12.10 Division of stages in a diffusion plant

liquid centrifuges was that by Svedberg on the modern high-speed analytical centrifuge.

The suggestion that centrifugal fields were suitable for isotope separation was made by Lindemann and Aston[1] in 1919, and this was followed by experimental attempts to realise the separation. Mulliken[24] introduced the evaporative centrifuge, in which a small amount of liquid is introduced into the centrifuge; this forms a layer on the periphery, which is removed through a pipe along the axis during the spinning of the rotor vapour. Beams[25] developed vaccum chamber centrifuges that were free from vibration and thermally isolated to eliminate convection currents, these being the cause of the failure of the earlier experiments. With Skarstrom and Armstead[26] he produced a change of ^{35}Cl and ^{37}Cl in carbon tetrachloride. The evaporative centrifuge was soon replaced by the concurrent type, in which the isotopic mixture to be separated is introduced continuously into one end of a rotating tube, and is collected as two separate fractions at the other end. The tubular centrifuge was pioneered by Beams in the late 1930s and was further developed during the 1940s as a proposed method of separating uranium isotopes for nuclear weapons. The concurrent centrifuge has two major disadvantages which preclude its use in an economic enrichment plant. Firstly, the centrifugal field establishes an exponential pressure profile in the gas, and the inner enriched gas stream is at a low pressure and is difficult to remove.

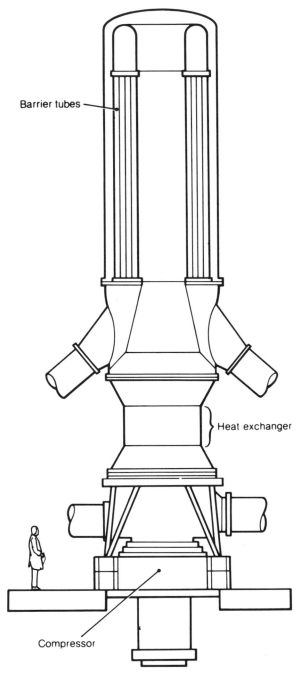

Barrier tubes

Heat exchanger

Compressor

Fig. 12.11 Integrated compressor, heat exchanger and diffuser unit in French diffusion plant

Secondly, the degree of enrichment achieved is relatively small. These developments were followed by the countercurrent machine, in which the gas at the periphery, depleted in the light isotope, is swept to one end of the rotor, whilst the enriched gas in the central zone is swept to the other. In this way, an isotopic concentration difference is produced along the length of the rotor, so that both the enriched and depleted fractions can be taken off near the periphery where the pressure is highest. The enrichment factor is increased many times by the counter-current flow and is proportional to the length of the rotor. Work was carried out by Beams[27] in the USA and Groth[28] in Germany. However, the speed of the machines was limited by the strength of the available rotor materials and the power consumption in the bearings was high.

In the years between 1957 and 1960, Zippe[29] and others published details of the design of a light-weight centrifuge, which was fitted at the lower end with a central pivot operating in an oil-lubricated cup. A system of centralizing magnets provided the non-contacting upper bearing and this provided some lift (Fig. 12.12).

Since then, speeds have increased because of the use of stronger materials, and an economic uranium enrichment plant based on centrifuges has become feasible.

3.1 Maximum output from a centrifuge

It has been shown previously that the separative work of a device which processes L gram-mol at a fraction N, producing a fraction θL of product and a fraction $(1-\theta)L$ of waste is given by

$$\delta U = \frac{\theta}{1-\theta} L \frac{\epsilon^2}{2} \text{ mol/s,} \tag{32}$$

where θ is the cut, a is the enrichment factor in the product, which depends on the cut, and $\epsilon = a - 1$. This has a maximum at $\theta = \frac{1}{2}$ when

$$\delta U = \frac{L}{2} \varepsilon^2. \tag{33}$$

This can be rewritten as

$$\delta U = \frac{L}{8} (2\varepsilon)^2, \tag{34}$$

where 2ϵ is the enrichment factor between the product and the waste.

In the centrifuge

$$\frac{N}{1-N} = \exp \left[\frac{\Delta M \omega^2 r^2}{2RT} \right], \tag{35}$$

where,

   N is the fraction of the lighter isotope,
   ω is the angular velocity,

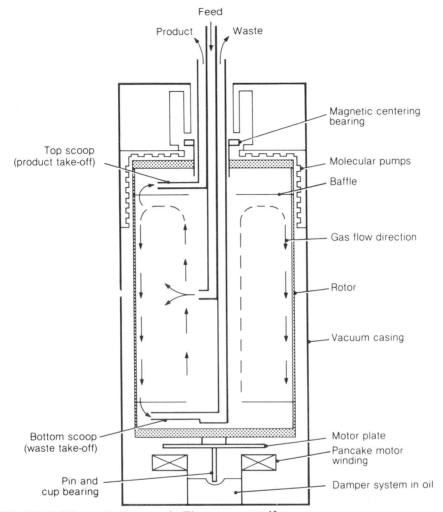

Fig. 12.12 Schematic diagram of a Zippe type centrifuge

r is the distance from the centre,

R is the gas constant,

T is the temperature, and

ΔM is the mass difference between the two molecular species.

Then

$$\frac{\left(\dfrac{N}{1-N}\right)_{r+\mathrm{d}r}}{\left(\dfrac{N}{1-N}\right)_{r}} \approx \exp\left[\frac{\Delta M\omega^2 r\,\mathrm{d}r}{RT}\right] \tag{36}$$

$$\approx 1 + \frac{M\omega^2 r\,\mathrm{d}r}{RT},$$

and

$$2\varepsilon = \frac{\Delta M\,\omega^2 r\,\mathrm{d}r}{RT}. \tag{37}$$

The flow of light isotope is given by

$$\tfrac{1}{2}L[N(r) - N(r + \mathrm{d}r)] = \tfrac{1}{2}L(2\varepsilon)N(1 - N) \tag{38}$$

and this is equal to $\dfrac{\rho D}{M}\dfrac{\partial N}{\partial r}$ where ρ is the density and D the diffusion coefficient.

Eqn (33) can be re-arranged as follows

$$\delta U = \frac{2\varepsilon}{4} \frac{\rho D \dfrac{\partial N}{\partial r}}{N(1 - N)}, \tag{39}$$

where multiplication by M has taken place separative work is measured in grams per second.

It can easily be shown that

$$\frac{\dfrac{\partial N}{\partial r}}{N(1 - N)} = \frac{\Delta M \omega^2 r}{RT}. \tag{40}$$

Then

$$\delta U = \frac{1}{4}\left[\frac{\Delta M \omega^2 r}{RT}\right]^2 \mathrm{d}r. \tag{41}$$

Integrating over a machine of radius a and length Z gives

$$\delta U = \frac{2\pi}{4} Z \int_0^a r^3 \rho D \left[\frac{\Delta M \omega^2}{RT}\right]^2 \mathrm{d}r$$

$$= \frac{\pi}{2} \rho D Z \left[\frac{\Delta M \omega^2 a^2}{2RT}\right]^2. \tag{42}$$

As can be seen from eqn (42) there is a major advantage in achieving higher peripheral speeds, V, and longer lengths, as the output can in theory vary as the fourth power of V and as the length Z. Development therefore has been directed to higher speeds and longer lengths.

The speed is limited by the strength of the material. The strength must be greater than the stress

$$\rho V^2. \tag{43}$$

In Table 12.1[30] typical maximum peripheral speeds of thin-walled cylinders are given for various materials. However, rotors have to be operated continuously for many years and safety margins have to be included.

Vibration problems arise when the length of the rotor is increased. The main problem is that long rotors have natural frequencies of vibration lower than the operating frequency of rotation. The most important modes of vibration are the transverse or flexural ones and these are easily excited by

Table 12.1[30] Maximum peripheral speeds attainable with various materials

Material	Tensile strength (kg/cm^2)	Density (g/cm^3)	Approximate maximum peripheral speed (m/s)
Cr–N steel (1935)	13 000	8.0	400
Aluminium alloy	5 200	2.8	425
Titanium	9 200	4.6	440
Maraging steel	22 500	8.0	525
Glass fibre/resin	7 000	1.9	600
Carbon fibre/resin	8 500	1.8	685

rotor imbalance forces when the frequency of rotation coincides with these natural frequencies or critical speeds.

Table 12.2[31] gives the critical speed for aluminium rotors designed to operate at 350 m s^{-1} as a function of length/diameter ratio L/D.

Two methods[30] have been described for the practical realization of a supercritical machine. The critical speeds can be negotiated successfully if heavy-duty bearings are available, together with suitable dampers to dissipate the energy, which are similar to those used with supercritical shafts in electric generators. The other approach is to connect a series of short rotor tubes with flexible bellows, which lowers the bending frequencies of the overall rotor and so eases the bearing problems in traversing the critical rotational frequencies.

There are of course other dynamical problems. These include bearing resonances, the rigid-body modes of vibration of the rotor in its suspension system, rotor precession at speed, and external vibrations.

3.2 Flow in centrifuge

In eqn (41) the maximum separative power of an elementary element is given. The matching of the flows in each counter-current element has not been considered in eqn (42) which deals only with elementary effects. The maximum output is for an idealized system, and the predicted LV^4 variation of separative power with length and peripheral speed may not be realized in practice.

Table 12.2[31] Critical speed for aluminium rotors

L/D	Critical speeds (m s^{-1})			
	1st	2nd	3rd	4th
7	400			
11.6	145	400		
16.3	74	204	400	
21	45	123	242	400
25.5	30	83	162	269

To achieve a longitudinal separation of the isotopes in the countercurrent machine, it is necessary to have a flow profile that results in the lighter isotope, which is concentrated towards the middle of the machine, being swept to one end, and the heavier, which is concentrated nearer to the outer wall, being swept to the other. A profile to achieve this requires a velocity inversion point near the wall and high velocities in the interior, where the density is low.

The fraction of the lighter isotope N satisfies a partial differential equation. As approximation to this equation has been derived by Cohen[3], who makes the following assumptions:

(1) isothermal flow, which implies that the product of the diffusion coefficient and pressure is constant.
(2) non-decaying axial flow, and
(3) solid-body rotation.

Cohen's method depends essentially on the fact that the radial gradient $\partial N/\partial r$ of the concentration of the lighter isotope N is of the order of

$$N(1-N)\frac{\Delta M}{2RT}w^2 r \tag{44}$$

using the previous notation. This is a small quantity. Therefore, the variation of N with respect to r is small, as compared to the variation of p or pw, where p is the pressure and w the axial velocity. With these approximations Cohen derived the following equation for the fraction N averaged over the radial distribution

$$\frac{\partial}{\partial z}C_5\frac{\partial N}{\partial z}-\frac{\partial}{\partial z}[PN+C_1 N(1-N)]=0, \tag{45}$$

where P is the product flow. Normally C_5 and C_1 are assumed to be independent of z and are given by

$$C_1=\frac{2\pi}{RT}\frac{\Delta M\omega^2}{RT}\int_0^a r\,dr\int_0^r pwr\,dr, \tag{46}$$

and

$$C_5=\frac{2}{RT}Dp\frac{a^2}{2}+\frac{2}{RT}\times\frac{1}{Dp}\int_0^a\frac{dr}{r}\left[\int_0^r pwr\,dr\right]^2. \tag{47}$$

It can be seen that the fraction distribution is dependent only on the integral of the mass-flow profile, which means that approximations to the flow field give reasonable fraction distributions.

A solution of the above equation shows that the influence of the mass-flow profile on the efficiency of the separation can be described in terms of a quantity called the profile efficiency, given by

$$\frac{4}{a^2}\frac{\left[\int_0^a r\left\{\int_0^r pwr\,dr\right\}dr\right]^2}{\left[\int_0^a\frac{1}{r}\left\{\int_0^r pwr\,dr\right\}^2 dr\right]}. \tag{48}$$

It is a simple variational problem to find the function $w(r)$ which optimizes this expression. This leads to pw being a constant and independent of r. Now, it is also necessary to fulfil a condition on $\int pwr \, dr$ as this is proportional to the output rate P, and since this is a small quantity, the integral of pw should be very small. If a variational condition is written down subject to this integral vanishing, it is found that the best arrangement is to leave all the return flow in a thin layer along the wall.

There are other factors which affect the efficiency of the machine and reduce the separative work output from the Dirac upper limit. Consider a three-pole machine, i.e as shown in Fig. 12.12, where the feed enters in approximately the mid-plane and the product and waste are taken from the ends.

Let P be the product flow of enrichment N_P, and let W be the waste flow of enrichment N_W. Let the feed be introduced at $Z = Z_F$. Then from the definition given previously, the amount of separative work produced in a slab between Z and $Z + dZ$ is equal to

$$\delta U = P(N_P - N) \frac{dN}{N^2(1 - N)^2},$$ (49)

for the product end of the machine, and

$$\delta U = W(N - N_W) \frac{dN}{N^2(1 - N)^2},$$ (50)

for the waste end.

A local efficiency $E(Z)$ can be defined as

$$E(Z) = \frac{\delta U}{\delta z} \Big/ \frac{\delta U_D}{\delta z},$$ (51)

where δU_D is the Dirac upper limit.

In Fig. 12.13 the efficiency $E(Z)$ is plotted for a typical centrifuge together with the profile efficiency. It can be seen that the separative work is zero at either end of the machine, and care must be exercised in positioning the feed point in order to avoid a mismatch between the feed concentration and the concentration of the gas in the machine.

The maximum efficiency which can be obtained allowing for the end effects is 0.814 with constant flow. This was derived first by Los.[32]

For each value of the feed there is an optimum counter-current flow. If the countercurrent is too large, the axial concentration difference will increase only slowly over the machine and an optimum difference will not be reached.

Although the Cohen theory is approximate, it gives a fair description of the factors affecting the separation in the centrifuge. A more accurate description is obtained from a numerical solution of the partial differential equation describing the fraction N.

The Dirac formula predicts that the maximum separative power of a centrifuge is proportional to the fourth power of the rotational velocity and

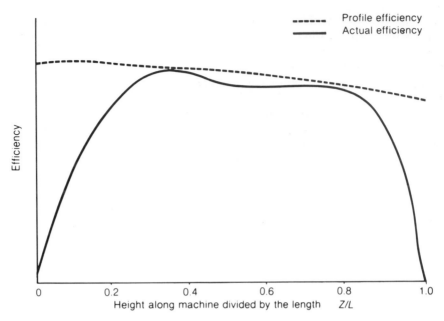

Fig. 12.13 Variation of the efficiency along the length of a centrifuge

to the length. However, at very high speed the interior of the machine is evacuated, and the mean free path of the molecules is of the order of the radius of the machine, which is no longer separating in this region. Raetz[33] has calculated the radial pressure gradient across the machine as a function of its speed. In the inner region of the machine, with a pressure of 10^{-3} of that at the wall, it is assumed that no separation occurs in the inner region of the machine bounded by this radius. The separative power is then less than that given by the Dirac formula. It can be expressed by a formula as follows:

$$\frac{\pi}{2}\frac{DP}{RT}\left(\frac{\Delta M}{2RT}\right)^2 LA\, V^n. \tag{52}$$

The constants A and n are determined such that at a certain speed the value and the gradient with respect to V are exact. Table 12.3 from Raetz[33] shows the local velocity exponent of the optimum separative power as a function of speed V. At the higher speeds the exponent is nearer to two. A constant wall

Table 12.3 Variation of the velocity exponent of the separative power with speed

V (m/s)	Velocity exponent, n
313	4
344	3.4
438	2.7
563	2.4
751	2.2

pressure is assumed in the calculation. If the hold-up is constant, i.e. pressure increases with increasing speed then it does not decrease so rapidly.

The counter-current flow in the centrifuge can be generated by thermal or mechanical means.

In the case of thermal drive, the radial pressure gradient becomes flatter in the areas of higher temperatures. Thus, along the wall the gas moves from the cold to the hot end, and in the centre from the hot to the cold. If there are temperature jumps at the two ends, the gas will flow in the centre from the hot to the cold end and then into a boundary layer at the end-cap, returning within a thin layer on the wall and again into a boundary-layer flow at the hot end. This latter type of flow was analysed by Martin[34] and Whipple.[35]

In the high-pressure case the stream function in the main body of the machine is proportional to

$$\rho^{1/2} \tag{53}$$

where ρ is the density of the gas. This is not a very efficient profile at high speeds, where the gas is concentrated nearer and nearer to the wall. Neglecting the evacuation effect, the efficiency varies as V^2 rather than V^4. However, a temperature gradient on the wall enables a counter-current flow to be maintained along a centrifuge.

A counter-current can also be generated by disturbing the rotational velocity. A deceleration will flatten the density distribution and an acceleration will steepen it.

A deceleration can be achieved by a scoop and this generates a counter-current similar to the thermal component. However, the counter-current generated by the scoops can be channelled to the interior by the use of a scoop and a baffle, and a better profile established, although the ideal Cohen profile of ρ w = constant cannot be achieved. The profile generated by the mechanical means will change as the flow proceeds along the machine. Eventually the profile will be similar to the thermal profile. The length over which the profiles change depends on the operating pressure of the machine and also on its diameter.

To describe the methods by which the flow pattern can be calculated is beyond the scope of the present chapter. Briefly, the flow pattern can be obtained by a numerical solution of the non-linear Navier–Stokes equations. A solution of the linearized Navier–Stokes equations can be obtained semi-analytically by an expansion in terms of Steinbeck or Parker eigenfunctions plus a simulation of the boundary layer flows, or, again, a numerical solution can be obtained.

Summarizing, it can be seen that the prediction that the separative power given by the Dirac formula, which is proportional to LV^4, has to be modified to allow for various factors, for example, the evacuation effect and the type of flow pattern that can be generated in the centrifuge. However, optimization studies indicate that there is an advantage in proceeding to higher speeds and longer lengths. The current price of separative work from centrifugation is roughly equal to that quoted for the US diffusion plants, but taking into account the potential from operation at higher speed and longer lengths, the cost eventually should be below that of gaseous diffusion.

Figures 12.14 and 12.15 are photographs of the cascade halls at the Capenhurst plant and at the Almelo plant in Holland.

Figure 12.16 shows the Component Test Facility at Oak Ridge. From this it will be seen that the US machines are large. However, von Halle in his paper[21] discusses a model gas centrifuge of modest size—1 m in length, 10 cm in radius—rotating with a peripheral velocity of 450 m/s. A centrifuge cascade with a capacity of 8.75 million SWU/y composed of these centrifuges, producing reactor-grade uranium containing 4 per cent ^{235}U and 0.25 per cent tails, would be as shown in Fig. 12.17. The power required to run such a plant would be 100 MW, i.e. 4 per cent of that required to run a diffusion plant of the same size.

The cascade would be divided into banks of centrifuges with many machines in parallel. The output from his model machine is given as 4.136 SWU/y, which is small.

4 Other aerodynamic methods

Details of experimental work up to the pilot stage have been published in the literature on two processes: the Becker jet nozzle process[37] and the South African stationary wall centrifuge.[38]

4.1 **The jet nozzle process**

The first experimental work on jet separation processes was carried out by P.A. Tahourdin[36] at the Clarendon Laboratory during the war. This was based on an idea of P.A.M. Dirac, who worked out the theory of the process. The jet process is also discussed in London's book.[2]

The Becker jet nozzle, which represents a further development of these ideas, is based on the mass dependence of the centrifugal force in a fast, curved flow of UF_6. The separation unit is as shown in Fig. 12.18.

This process has been developed for over ten years by Becker at Karlsruhe. Since 1970, the German Company Steag has been involved in the technological development and commercial exploitation of the nozzle process. A test loop was built at Essen operating with a hydrogen–UF_6 mixture having a 95 per cent molecular fraction of hydrogen. The mixture expands into the curved duct and the flow is split into a higher and a heavier fraction by means of a skimmer. The optimum pressure of a commercial separation nozzle with a radius of curvature of 0.1 mm is of the order 13 kPa. The effective velocity of UF_6 molecules is increased to three times the sonic velocity. A separation factor of 1.02 is obtained, and this factor is higher than would be predicted theoretically. For the operating pressures, the ratio of the width of the nozzle to the mean free path is 50.

Fig. 12.14 Cascade hall, Capenhurst

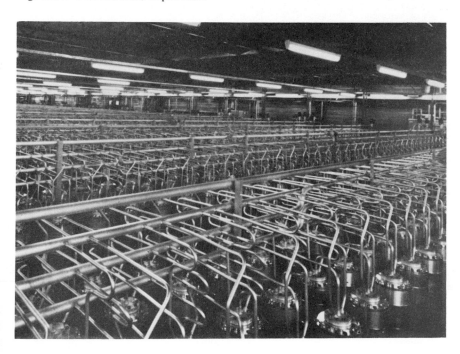

Fig. 12.15 Cascade hall, Almelo

Fig. 12.16 Component test facility, Oak Ridge, Tennessee

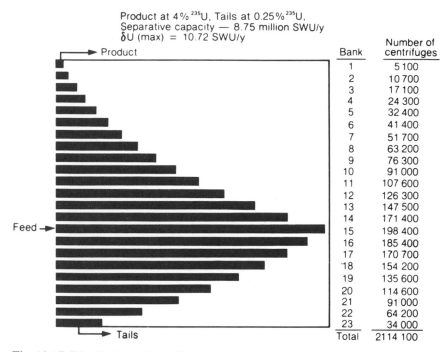

Product at 4% ^{235}U, Tails at 0.25% ^{235}U,
Separative capacity — 8.75 million SWU/y
δU (max) = 10.72 SWU/y

Bank	Number of centrifuges
1	5 100
2	10 700
3	17 100
4	24 300
5	32 400
6	41 400
7	51 700
8	63 200
9	76 300
10	91 000
11	107 600
12	126 300
13	147 500
14	171 400
15	198 400
16	185 400
17	170 700
18	154 200
19	135 600
20	114 600
21	91 000
22	64 200
23	34 000
Total	2 114 100

Fig. 12.17 Distribution of centrifuges in a 8.75 million SWU/y plant

The cut is low, about ¼, and there is a problem in the handling of hydrogen as a carrier gas. However, reasonable solutions have been found to the main technical problems to permit the construction of a full-size plant. There is only one critical element, i.e. the nozzle which is static and un-stressed. The energy consumption is higher than that in a diffusion plant, i.e. 4000 kW h/SWU. However, with $\epsilon = 0.02$, there is no reason why this cannot be reduced, see ref. (19). Variants of the process which might reduce power consumption have been described. If opposed jets were used to deflect one another, wall-friction and hence power consumption should be lessened. Also, a non-uniform distribution of the mixture composition across the nozzle inlet could be used. The power consumption might be reduced to 2700 kW h/SWU, which is comparable with that for a diffusion plant.

4.2 The South African process

Some details of this process were released at the Paris conference on Nuclear Energy Maturity, 1975.[38] In this process, a mixture of hydrogen and Hex is compressed to 6 bar and allowed to swirl in a separating element which is, in effect, a high-performance stationary-walled centrifuge. Details of this element have not been disclosed. The separation factor is higher than that

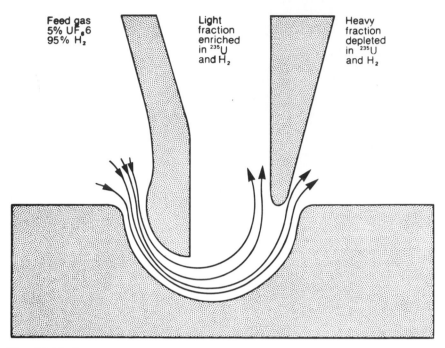

Fig. 12.18 Cross-section of the separation nozzle system used in the commercial implementation of the separation nozzle process

for the jet nozzle process and varies between 1.025 and 1.030, but the cut is small.

The Helikon technique integrates the jumped cascade ($\theta = \frac{1}{4}$) with close coupling of the compressor, cooler, and separating element in one housing to minimize duct loss. One compressor is used to transmit several streams of different isotopic composition without there being significant mixing between them. An important feature of the separation elements is that they can be used also to yield high-purity hydrogen in a single step when fed with material of approximately the enriched stream composition. This feature is incorporated in the Helikon technique to make gas separation by freezing unncecessary.

Apparently the plant has a low uranium inventory, and this results in a short cascade equilbrium time, of the order of 16 hours for a commercial plant enriching uranium up to 3 per cent ^{235}U.

The theoretical lower limit to the specific energy consumption of the separation element is about 300 kW h/kg SWU as obtained from the eqn (20). The minimum figure which has been obtained with laboratory separating elements is about 1800 kW h/kg SWU, disregarding all system inefficiencies. It is argued that there is scope for a large development potential, and that the power consumption should fall below that of diffusion.

5 Plasma processes

5.1 **Plasma centrifuges**

The use of electromagnetic forces to generate high rotational velocities in a plasma-ionized state of molecular or atomic species was discussed in some detail at the British Nuclear Energy Society Conference on Uranium Isotope Separation.[39] Very high rotational velocities of the order of 10^5 m/s can, in theory, be obtained. However, there are factors, such as high temperature, that offset the increase in separative power caused by the higher rotational speed. Nathrath *et al*[39] reported experimental work using uranium. They have shown that stable plasmas could be established using both U^+ and UF_6^+ ions, and that these plasmas can be rotated, but none of the experimental work had reached a stage where continuous separation of isotopes had been attempted.

The impression given by the various authors was that the work was still at an early stage. The exploration of the theory to very high rotational velocities and varying degrees of ionization had yet to be validated. The disadvantageous effects of the necessarily high temperatures appeared to be inadequately understood, and in general the basic physics of the processes likely to be involved in rotating plasma centrifuges was more complicated than had been originally envisaged.

5.2 **Other plasma-based methods**

At the 1977 Salzburg Conference, Vanstrum and Levine[40] described two other plasma-based processes. In one of these, the Dawson process, a plasma of uranium atoms within a strong uniform magnetic field is exposed to a low-energy radio frequency wave, resonant with the cyclotron frequency of the ^{235}U ions. The rotation thereby imparted preferentially to the ^{235}U ions enables the ^{235}U to be separated from the ^{238}U by properly placed collection plates. The second process involves the production of a UF_6 plasma by chemi-ionization: UF_6 molecules are accelerated by expansion with an inert carrier gas through a supersonic jet. A cross beam of alkali metal molecules results in the formation of Na^+ or Cs^+ and UF_6^-. A radio frequency quadrupole mass filter deflects the $^{238}UF_6$ out of the plasma beam, permitting the separating of the two beams on separate baffles cooled by liquid nitrogen. The two processes are shown schematically. Figs 12.19 and 12.20.

6 Chemical exchange methods

Chemical exchange processes are reviewed in London's book.[2] Recently they have been considered in some detail in the report of the Uranium

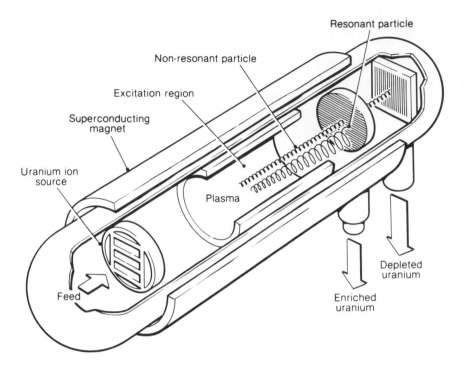

Fig. 12.19 Dawson plasma isotope separation process using ion cyclotron resonance principle

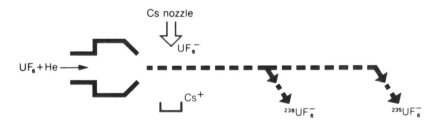

Fig. 12.20 Schematic of prototype chemi-ionization separator

Isotope Separation Review *ad hoc* Committee.[19] The Committee recommended that further work should be carried out on chemical exchange reactions employing

 (1) distillation,
 (2) gas–liquid absorption,
 (3) liquid–liquid extraction, and
 (4) gas–solid chromatography.

It is also recommended that a limited amount of work should be carried out on the Redox ion-exchange process. This process, involving a change in valence state, has been investigated in detail in France and Japan.

The principle of chemical exchange is that a reversible chemical equalibrium, normally taking place between two separate phases, gives a small isotope-separation factor, and because it is reversible only small energy requirements arise at each stage. However, the counter-current flow of uranium in the two phases has to be totally refluxed at each end of the cascade, either by very large chemical-reaction or by electrochemical changes of uranium valency.

The factors which are important in keeping down plant size and equilibrium time as can be seen from eqns (15) and (16) in §1, are

(1) a large single-stage separation factor,
(2) a high molecular density of uranium in each phase,
(3) small single-stage process equipment,
(4) rapid chemical exchange kinetics, and
(5) a simple method of providing chemical, or electrochemical, or physical reflux of uranium at each end of the cascade.

The oxidation–reduction of tetravalent uranium and hexavalent uranium in an aqueous solution has been studied in great detail as mentioned earlier. Other processes such as the oxidation–reduction reaction of trivalent and tetravalent uranium have also been studied. In the case of the first redox (oxidation–reduction) reaction, tetravalent uranium and hexavalent uranium are combined to yeild the following isotope exchange equilibrium

$$^{238}U^{4+} + \ ^{235}U^{6+} \rightleftharpoons \ ^{238}U^{6+} + \ ^{235}U^{4+}$$

The equilibrium constant, K, is 1.0013 according to the results of several determinations. In solution, hexavalent uranium is bonded to two oxygen atoms to give the uranyl ion $U(VI)O_2^{2+}$. Uranium isotopes can be separated when hexavalent uranium is separated from tetravalent uranium through chromatography, electrophoresis, and solvent extraction.

Ion exchange between the tetravalent U^{4+} and hexavalent UO_2^2 in aqueous and organic systems has been examined at Harwell. The hexavalent state distributes preferentially into the organic phase, thus allowing the single-stage separation to be cascaded. Experiments were carried out with a wide range of organic solvents. Polymerization of U^{4+} species occurred in the aqueous phase at high pH's required for rapid isotope exchange. These polymers absorb at the interface of the solvent droplets and greatly reduce the rate of mass transfer of ionic species between aqueous and organic phases.

The information released in a French paper[41] at Salzburg in 1977 has renewed interest in chemical exchange methods. A complete specification by the Commissariat à l'Energie Atomique was published on 16th March 1977.[42] The invention makes use of the observation that it is possible to form solutions containing U^{3+} in which this normally unstable valence state can be preserved in a metastable manner, even in acid solution, for long periods. These times are sufficient to perform isotopic exchange under industrial conditions and are achieved when these solutions are now allowed to contact conducting bodies and when they are substantially free, apart from the uranium, of metal ions of groups III to VIII of the Periodic Table.

Many examples are quoted in the patent specification of the magnitude of the separation for different conditions and organic solvents.

Work has also been reported in other countries. An Australian paper[43] by Hardy at the BNES Conference in 1975 described the results of experiments using ion exchange resin.

In Japan the chemical exchange process has been studied in some detail, and recently there have been a number of publications[44,45] giving details of their work on isotope separation by redox chromatography of tetravalent and hexavalent uranium. Chromatography was considered to be the most suitable separation means by which several hundred theoretical separation stages per metre of a simple packed column can be provided. However, in order to make the process commercially feasible an extensive programme of work on suitable adsorbents had to be undertaken.

The redox chromatography is carried out in a column packed with an adsorbent, as shown in Fig. 12.20. An oxidation agent is fed into the column through the selector valve at the top until it begins to flow through the lower valve, and the supply of uranium solution is then begun. The U^{4+} ion in the solution is oxidized to UO_2^{2+} by the oxidation agent and absorbed, and a boundary is formed between the oxidation agent and UO_2^{2+} adsorption zones. If the rates of adsorption and oxidation are high enough, the boundary will be narrow and clearly defined. When the supply of uranium ion sufficient to allow formation of the desired UO_2^{2+} adsorption zone has been completed, then a reduction agent is fed to the top of the column to reduce the UO_2^{2+} in the upper part of the adsorption zones to U^{4+}. Thus, uranium ions are desorbed in the form of U^{4+} and a boundary is formed between the reduction agent zone and the UO_2^{2+} adsorption zone.

With continued feed of the reducing agent, the UO_2^{2+} adsorption zone descends in the column, and the isotope exchange reaction occurs numerous times between the adsorbed and solute uranium ions. When these conditions are maintained for a sufficient period, the concentration of uranium at various points throughout the column follows the trapezoidal curve as shown in Fig. 12.21, and the isotopic distribution follows the S-shaped curve.

The time required to achieve the equilibria, the stage time, is very important in this process. The approximate stage time is generally assumed to depend on, the contacting method, e.g. liquid–liquid several minutes; gas–liquid one minute; and liquid–solid one minute. However, the Japanese work has resulted in a stage time of less than one second. Details are given in their patent.

In Fig. 12.22 a system is illustrated consisting of a module of four columns for uranium enrichment, plus one oxidation and one reduction column.

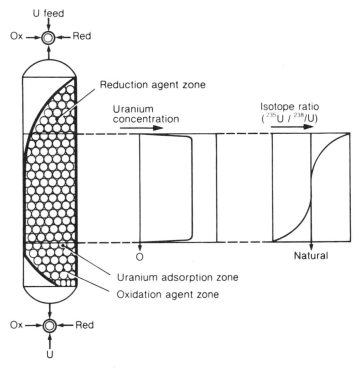

Fig. 12.21 Isotope distribution in a unit of the Japanese chemical exchange process

Each uranium enrichment column is equipped with rotating selector valves at the top and bottom; supply lines for feed uranium, reduction agent, and oxidation agent; and discharge lines for enriched and depleted uranium, deactivated reduction agent, and deactivated oxidation agent. Deactivated redox agents are recycled through oxidation and reduction columns for reactivation by O_2 and H_2, and then back to the uranium enrichment columns. In Fig. 12.22 the reduction agent zone in column 4 is being replaced by the oxidation agent zone as a process of regenerating a once-used column.

The data quoted in ref. (44) indicates that the uranium inventories are not large. For example, it is stated that an initial load of 26 tonnes would yield 1000 tonnes SWU/y enrichment if the separation factor was 10^{-3} and the stage time was 0.1 s.

It is claimed that highly enriched uranium cannot be easily produced using chemical separation methods, owing to time constants which are large compared with those of conventional means. It is stated that it would take thirty years to make highly-enriched uranium for nuclear weapons by the French chemical exchange process. In the case of the Japanese process, calculations indicate that it would require five to eight years to make weapons-grade material. It is also stated that difficulties would also arise

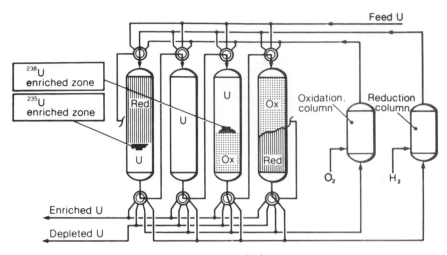

Fig. 12.22 Coupling of units in Japanese chemical process

from problems of criticality posed by the presence of highly enriched uranium in aqueous solutions, although it would appear reasonable to expect that these might be solved—at a price. The long time necessary for the production of weapons-grade material can be reduced by the clandestine use of some reactor enriched material (3 per cent enrichment) in another much smaller enrichment plant (however inefficient) which could not be detected. The bulk of the separation effort is used in enriching the uranium from natural to reactor-grade enrichment (≈ 3 per cent).

7 Laser isotope separation

7.1 Introduction

Isotope shifts occur in atomic and molecular spectra to a varying degree determined by mass differences of the isotopes, nuclear spin, and variations in the nuclear volume. These shifts form the basis of the photochemical approach to isotope separation. The subject has been reviewed by Farar and Smith,[45] who consider the applications of these methods to uranium isotope separation, review the work carried out in the United States during and after the war, and also give a survey of the literature up to 1957. During the war the work was unsuccessful when compared with other methods, i.e. gaseous diffusion and electromagnetic separation.

The development of the laser has increased the interest in this method, and many papers have appeared discussing the use of lasers in uranium isotope separation. Wavelength-tunable lasers have opened up possibilities of the separation of other isotopes, and there is a lengthening list of successful experiments. For example, the separation of boron, hydrogen, nitrogen, and chlorine isotopes has been achieved by photodissociation, calcium isotopes by photoionization, and barium by atomic beam deflection.[47]

Isotope shifts have been identified for uranium in both atomic and molecular forms, and these form the basis for the atomic and molecular routes.

In the atomic route, selective excitations of ^{235}U in a low-pressure vapour at a temperature over 2300 K is obtained using tunable lasers, thereby producing ionized ^{235}U species, which are separated electromagnetically. There are variants of this scheme.

In the molecular route selective infra-red absorption in $^{235}UF_6$ gas is followed by further irradiation at infra-red or ultraviolet frequencies, resulting either in dissociation of the excited molecules or their participation in a subsequent chemical reaction which the unexcited ones do not.

These processes have been reviewed recently in a paper by Robinson and Jensen.[48] It is a statement of the work carried out in the US, including that undertaken at the Los Alamos Scientific Laboratory and the Lawrence Livermore Laboratory. It is pointed out that work at these laboratories is classified, and only general information can be given.

7.2 The atomic route

In this route a multistep photoionization process is performed in uranium vapour with tunable-dye lasers, and the ions are extracted electromagnetically.

A technique for the generation of uranium vapour at high temperatures (≈ 2300 K) is described in a patent by Jersey Nuclear AVCO Isotopes Inc.[49] A method has been developed using a vessel containing uranium with an electron-beam heating mechanism. This allows heat deposition exactly along the line where vaporization is required, and the hot uranium is contained by cooler uranium around the outside. It is claimed that the combination of the electron beam with its controllable intense heating, together with the control of the convective losses, has provided a high-efficiency method for the evaporation of uranium. However, no information is given on the actual energy used in the method.

There exists a low-lying (620 cm^{-1}) state of the uranium atom, which may be populated by the electron beam, thus complicating the excitation process.

7.2.1 *Spectroscopy of uranium*
There are several hundred atomic lines in the absorption spectrum in the visible and near ultraviolet range that exhibit isotope shifts (^{238}U–^{235}U) of magnitude up to 10 pm, which is sufficient to allow resolution by conventional spectrometers. A tunable laser can be used to make a selective change in the energy level of the ^{235}U isotope to bring it to an excited state (≈ 2 eV above the ground state).

A complication arises in the excitation of the ^{235}U nucleus owing to small interactions between the electrons and the shape of the nucleus resulting in hyperfine splittings. These are illustrated in Fig. 12.23. It can be seen that hyperfine splitting in ^{235}U excitation curve produces additional broadening, and adds to the complication of the selective excitation of this atom.

7.2.2 Lasers for the atomic route
Tunable lasers or laser systems are available for most of the relevant spectral region. In the visible range, the dye laser has continuous tuning, high stability, and power. The laser line-width can be reduced to about 100 fm.

All these facilities can be extended into the ultraviolet, although with considerably reduced power, by using frequency-doubling techniques. By choosing laser pumps well suited to the dye laser absorptions, a high efficiency is possible (1 per cent) even for narrow-output line-widths.

7.2.3 Ionization process
In many schemes more than two steps are required to achieve the ionization of the uranium atoms, i.e. to furnish 6.2 eV of energy. This is because efficient tunable lasers are availabe for steps up to 2 eV, but above this energy they tend to become less efficient. The overall selectivity is the product of the selectivity of individual steps. The first steps are resonant and the selectivity will be high. However, the last step is one from an excited level into the ionization continuum. The transition is from a bound to an unbound level, and consequently the cross-section is much lower, but this can be compensated for by the use of a large efficient laser, e.g. a CO_2 laser. The problems in the last step can also be eased by the identification of a

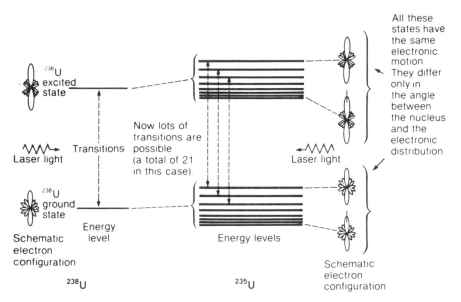

Fig. 12.23 Hyperfine interaction between the nucleus and electrons of uranium

suitable autoionizing level in the continuum with an enhancement of the cross-section.

Collectors have been designed[50] to obtain the selectively ionized uranium. The collector is pulsed directly after each laser pulse, and the current pulse with magnetic field creates a circumferential electric field in the direction normal to the alignment of the collector plates. There remains the problem of removing metal from the collector plates for processing.

7.2.4 *Conclusions*

The obstacles in the way of the atomic route are

    (1) the corrosiveness of uranium vapour,

    (2) the high temperatures producing thermally excited atoms in the beam.

    (3) charge exchange between ^{235}U ions and ^{238}U atoms,

    (4) line broadening caused by hyperfine splitting,

    (5) the unreliable performance of tuned lasers with a high repetition rate, and

    (6) uranium metal removal.

Estimates have been made of the charge exchange cross-section between ^{235}U ions and ^{238}U atoms, and these show that it is not so large as to limit the beam density to an uneconomic value.

Despite all the other apparent difficulties, Jersey Nuclear–Avco Isotopes Inc. are said to have developed a commercially competitive separation process, and it is believed that the Lawrence Livermore Laboratory is making dramatic progress towards the development of a large-scale process.

The attractiveness of the proposal can be seen from the fact that 1 watt of laser power absorbed gives no less than 1.4 ^{235}U per hour.

7.3 **The molecular route**

This route is based on the selective excitation of one isotopic form of a molecular compound with a resulting alteration in the chemical properties of this compound. As examples, there could be selective photodissociation or a selective photochemical reaction with some other chemical compound.

It should be pointed out that gas-phase processes are being dealt with and not those of liquids or solids, where the isotope effect is masked by broadening effects of absorption lines in the condensed phase. As previously stated, uranium hexafluoride has the highest vapour pressure of any uranium compound, and it is for this reason that it is used in centrifuge and diffusion plants. It is then the most obvious choice of gas here also. However, there has been interest in other uranium compounds, because of the difficulty in producing powerful lasers to excite preferentially the $^{235}UF_6$ species. Work has been reported[51] on the multiphoton dissociation of hexafluoroacetyl acetonate. Transitions of the 10.6 m CO_2 laser are in resonance with the asymmetric O–U–O stretch of the uranyl half, a vibrational mode in which the frequency is sensitive to the mass of the uranium and oxygen isotopes. Unimolecular dissociation is observed mass-spectrometrically at low energy fluence (less than 0.25 mJ/cm²) with no evidence of an energy fluence threshold. The dissociation yield is observed to increase nearly linearly with increasing energy fluence, before approaching 100 per cent dissociation at an energy fluence of about 100 mJ/cm². Both the oxygen isotope selectively and the uranium isotope selectively measured in these photodissociation experiments are very nearly equal to those predicted by the ratio of the linear absorption cross-sections for the respective isotopes. Essentially complete selectivity is observed for the oxygen isotopes, whereas a selectivity of only 1.25 is measured for the uranium isotopes. Interest centres on higher enrichment factors than 1.25, and the remainder of the discussion will relate to methods of achieving this in UF_6, assuming that the necessary lasers can be developed to excite specific vibrational modes.

7.3.1 *Spectroscopy of UF₆*
The isotope shifts which occur in the electronic states of atoms are associated with nuclear spin and nuclear volume. In molecules the effects occur as a result of variations in the isotope mass as in the case of the molecular hexafluoracetyl acetonate; UF_6 has a large number of molecular vibrational modes, but only those in which the centre of symmetry of the molecule is disturbed show isotope shifts, and these vary between 0.16 cm⁻¹ and 0.65 cm⁻¹. The ν_3 mode, as shown in Fig. 12.24, gives an isotope shift of about 0.6 cm⁻¹ between $^{235}UF_6$ and $^{238}UF_6$.

The isotope shifts are masked by the wide spread of rotational structure (≈ 10 cm⁻¹), and in UF_6 gas at room temperature each band in the infra-red

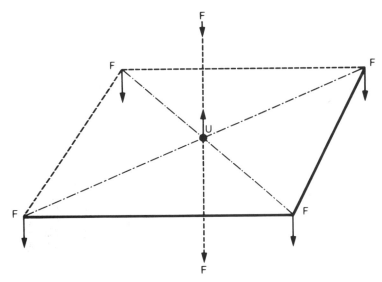

Fig. 12.24 ν_3 vibration mode of UF_6 molecule

spectrum is in reality a superposition of vibrational transitions arising from thermally populated hot bands. A consequence of the large number of low frequency vibrations is the small population of molecules, 0.4 per cent, in the vibrational ground state. The net result is that the density of rotation–vibration lines is so high that individual transitions overlap, and this prevents both resolution of the rotational structure and also isotopically selective excitation.

The hot bands for UF_6 can be eliminated by cooling the gas. Dynamic cooling can be achieved in a nozzle by the expansion of UF_6 mixed with an inert gas such as nitrogen. The nitrogen cools very rapidly and by collisions causes vibrational deactivation of the UF_6 molecules. The gas mixture has to be cooled to less than 80 K before half the UF_6 molecules are at the ground vibrational level.

Robinson and Jensen show the spectrum of the Q branch of the ν_3 mode of UF_6 at a temperature of approximately 50 K, as illustrated in Fig. 12.25.

The region of the Q branch was measured with a semiconductor laser, and there is nearly complete resolution between the ^{235}U and the ^{238}U features of the spectrum. Details of the infrared spectroscopy of supercooled gases have been reported elsewhere in the literature.

UF_6 exhibits no electronic absorption spectrum in the visible region, but there is a weak transition with some vibrational structure in the near ultra-violet (350–400 nm) followed by much stronger transitions at shorter wave-length (≈ 330 nm).

7.3.2 *Multiphoton excitation*

One of the most exciting developments in the field of laser isotope separation has been the discovery of isotopically selective multiphoton dissociation of polyatomic molecules in an intense infra-red laser beam.[52] In this process molecules with a vibrational absorption band close to the laser frequency

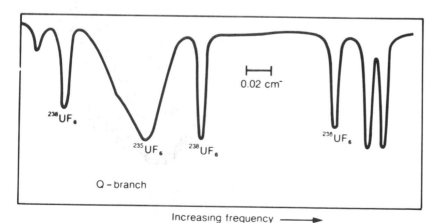

Fig. 12.25 High resolution scan of flow cooled UF_6

rapidly absorb in excess of thirty quanta and dissociate without the aid of collisions. In the case of the irradiation of sulphur hexafluoride it is isotopically selective, and large enrichment factors of particular isotopes can be obtained. Isotopes of a number of other elements have been enriched in this way on a laboratory scale. Recently Russian work[50] has been reported on the selective dissociation of $^{33}SF_6$ in the natural isotope mixture in the temperature range 123–293 K. Gas cooling makes it possible to achieve separation ratios of ^{33}S to ^{32}S as high as 5 in the dissociation products in an optimum temperature range of 170–190 K. The peak energy flux in these experiments was of the order MW/cm². It was thought originally that the threshold for multiphoton effects was at a much higher flux in the GW/cm² range.[53]

A complete analysis of the multiphoton effect has yet to be published. It is a complex theoretical problem and among the effects which have to be considered are anharmonicity, Coriolis coupling, and octahedral symmetric anharmonicity.[54]

The degree of multiphoton excitation is critically dependent on the choice of anharmonic constants. The correct values for the SF_6 constants have been obtained recently from an analysis of high-resolution spectra of the 0–3 ν_3 overtone band.[54] The constants for UF_6 are different, and therefore the results for SF_6 are only a guide to the multiphoton excitation in UF_6. However, it can be said that the multiphoton excitation proceeds at modest fluences. The degree of isotopic selectivity in UF_6 has not been disclosed; it is lower than that in SF_6 because of the smaller isotope shift.

The implications of the multiphoton process are discussed by Robinson and Jensen.[48] It is pointed out that the molecules which are placed in highly excited vibrational states are often more chemically reactive than those in their initial-state distribution. Thus preferential excitation of one isotope to such states can yield chemical products which are highly enriched. However, in order to examine the potential of the chemical reaction scheme, details of the relaxation and excitation process which take place through collisions following laser excitation must be known. In order to prevent the initial isotopic selectivity from being scrambled rapidly, the non-uranium reaction partner should be present in large excess.

7.3.3 *Laser requirements*
To excite the ν_3 vibration in $^{235}UF_6$, a laser of approximately 16 μm wavelength is required. If the infra-red-excited molecule is to be dissociated by ultraviolet relying on vibronic effects, i.e. coupling between the u.v. and the vibrations, a powerful laser of wavelength of range 300 nm is required. If dissociation is to be achieved by multiphoton effects, then a powerful non-selective laser could be used after irradiation by the 16 μm laser tuned to the $^{235}UF_6$ molecule.

The laser repetition rate is determined by the speed at which the feed enters a stage—the laser is pulsed each time a new volume of feed gas enters the irradiation volume.

7.4 Economics

The discussion in the previous paragraphs is of a general nature, and consequently it is difficult to compare the potential economics of laser isotope separation with that, for example, of diffusion. Howeer, the high separation factor in the laser separation process introduces differences in the design of a cascade of elements. The cascade could be designed to produce a waste containing a small concentration of ^{235}U, and the most efficient operation of a stage is not necessarily the one where $\alpha = \beta$ (see §1)

The most economic application of a laser isotope-separation plant would be to strip the maximum amount of ^{235}U from the tails of a diffusion plant, which might otherwise be argued to be of little value.

8 Conclusions

A review has been given of some of the methods of uranium enrichment. The views expressed are obviously influenced by the suitability of various methods for the UK enrichment programme, although an attempt has been made to give a more global picture.

The enrichment scene is still dominated by the gaseous diffusion process althugh centrifugation is proving attractive to countries not having access to blocks of cheap power.

In the future there will be incentive to strip the waste to a lower concentration of ^{235}U, and for this reason there seems to be considerable interest in the laser isotope-separation technique.

The future demand for enrichment capacity was estimated in the context of the Interntional Nuclear Fuel Cycle Evaluation, and projections from the report of INFCE Working Group 2 are shown in Table 12.4. The corresponding scale of projected nuclear power growth is shown in Table 12.5.

Table 12.4 Installed nuclear power capacity projections in WOCA† countries

Year	High projection‡	Low projection‡
1985	274	245
1990	462	373
1995	770	550
2000	1200	850
2005	1650	1100
2010	2150	1300
2015	2700	1450
2020	3350	1650
2025	3900	1800

Note
The projections presented here are recognized as not being limiting bounds, but representative projections. Both higher and lower capacities are possible, depending on decisions and events occurring in the future.
† World Outside Centrally Planned Economies (CPE) Area
‡ data beyond 2000 are rounded to the nearest 50 GW(e)

Table 12.5 Projected annual separative work demand (MSWU/Y at 0.20% tails assay)

Strategy	1980	1990	2000	2010	2020	2025
H1a	19	53	128	232	365	427
b	19	54	130	194	303	353
H2a	19	54	123	124	95	61
b	19	54	123	146	160	152
H3a	19	54	123	168	216	227
b	19	54	122	127	96	60
H4a	19	52	106	187	289	335
b	19	52	94	129	195	224
H5a	19	54	123	125	95	61
b	19	54	138	272	333	335
L1a	16	40	88	134	175	192
b	17	42	90	111	145	158
L2a	17	42	85	79	58	38
b	17	42	85	91	87	77
L3a	17	41	85	92	84	71
b	17	41	84	84	64	44
L4a	17	40	72	106	136	148
b	17	40	65	74	90	98
L5a	17	42	88	79	55	38
b	17	42	94	147	153	147

Notes
H is high nuclear growth projection
L is low nuclear growth projection
1–5 are illustrative strategies
a is current technical characteristics
b is improved technical characteristics

Appendix A

In eqn (5) the separative power is defined as

$$\frac{L \, \Delta S}{RN(1-N)},$$
(54)

where the entropy S is given by

$$S(N) = R\{N \ln N + (1-N) \ln (1-N)\}.$$
(55)

The separation device has a feed of L gram-mol and produces a product θL at a fraction $N + \Delta N_1$ and a waste $(1 - \theta) L$ at a concentration $N - \Delta N_2$, where θ is the cut.

The total change of entropy is

$$L \, \Delta S = \theta L \, S(N + \Delta N_1) + (1 - \theta) L \, S(N - \Delta N_2) - L \, S(N).$$
(56)

If ΔN_1 and ΔN_2 are small expanding $S(N + \Delta N_1)$ and $S(N - \Delta N_2)$ gives

$$L \, \Delta S = \{\theta L + (1 - \theta)L - L\}S(N) + \{\theta L \, \Delta N_1 + (1 - \theta)L \, \Delta N_2\}S'(N)$$
$$+ \tfrac{1}{2}\{\theta L(\Delta N_1)^2 + (1 - \theta)L(\Delta N_2)^2\}S''(N).$$
(57)

The coefficients of $S(N)$ and $S'(N)$ vanish by conservation of mass and conservation of light isotope.
Then,

$$L \, \Delta S = \tfrac{1}{2}\{\theta L(\Delta N_1)^2 + (1 - \theta)L(\Delta N_2)^2\}S''(N).$$
(58)

From (55) by differentiation with respect to N

$$S''(N) = \frac{R}{N(1-N)}.$$

Then,

$$L \, \Delta S = \frac{R}{2N(1-N)} \{\theta L(\Delta N_1)^2 + (1 - \theta)L(\Delta N_2)^2\},$$
(59)

and from (54)

$$\delta U = \frac{L}{2N^2(1-N)^2} \{\theta(\Delta N_1)^2 + (1 - \theta)(\Delta N_2)^2\}.$$
(60)

From the conservation of mass,

$$\Delta N_1 \theta = (1 - \theta)\Delta N_2,$$

and

$$\delta U = \frac{L}{2N^2(1-N)^2} \frac{\theta}{1-\theta} (\Delta N_1)^2, \tag{61}$$

and, in terms of the enrichment factor,

$$\Delta N_1 = (\alpha - 1)N(1 - N_1) \approx \varepsilon N(1 - N).$$

If $N_1 \sim N$ eqn (61) then becomes

$$\delta U = \frac{\theta L}{2(1-\theta)} \varepsilon^2. \tag{62}$$

Appendix B

Flow through a membrane

This has been studied in great detail but nearly all of the work is classified. The most comprehensive unclassified publications are by R. Present[14] and Massignon[15]. Space does not allow for such a detailed description, but the results will be summarized.

The results of numerous experiments with pure gases have shown that flow through a porous membrane is similar in its essential features to flow through long capillaries, rather than to flow through orifices in a thin sheet. The nature of the flow depends on the ratio of the mean free path to the pore radius a. If the conditions are such that $\lambda \ll a$ the flow will be viscous and obey Poiseuille's law, which for circular capillaries is

$$J = \frac{\pi(p_F^2 - p_B^2)a^4}{16 \eta l \, RT} \tag{63}$$

If the conditions are such that $\lambda \gg a$, the flow will be molecular and given by Knudsen's law, which for capillaries is

$$J = \frac{8\pi a^3}{3l} \sqrt{\left(\frac{1}{2\pi RTM}\right)}(p_F - p_B), \tag{64}$$

where

J = molar flow per unit time down the capillary
a = radius of the capillary
η = coefficient of viscosity
l = length of the capillary
M = molecular weight of the gas
R = gas constant
T = absolute temperature
p_F = fore pressure
p_B = back pressure.

It has been shown in experiments covering a wide range of pressures, that the flow through a membrane consisting of short capillaries or packed bed barriers can be represented by a relationship which comprises the sum of the two component flows. Thus

$$J = \frac{A}{\sqrt{M}}(p_F - p_B) + \frac{B}{\eta}(p_F^2 - p_B^2), \qquad (65)$$

where A and B are constants which must be determined experimentally for any particular membrane. J is the flow rate per unit area. The permeability of a porous barrier is proportional to the porosity δ and is defined as

$$G = \delta \frac{J}{p_F - p_B}, \qquad (66)$$

where the porosity δ is the raio of the volume of the voids to the total volume.

The permeability of a long capillary does not show a linear relation with pressure as deduced from (65). The shape is as indicated in Fig. 14.5. The permeability initially drops before increasing linearly with pressure. The reason for the initial drop is the removal of long paths for molecular streaming when there are collisions in the capillary. Massigon[14] discusses the shape of the permeability curve for a composite barrier where the separating layer is mounted on a sturdy macroporous layer as shown in Fig. 14.6. The permeability is as shown in Fig. 14.7. However, if the holes in the macroporous layer are large enough, then the permeability is practically identical with that of the separating layer.

Membrane separation efficiency

As the pressures are raised on either side of an actual membrane the over-all concentration change ΔN across the membrane will decrease for the following reasons.

1. The existence of a finite back pressure, p_B, will cause enriched gas to diffuse back through the membrane.
2. The higher fore pressure, p_F, introduces some non separative viscous flow.

When the mambrane is operating under non-ideal conditions, the separation efficiency can be defined as the ratio of the actual concentration change N to the ideal fraction change N_{ideal}, i.e.

$$S = \frac{\Delta N}{\Delta N_{ideal}}. \qquad (67)$$

For any membrane the value of S will change with pressure, temperature, and gas.

A simple model problem is one in which it is assumed that the only non-separative flow is the Poisseuile flow. The equations describing the flow are then a generalization of eqn (65), i.e.,

$$J_{\mathrm{L}} = \frac{A}{\sqrt{M_{\mathrm{L}}}} (N_0 p_{\mathrm{F}} - N_1 p_{\mathrm{B}}) + \frac{B}{\eta} N_0 (p_{\mathrm{F}}^2 - p_{\mathrm{B}}^2), \tag{68}$$

and

$$J_{\mathrm{H}} = \frac{A}{\sqrt{M_{\mathrm{H}}}} \{(1 - N_0) p_{\mathrm{F}} - (1 - N_1) p_{\mathrm{B}}\} + \frac{1}{\eta} B (1 - N_0)(p_{\mathrm{F}}^2 - p_{\mathrm{B}}^2), \tag{69}$$

where J_{L} and J_{H} are the flows of the light and heavy components, N_0 and N_1 are the concentrations of the light components entering and leaving the membrane, and M_{L} and M_{H} are the molecular weights of the lighter and heavier species.

The concentration of the gas leaving the membrane is

$$N_1 = \frac{J_{\mathrm{L}}}{J_{\mathrm{L}} + J_{\mathrm{H}}}, \tag{70}$$

and combining this with eqn (67), (68), and (69) gives the following formula for the separation efficiency:

$$S = \frac{1 - r}{1 + \dfrac{(1 - r)(p_{\mathrm{F}} + p_{\mathrm{B}})/p_{\mathrm{C}}}{\alpha + N_1(\alpha - 1)}}, \tag{70a}$$

where $r = P_{\mathrm{B}}/P_{\mathrm{F}}$, $\alpha = \sqrt{(M_{\mathrm{H}}/M_{\mathrm{L}})}$ and P_{c} is a characteristic pressure for the membrane and is given by

$$p_{\mathrm{C}} = \frac{\mu \alpha}{B \sqrt{M_{\mathrm{H}}}}. \tag{71}$$

This expression further simplifies as $\alpha - 1 \ll 1$ to

$$S = \frac{1 - r}{1 + (1 - r)(p_{\mathrm{F}} + p_{\mathrm{B}})/p_{\mathrm{C}}}. \tag{72}$$

For the case of an ideal membrane with pure molecular flow,

$$B = 0 \tag{73}$$

and

$$S = 1 - r$$

while for a non-ideal membrane with zero back-pressure,

$$S = \frac{1}{1 + p_{\mathrm{F}}/p_{\mathrm{C}}}. \tag{74}$$

The above treatment is an over-simplification of the problem as the result of work carried out independently by Present and de Bethune and also by Bosanquet. Peasant and de Bethune considered the case in which part of the molecular flow is non-separative, whereas Bosanquet used the concepts of mean free path and collision frequency to derive expressions for relative persistence of velocity and momentum.

The results of these treatments were to reduce p_c below the value given by eqn (10) and S is then given by

$$S = \frac{\int_{x_B}^{x_F} \exp(1.430x + 0.1142x^2)\, dx}{x_F \exp(1.430x_F + 0.1142x_F^2)}. \tag{75}$$

where $x_F = p_F/p'_C$, $x_B = p_B/p'_C$, and p'_C is a new characteristic pressure.

For the extreme case of $x_B = 0$ (zero back-pressure), the above equation approximates to

$$S_0 = \frac{1}{1 + 0.662x_{50} + (0.338x_{50})^2}, \tag{76}$$

where $x_{50} = p_F/p_{50}$ and p_{50} is the characteristic pressure at which the membrane efficiency $S = 0.50$. This latter expression has been shown to correlate the results of several experiments.

Acknowledgements

I am indebted to many of my colleagues who have read the manuscript and have made helpful comments, in particular Dr Joanna Taylor, Dr P. Hodgkinson, Mr I. Heriot, Mr D. Aston, prof. Sir Rudolph Peierls, Mr D. Wort, Dr S. Whitley and Dr J. Ramsay.

The Figures in this chapter are reproduced by permission of the following authors and organizations: URENCO (Figs. 2, 14, 15), Dr M. D. Massignon and Springer Verlag (Figs. 6, 7), K. W. Sommerfeld (Fig. 9), American Institute of Chemical Engineers (Figs. 10, 17), Eurodel (Fig. 11), Department of Energy, USA (Fig. 16). Pergamon Press (Fig. 18), American Nuclear Society (Figs. 21, 22), Dr R. Jenson and Springer Verlag (Fig. 25).

References

1. LINDEMANN, A. and ASTON, F.W. *Phil. Mag.* **37**, 523–534 (1919).
2. LONDON, H. *Isotope separation*. Newnes, London (1961).
3. COHEN, K. *The theory of isotope separation as applied to the large scale production of* ^{235}U. McGraw-Hill Book Company, New York (1951).
4. BENEDICT, M. and PIGFORD, T.H. *Nuclear chemical engineering*. McGraw-Hill, New York (1957).
5. PRATT, H.R.C. *Countercurrent separation processes* Elsevier, Amsterdam (1967).
6. VILLANI, S. *Isotope separation*, Chap. 4. American Nuclear Society, Hinsdale (1976).
7. BRIGOLI, B. *Uranium enrichment-cascade theory*. Springer-Verlag, Berlin (1979).
8. GRANT, W.L., WANNERBURG, J.J. and HAARHOFF, P.C. The cascade techniques for the South African enrichment process. In *American Institute of Chemical Engineers Symposium Series* **73** (169) p. 20 (1977).
9. GRAHAM, T. (a) *Phil. Mag.* **136**, 573 (1846).
   (b) **153**, 385 (1863)
10. ASTON J.G. *Phil.Mag.*, **39**, 449 (1920).
11. HERTZ, G. (a) *Z. Phys.* **79**. 108, (1932).
    (b) *Z. Phys.*, **79**, 700 (1932).
12. SMYTHE, H.DE W. *Atomic energy for military purposes*. Princeton University Press (1945).
13. JAY, K.E.B. *Britain's atomic factories*. HMSO, London (1954).
14. PRESENT, R.D. and DE BETHUNE, A.J. Phys Rev **75** 1050–57 (1949).
15. MASSIGNON, D. *Uranium enrichment—gaseous diffusion*. Springer-Verlag, Berlin (1979).
16. ALDERTON, G.W. *J. Inst. Fuel*, **49** (401) 210–217 (1976).
17. DE BOER, J.H. *The dynamical character of adsorption*. Oxford University Press (1968).
18. HIGASHI, K., OYA, A., and OISHI, J. *J. Nucli. Sci. Technol.* **3**, 51 (1966).
19. BENEDICT, M., BERMAN A.S., BIGELEISEN, J., POWELL, J.E., SHACTER J., VANSTRUM P.R. *Report of Uranium Isotope Separation Review ad hoc Committee*. Oak Ridge Report *ORO*-694. U.S. Atomic Energy Commission, Washington (1972).
20. VON HALLE, E. The counter-current gas centrifuge for the enrichment of U-235. In *Proc. A.I.Ch.E. Symp. on Recent· Advances in Separation Techniques II*. American Institute of Chemical Engineers Symposium Series Vol. 76 (1980).
21. SHACTER, J.,VON HALLE, E., and HOGLUND, R.L. The gas centrifuge method. In *Encyclopaedia of chemical technology*, vol. 7. pp. 149–165 Interscience, New York (1965).
22. AVERY, D.G. and DAVIES, E. *Uranium enrichment by gas centrifuge*. Mills and Boon, London (1973).
23. OLANDER, D.R. *Adv. Nucl. Sci. Technol.* **6**, 105–174 (1972).
24. MULLIKEN, R.S. *J.Am. Chem. Soc.*,**44**, 1033 (1922).
25. (a) BEAMS, J.W. and HAYNES, F.B. The separation of isotopes by centrifuging Phys. Rev., 50, (5) (1936) pp. 491-492
    (b) BEAMS J.W. and MASKET A.V. Concentration of chlorine isotopes by centrifuging. Phys. Rev. *51* (5)(1937) p. 384 only
    (c) BEAMS J.W. A tubular vacuum-type centrifuge. Rev. Scient. Instrumn. 9 (1938) pp. 413–416.
    (d) BEAMS J.W. Rev. Mod. Phys. **10** pp. 245 (1938).
26. (a) BEAMS J.W. and SKARSTROM C. The concentration of isotopes by the evaporative centrifuge method. Phys. Rev. 56 (3) (1939) pp. 266-272.

(b) Amistead, F.C. and Beams J.W. Concentration of chlorine isotopes by centrifuging at dry ice temperature. Phys. Rev. *57* (4) (1940) pp: 359

27. Beams J.W. Snoddy, L.B. and Kulthau, A.R. Tests of the theory of isotope separation by centrifuging. In *Proc. 2nd United Nations Conf. on the peaceful uses of Atomic Energy, Geneva, September 1958*, vol. 4, pp. 428-434, United Nations, Geneva (1958).

28. Groth, W.E., Beyerle, K., Nann, E. and Welge, K.H. In *Proc. 2nd United Nations Conf. on the Peaceful Used of Atomic Energy, Geneva, September 1958*, vol. 4, pp. 439-446. United Nations, Geneva (1958).

29. Zippe, G. The development of short bowl ultra centrifuges, Report ORO 315, Research laboratories for the engineering sciences. University of Virginia, Charlottesville (1960).

30. Aston, D. *Endeavour*, *2*, 142–148 (1978).

31. Whitley, S. *Phys. Technol.* **10**, 26–33 (1979).

32. Ouwerkek, C. and Los, J. *In Proc. 3rd United Nations Int. Conf. on the Peaceful Uses of Atomic Energy*, p. 1637. United Nations, Geneva (1964).

33. Raetz, E. Uranium isotope separation in the gas centrifuge. Presented at VKI lecture series on aerodynamic separation of gases and isotopes 29 May-2 June, 1978. Von Karman Institute for Fluid Dynamics; Belgium (1978).

34. Martin, H. Z. Elektrochem. **54**, 120–129 (1950).

35. Whipple, R.T. Unpublished work (1962).

36. Tahourdin, P.A. *Final report on the jet separation method. BR 694*, Clarendon Laboratory, Oxford, (April 1953).

37. Becker, E.W. Berkhahn, W., Bley, P., Ehrfeld, U., Ehrfeld, W., and Knapp, U. Physics and Development potential of the separation nozzle process. In *Proc. Int. Conf. on Uranium Isotope Separation. London (1975)*. British Nuclear Energy Society. 1975. pp. 1–7 Session I paper 1.

38. Roux A.J.A. and Grant, W.L. Uranium enrichment in South Africa. In *Proc. Conf. on Nuclear Energy Maturity, Paris, April 1975*. Pergamon Press, Oxford (1975) pp. 39–43.

39. Nathrath, N., Kress, H., McClure, J., Mück, G., Simon, M., and Dubbet, H. Isotope separation in rotating plasmas. In *Proc. Uranium Isotope Conference, London, 1975*. British Energy Society, London (1975), Session II, Paper 8.

40. Vanstrum, P.R. and Levin, S.A. New processes for uranium isotope separation. IAEA-CN-36/12. In *Proc. Int. Conf. on Nuclear Power and its Fuel Cycle* vol. 3 IAEA, Vienna (1977). pp. 215-226.

41. Unpublished Statement by Giraud, M. at IAEA Salzburg Meeting (1977).

42. French Patent Specification 1467 174. Isotopic enrichment of uranium with respect to an isotope. Commissariat à l'Energie Atomique (February 1974).

43. Hardy, C.J. Recent experimental and assessment studies of uranium by ion exchange. In *Proc. Uranium Isotope Conference, London 1975*. British Nuclear Energy Society, London (1975) Session II, Paper 12.

44. Seko, M., Miyake, T and Inada, K. Uranium isotope environment by chemical method. *Nucl. Tech.*, **50** (2), 178 (1980).

45. Dutch Patent 7609046, filed (August 1976). *Procedure for the separation of uranium isotopes by acclerated isotope exchange reaction. Asahi chemical industry Co., Ltd.*

46. Farrar, R.L. and Smith, D.F. Photochemical isotope separation as applied to uranium. Oak Ridge Rep. K-3054 rev I. National Technical Information Service, Springfield. (1972).

47. Nebenzahl, I. Some technical aspects of the laser isotope separation method. In *Proc. Int. Conf. on Uranium Isotope separation, London, 1975*. British Nuclear Energy Society, London (1975) Session I, Paper 5.

48. Robinson, C.P. and Jensen, R.J. Laser methods of uranium isotope separation.

In *Villani, S. (ed.). Uranium enrichment. Topics in applied physics*, vol. 35. pp. 269–290 Springer-Verlag, Berlin (1979).

49. LEVY, R.H. JANES, G.S. Method of and apparatus for the separation of isotopes. *US Patent* 3 772 519 (1973).
50. KANTROVITZ, A. *Wide angle isotope separator. US Patent* 3 940 615 (1976).
51. COX, D.M., HALL, R.B., HORSLEY, J.A. KRAMER, G.M., RABINOWITZ, P., and KALDER, A. *Science*, **205**, 390–394 (1979).
52. AMBARTSUMYAN, R.V., GOROKOV, YU.A., LETOKHOV, V.S. MAKARV, G.N., and PUROTSKY, A.A. *Zh ETC*, **11**, 440 (1976).
53. BARANOV, V.Yu. VELIKHOV, E.P., KOLOMIISKY, Yu.R., LETOKHOV, V.S., NIZYEV, V.G., PISMENNY, V.D., and RYABOV, E.A. *Quantum Electron* **6** no. 5, (1979).
54. PINE, A.S. and ROBIETTE, A.G. *J. Mol. Spectroscopy* (to be published).
55. INTERNATIONAL NUCLEAR FUEL CYCLE EVALUATION *Enrichment availability, Report of Working Group 2. INFCE/PC/2/2*. IAEA, Vienna (1980).

13

Fuel design and fabrication

D. O. PICKMAN

Fuel elements are at the heart of a nuclear reactor, the heat resulting from nuclear fission is generated in the fuel material they contain and conducted into the coolant medium that flows past them.

It is a prime requirement that a fuel element generates its design heat output. This is determined by the quantity of fuel material and its specific heat rating, a function of the material composition, the uranium enrichment, and the neutron flux. A fuel pin, the key component of a fuel element has the fuel material sealed into it, and the cladding material is the primary barrier to the release of radioactive fission products.

Fuel elements have long lives compared with conventional types of fuel, and they remain in reactor for periods varying from about one year (fast reactors) to seven years (Magnox reactors). The rod/oxide types used in the predominant light water reactors are commonly in the reactor for between three and five years. Damage to materials caused by the intense neutron bombardment includes embrittlement, dimensional changes, and swelling. In selecting suitable materials and designs, these changes must be allowed for, as well as compatibility of materials, corrosion, and neutron absorption.

The common fuel materials, uranium metal and uranium dioxide, are produced from the impure U_3O_8 coming from the mines by converting to uranyl nitrate, solvent extraction, thermal de-nitration to UO_3, hydrogen reduction to UO_2, reaction with HF to yield UF_4 (for metal production), and further reaction with F to UF_6 for UO_2 production. The metal is made by magnesium reduction of the UF_4, and UO_2 by reduction of UF_6 by steam and H_2. Where necessary, enrichment is carried out at the UF_6 stage, either by gaseous diffusion through membranes or by the centrifuge process.

Fuel cladding and other metal components are all made by conventional metal-working techniques, and a variety of operations including high-quality precision welding are used in the assembly operation.

High standards of integrity and reliability are required in the manufacture of fuel elements, and well-established quality assurance (QA) schemes are used to ensure that the design intent and required manufacturing standards are achieved.

Contents

1 Introduction

A fuel element is a replaceable component of a reactor core which contains the fissile material, the fuel, in which the heat of nuclear fission is generated. It may comprise a single rod, as in a Magnox reactor, Fig. 13.1, or an array of pins, as in a commercial advanced gas-cooled reactor (CAGR), Fig. 13.2, or a light water reactor (LWR), Fig. 13.3. There are other types; some comprising an assembly of flat or curved plates containing the fuel material are used in naval reactors and materials-test reactors. The fuel for high-temperature reactors (HTRs) consists essentially of small spheres of UO_2 coated with various materials to seal in the fission products. These spheres may be arranged in a bed, as in the pebble-bed version, or embodied in various geometrical forms embedded in a graphite matrix.

The pin-cluster fuel elements used in the majority of power reactors consist of an array of pins together with a support structure, to hold them in the required position for reactor physics and heat transfer reasons, plus certain ancillary features to permit handling, and to maintain stability and freedom from damage. The fuel material in the pins is almost universally a compound of uranium or uranium–plutonium, the principal one being uranium dioxide, UO_2.

Both U and U–Pu can be used as the fuel material in thermal reactors (moderated) and fast reactors (un-moderated), but the usual practice is to

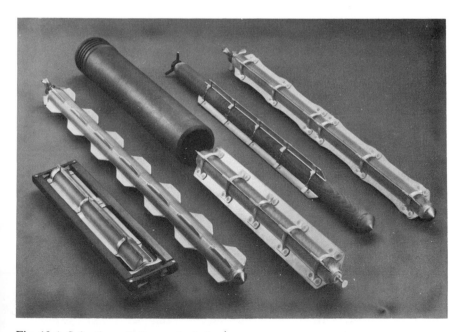

Fig. 13.1 Selection of Magnox fuel elements

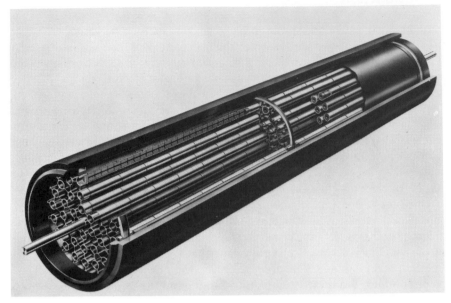

Fig. 13.2 Cut-away CAGR fuel element

use U fuels in thermal reactors and U–Pu in fast reactors. Plutonium recycle in thermal reactors has proved to be of limited interest and serious programmes are under way only in Japan and Belgium. In other countries where reprocessing is practised or planned, the intention is to use the Pu in fast breeder reactors. The fuel material is hermetically sealed into a container, (the can or cladding) generally a round tube, which, in gas-cooled reactors, has an extended external heat transfer surface. The material varies depending on the coolant medium employed and the temperature of operation, the most common cladding materials being stainless steel and zirconium alloys. The Magnox reactors are unique in having only a single rod, the fuel material being a metallic uranium alloy and the cladding material a magnesium alloy with a substantial extended heat transfer surface.

The fabrication of fuel elements involves a variety of conventional chemical and engineering processes, such as solvent extraction, sintering, tube reduction, welding, brazing, casting, and machining, none of which is unique to the nuclear fuel industry. Because the consequence of a fuel element failure in service, such as severe distortion or rupture of the fuel cladding, involves a heavy financial penalty, there is a requirement for high quality and reliability, but again the standards demanded are not unique.

Unusual aspects of the fuel element manufacturing industry are the radioactivity and toxicity of some of the fuel materials used, which require stringent precautions. In the case of plutonium-bearing fuels, mostly used in fast reactors (FRs), this requires a number of the manufacturing operations

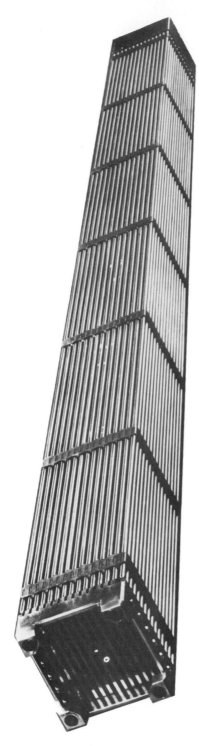

Fig. 13.3 PWR fuel element

to be done under completely remote-handling conditions. Most of the fuels for today's power reactors present no more difficulty or inconvenience because of the nature of the fuel material than is encountered in sterile conditions working in the drug industry, or even in the clean-conditions working required in high-precision engineering.

It is a peculiarity of fuel elements that feedom from certain impurities is required. Some elements, such as boron, hafnium, and cadmium, have the property of absorbing neutrons to a marked extent and must be rigorously excluded; others, such as fluorine, chlorine, and hydrogen can be involved in attack on the cladding or structure, sometimes assisted by the radiation field within the reactor core and must be maintained below specified limits.

2 Design and performance requirements

A certain heat output is required of the fuel elements in a reactor core, and the designer is particularly concerned with the maximum heat rating at which the fuel has to operate, since this is generally the most onerous condition. There are systematic variations in heat rating, both axially and radially, caused by neutron leakage from the core, which determine the maximum heat rating. A constraint on maximum permissible heat rating is the consideration of how the fuel elements will behave under abnormal or accident conditions, and in general terms design limits are set by a combination of material property limitations in normal steady running and abnormal conditions.

The designer is also required to ensure that his design and specification, in terms of engineering limits and material specifications, is attainable at the necessary quality level in large-scale production, bearing in mind radiological and toxicity limitations.

Some of the more important requirements are now discussed.

2.1 Fission-product retention

A range of fission products, some solid, some gaseous, and some volatile at operating temperature, are produced in the fuel material as a consequence of the fissioning of uranium and plutonium atoms. Many of these fission products are highly radioactive, and it is necessary to prevent their release either to operating areas within the generating station or off-site. In all reactor designs there are a number of barriers to such release, the first of which is the fuel cladding. The cladding must be strong enough to accommodate the internal pressure build-up from any release of gaseous and volatile fission products, and chemically compatible not only with the fuel material itself and with the reactor coolant, but also with the fission products. The internal pressure generated can be a problem: it must not be allowed to cause significant creep of the cladding away from the fuel or temperatures will rise and more fission products will be released. The design intent in most reactors is that internal pressure will not exceed reactor coolant pressure, but the sodium-cooled fast reactor (LMFR) with its unpressurized coolant circuit is an exception. The problem here has to be solved by the choice of high-strength, creep-resistant cladding materials. The internal gas pressure is also of importance as a potential cause of uncontrolled swelling of the cladding in foreseeable accident situations. The maximum pressure can be controlled by the inclusion of a gas storage plenum at the ends of the fuel pins in designs where the pins are full core length, but in designs such as CAGR, with segmented pins, this is not permissible. However, the hollow UO_2 pellets used provide some gas storage volume and beyond that the heat rating must be limited to control the amount of gaseous fission products released from the fuel material. The uranium metal fuel used in Magnox reactors releases very little of the gaseous fission products, so there is no corresponding problem in these reactors.

A further source of activity release, other than fission products, is active species such as ^{60}Co produced by neutron bombardment of the cladding material, which can be transferred to the coolant in the form of corrosion products. This source of activity release from the fuel elements cannot be entirely suppressed, but must be minimized by choice of suitable cladding materials.

2.2 Dimensional changes

The accommodation of thermal expansion is not an unusual engineering requirement, but it arises in an acute form in fuel elements because of the very steep temperature gradients in the fuel material, which lead to major differential expansions both axially and radially between fuel and cladding. The axial differential expansion problem is minimized in UO_2–pellet–fuelled pins by *dishing* of the pellet ends so that fuel expansion is determined by the outer regions at a lower temperature. When relatively weak cladding is used, as in CAGRs, it is also necessary to lock the fuel and cladding together by pressurizing into grooves in the fuel. This process is also used with Magnox fuel rods.

The differential thermal expansion could be minimized by the choice of high-conductivity fuel materials, by the best possible matching of material expansion coefficients, and by ensuring that the temperature drop across the gap between fuel material and cladding is reduced as far as possible. Because of other over-riding requirements, only limited success can be achieved in meeting these criteria, and the final designs are in all cases compromises with a need for close control on clearances and tolerances. A good example of the difficulties can be seen in the cases of LWR and CAGR fuel pins. In both cases UO_2, with a low thermal conductivity, has major advantages over other fuels. While in the LWR case zirconium alloy cladding, with an expansion coefficient about half that of UO_2, has obvious advantages over stainless steel, in the CAGR case the reverse is true: here stainless steel, with an expansion coefficient almost twice that of UO_2, is preferred, because at the higher cladding temperatures (700 °C) zirconium alloys undergo severe oxidation and embrittlement.

Gaps between fuel and cladding are generally kept to within the range 0.05–0.25 mm and filled with helium; in the case of LWR pins, where a significant gap may be maintained for a long time because of the strong cladding, the He is generally at an elevated pressure. This reduces the dilution by released fission-product gases, so minimizing the strong adverse effect which they have on the gas gap conductivity.

Radiation-induced dimensional changes occur both in the fuel and cladding. In UO_2 for example, the solid fission products, together with that proportion of the gaseous fission products retained, cause a swelling which will eventually appear as a strain in the cladding unless space for accommodation is provided. It is usual to provide some free volume in the form of distributed porosity in the UO_2 and macroscopically by dishing the end faces of the UO_2 pellets. If the porosity is too fine (5μm), however, the UO_2 readily sinters up, causing a shrinkage of the pellet. A further effect found in the high-temperature regions of pellets is a migration of irradiation-induced voidage to the centre of the pellet, leading to an expansion. There are further complications in that UO_2 pellets crack under the thermal stresses, and there is evidence that fragments move outwards to reduce the gap between fuel and cladding.

Radiation causes dimensional changes in some cladding materials, either a change in shape, such as an increase in length of zirconium alloy cladding tube, or an increase in volume. The latter effect, known as neutron-induced voidage (NIV), is peculiar to the high fast neutron dose received by materials in fast breeder reactors. Both effects can be substantial and life-limiting.

Dimensional changes are also influenced by changes in material properties under irradiation. It is characteristic of metals that their strength properties (yield and tensile strength) increase and ductility reduces. Creep behaviour on the other hand, which is a most important property in determining fuel element behaviour, may be strongly influenced, as with the major reduction in creep strength of zirconium alloys, or little affected, as with stainless steel cladding in the low fast neutron fluxes in thermal reactors.

The fuel element designer needs to have basic data on all these dimensional change effects and on the mechanical properties under relevant neutron flux and dose conditions.

2.3 Burn-up

Reactors are designed to a specified burn-up figure which may be expressed as a percentage of heavy atoms (uranium or plutonium) fissioned, or as a heat output per unit weight of fuel, frequently expressed as MW d/t U. By definition, natural uranium fuel is limited in attainable burn-up, and the higher the burn-up required, the larger must be the initial enrichment in fissile U or Pu atoms. From a fuel designer's point of view, burn-up is important in many respects. For any given reactor type, a higher burn-up means greater fission-product generation, and additionally leads to changes in material properties, to dimensional changes, and so on. The drive to get maximum utilization out of uranium resources and the high cost of fuel reprocesing are both factors leading designers to seek higher burn-up in nuclear fuel. There are certain potential life-limiting features which have to be evaluated, notably cladding corrosion, fission-product gas release, UO_2 swelling, and NIV swelling in fast reactor fuel components. Each of these factors can lead to an increasing incidence of leaking fuel pins, which represents a 'wear-out' situation, unlike the possible occurrence of leaking fuel pins at an earlier stage because of faulty design, manufacture, or operation.

2.4 Neutron economy

The economics of nuclear electricity generation are dependent on the most efficient utilization of neutrons in the reactor core. From the fuel element viewpoint this means minimizing the parasitic capture of neutrons by the materials of construction or by impurities in the fuel. Materials vary widely in their neutron absorption characteristics, usually expressed as capture cross-sections in a unit of area known as a barn (10^{-28} m^2)—a convention discussed in more detail in Chapter 1. Capture cross-sections of some materials of interest for thermal neutrons (2200 m/s) are magnesium, 0.069; zirconium, 0.185; stainless steel, 2.94; hafnium, 105; and boron, 755.

The fuel designer will choose the lowest neutron cross-section material which satisfies his other essential requirements and will want to minimize the quantity of such material in the core. The freedom of choice available is very limited, bearing in mind temperature of operation, corrosion, and strength, but it is clear why in LWRs zirconium alloys are preferred to stainless steel. Such alloys are unsuitable for CAGRs because of poor oxidation resistance at working temperature. It is also clear why common impurities, such as boron in stainless steels and hafnium in Zr, have to be reduced to low levels in nuclear applications.

2.5 Radioactivity and toxicity

In the manufacture of uranium-bearing fuels the problems posed by radioactivity and toxicity (by ingestion or inhalation) are minor, and not dominant in deciding in what form to use the fuel material, or how to fabricate it into that form. There is, however, a preference for clean processes not leading to dusty atmospheres or accumulation of fuel material in crevices or other parts of the plant. The increasing activity level associated with recycled uranium will make it more attractive to move towards remote-handling methods and to processes particularly suited to them.

In the manufacture of plutonium-bearing fuels for fast reactors, the high Pu content has already led to the need for remote operaton in fully contained facilities. In such conditions the use of fine powders, which lead to dusty atmospheres, is unattractive and has led to increased interest in gel precipitation· processes by which small spheres of (U–PuO₂) can be produced directly from solution. These are further discussed later in this chapter (§4.4). After debonding and sintering, such spheres are a suitable clean feed material for filling cans by means of a vibratory compaction process known as Vipak. As an alternative, methods of compacting such spheres directly into pellets are currently being examined. From a performance point of view pellet fuel is preferred because of its better thermal conductivity, but it may be necessary to sacrifice some performance and use Vipak fuel if dust-free processing is to be achieved.

2.6 Fuel pin assemblies

One of the prime functions of the assembly of pins used in most modern designs of reactor is to preserve the integrity of the fuel pins throughout their irradiation life, while also preserving the regular design spacing. Spacing is usually controlled by a series of axially separated spacer-grids attached to some full-length structural members. A little-used alternative is a helical wire wrapping around each fuel pin, the assembly being then bundled together either by external bands or within a shroud-tube. The use of external shroud-tubes introduces a lot of parastitic material, and control of the fit of the pins to avoid vibration and fretting damage is difficult.

Increasing attention is being devoted to dismantling of irradiated fuel assemblies prior to chemical reprocessing, but there is no hard evidence to favour any particular design style at the present time. In the case of fast reactor fuel, the NIV swelling of cladding could make pin extraction from gridded assemblies difficult. If experience shows this to be the case, more attention may be have to be given to alternative concepts such as wire wrapping, which is more attractive than with the other common fuels because an external shroud is already used for other reasons.

3 Fuel element design

The fuel element designer has the complex task of meeting the various requirements of neutron economy, heat output, heat transfer, corrosion resistance, and burn-up, taking account of known irradiation effects in fuel and structural materials. He must know and allow for the stresses and strains that will arise from differential expansion effects, resulting from the large temperature gradients in the fuel material, from the coolant pressure and flow, and from fission gas release. Fission-product compatibility with the cladding, specific effects arising from operational manoeuvres, such as power changes, and behaviour under a whole range of probable and improbable fault conditions must all be taken into account.

Inevitably, knowledge of the demands made on fuel elements and the behaviour of the materials and structures has been an evolutionary process that is not yet complete. Fuel designs have changed as a result of operational experience as well as in response to pressure for improved performance, improved economy and cheaper fabrication. In the LWR fuel area, for example, there is currently strong pressure to introduce designs that will reliably achieve a 50 per cent increase in burn-up (to 45 000 MW d/t U), and will give the reactor operators complete operational flexibility with respect to power changing, that will survive anticipated faults without damage and will retain a coolable geometry under the most improbable accident conditions.

3.1 Uranium–Magnox fuel elements

The Magnox reactor was used for the first generation of British power reactors. A similar type has been built in France, and there are single reactors of the same type in Japan and Italy. The reactor coolant is CO_2 and the fuel elements occupy vertical cylindrical channels in the graphite moderator block. These reactors use natural uranium fuel and so require low-neutron-capture materials within the core.

A uranium alloy with small amounts of iron and aluminium is used as the fuel material in these reactors with a magnesium alloy as cladding. Magnesium has the lowest thermal neutron capture cross-section of the common metals (0.069 barns), but its oxidation rate in CO_2 is high and a special alloy, Magnox AL80, containing 0.8 per cent A1 and 0.003 per cent Be was developed at Harwell to give improved oxidation resistance.

The fuel element designs[1] have all used short (c. 1 m) single rods (Fig. 13.1) stacked 6 to 12 high in the moderator channels. Uranium metal undergoes a phase change at 662 °C with a significant volume change, and this represents the maximum permissible limit on uranium centre temperature. However, uranium metal has a good thermal conductivity, and it is possible to use rods of about 25 mm diameter at the attainable heat ratings.

Because of the poor heat transfer performance of CO_2, and the need to keep maximum cladding temperatures below about 450 °C, extended heat transfer surfaces are needed. Two principal types now used in the UK reactors are known as helical and herringbone. An early type used in the Calder Hall and Chapelcross reactors had transverse finning with a central brace assembly to prevent bowing.

The uranium rods are sealed into the cladding tubes with screwed and welded cup-shaped end caps. Conical end fittings are screwed to the end caps, the male on the bottom end, the female on the top.

The final design feature is near-channel-diameter braces, or lugs, on the herringbone-type cans, to support the elements centrally in the channel and prevent excessive bowing, which would otherwise arise from the compressive loading on the elements in combintion with the low irradiation creep strength of uranium in the low-temperature positions. In the helical elements these braces fulfil another important function, which is to separate the CO_2 flow into four quadrants, so producing a spiralling flow in each quadrant and improved heat transfer coefficients.

Uranium shows pronounced irradiation growth at low temperatures if there is a preferred crystallographic orientation of grains. It will also develop a rough irregular surface if the grain size is large. Strains resulting from these effects can cause cavitation creep failures in the cladding if the grain size and orientation are not well controlled with a minimum number of grains across the wall thickness being retained.

The initial irradiation trials of helical elements showed up a dynamic stability problem. Pressure differentials between the four quadrants caused the elements to rattle within the clearance of the support arms on the top end fittings, the bumping from one support arm to the next caused flexing of the heavy rods, and the resulting strains in the cladding led to early fatigue failures. This problem was solved by attaching a spring stabilizer to one of the support arms. The later herringbone can design was inherently much more stable and the stabilizer was found to be unnecessary.

Under thermal cycling conditions uranium bars may ratchet within Magnox cans because of their differential thermal expansion. This process results in a gap forming at one, or both, ends and a rapid creep failure of the unsupported can. This problem was solved by machining an array of anti-ratchetting grooves into the uranium bars, concentrated close to each end, and pressurizing the can on to the rod.

In two of the Magnox reactors, Berkeley and Hunterston, the uranium rods are independently supported, between graphite struts and in a graphite sleeve respectively, the rod resting on a zirconium alloy strut. The Tokai Mura reactor has a tubular U fuel rod which is also supported in a graphite sleeve. The French reactors use a different U alloy (Si, Cr, Al) and tubular fuel rods, while Bugey, the last of the series, has a larger-diameter tubular rod with external and internal coolant flow. The cladding alloy used in France is Mg with 0.6 per cent Zr, which requires a barrier layer on the bore to prevent outward diffusion of Pu.

Design parameters for typical Magnox fuel elements are given in Table 13.1.

3.2 AGR fuel elements

Higher heat ratings and improved thermal efficiency can be achieved in CO_2-cooled, graphite-moderated reactors by a greater sub-division of the fuel and higher coolant pressures and temperatures. Such a reactor type is the advanced gas-cooled reactor (AGR), which has been built as the second generation of reactors in the UK nuclear power programme, and two more stations of this type have recently been ordered at Torness and Heysham.

Table 13.1 Fuel element design parameters

Reactor	U–MAG NOX	AGR	PWR	BWR	SGHWR	PHWR	FR	RBMK
fuel material and type	U Rod	UO$_2$ hollow pellet	UO$_2$ pellet	UO$_2$ pellet	UO$_2$ pellet	UO$_2$ pellet	UO$_2$-PuO$_2$ hollow pellet & Vipak	UO$_2$ pellet
Fuel diameter (mm)	28	14.5/6.3	8.2	10.57	10.55	12.2	5.03/1.7	11.5
Cladding material	Magnox A180	Stainless steel	Zircaloy 4	Zircaloy 2	Zircaloy 2	Zircaloy 4	Stainless steel	Zircaloy 2
Pin diameter (mm)		15.2	9.5	12.52	12.19	13.08	5.84	13.6
Cladding thickness (mm)	2.1	0.38	0.62	0.86	0.65	0.4	0.38	0.83
No. of pins per element	1	36	264	63	57	37	325	18
No. of elements per channel	8	8	1	1	1	12	1	2
Nominal peak rating W/g	2.6	26	76	56	43	37	243	38
Nominal peak rating kW/m	30	35	42	44	41	46	45	41
Mean core rating W/g	1.6	14	33	23	17	18.5	108	—
Mean core rating kW/m	18.5	18	18	18	16	23	20	—
Nominal burn-up MW d/t U	5000	18 000	30 000	25 000	21 500	8500	83 000	19 500

The higher coolant temperature requires a change of fuel and cladding materials, and all AGRs use UO_2 fuel and stainless steel cladding. These changes impair the neutron economy and enrichment is therefore required, giving a much increased burn-up (18 000–20 000 MW d/t U channel average) compared with the Magnox reactors (up to 5000 Mw d/t U).

The AGR designers adopted a rod bundle, with 36 small-diameter pins arranged in a wide-spaced array, Fig. 13.2. With the good high-temperature properties of stainless steel, only a surface roughening in the form of machined transverse ribs is required to give adequate heat transfer. AGR fuel, like that in Magnox reactors. is divided into separate elements about 1m long tor ease of charge and discharge in circular channels in the graphite moaerator block. The elements are all loaded and unloaded as one unit supported on a central tie bar. The individual fuel pins are supported by and attached to skeletal machined end grids with an intermediate support brace. This whole array is in turn supported within a double skin graphite sleeve, which serves to insulate the moderator from the hot coolant. In addition a purge flow of cool CO_2 around the outer sleeve helps to keep the graphite moderator cool.

In order to avoid a mass of absorber within the core, AGR fuel pins have thin-walled, cup-shaped end caps. After welding the pins are pressurized into anti-stacking grooves in some of the UO_2 pellets to avoid gaps opening up in the pellet stack as a result of differential movements. The relative weakness of the thin AGR cladding will cause it to creep rapidly during this operation, which imposes rigorous restrictions on the use of chipped pellets if local clad overstrain or even rupture is to be avoided.

Because AGR fuel stringers are segmented, it is not economic to store the fission gases (mainly krypton and xenon) released from the UO_2 in an end plenum, as is commonly done in water reactors. This puts a restriction on peak rating, to minimize gas release, but the use of hollow UO_2 pellets has now been adopted, so providing a central gas storage volume, as well as reducing the amount of gas actually released. The use of hollow pellets has also overcome problems of defects caused by power cycling and power ramping, which could arise because of the rapid creep-down of the stainless steel cladding and its embrittlement by helium bubbles formed by the transmutation of boron present in the steel as an impurity. The UO_2 pellets also have a small dish of truncated-cone form on each end to accomodate differential axial expansion.

The fuel element designs in all the commercial AGRs (CAGR)[2] are essentially the same, differing only in certain items of peripheral hardware. No other country has adopted this reactor type, but the French built a CO_2-cooled, D_2O-moderated reactor (EL4), whose fuel, although originally intended to have beryllium or zirconium cladding, is not unlike that in AGR. It is interesting to note that there was an intention in the UK to use beryllium cladding for AGRs, since it has a very low thermal neutron capture cross-section (0.01 barns), but this ran into development problems, mainly severe embrittlement and corrosion, and was abandoned because of limited development potential.

Design parameters of a typical CAGR fuel stringer are given in Table 13.1

3.3 High-temperature reactor fuel elements

A logical continuation of the gas-cooled reactor line is a reactor with a very high outlet gas temperature. The higher gas temperatures allow increased thermal efficiency for electricity generation using the steam cycle. Alternatively it can be exploited in conjunction with a closed-loop gas turbine generator or in the supply of hot gas directly to certain industries. Such reactors have been built and are operating in the United States and in Germany, while interest is being taken in their development in Japan to supply process heat for steel-making. There are at the moment, however, no firm plans for the construction of large commercial HTRs (high-temperature reactors).

Fuel elements for HTRs have varied widely in design and are very different from those used in AGRs and LWRs. Because of the high outlet gas temperatures involved (750–900 °C) and more particularly the high fuel surface temperatures (900–1100 °C), the conventional type of metal-clad UO_2 fuel cannot be used, even in helium-cooled reactors, which are the preferred type.[3]

Two basically different core designs have emerged, generally referred to as *prismatic* and *pebble bed*. The fuel is a coated-particle type, having a spherical UO_2 or UO_2–ThO_2 kernel coated in layers of pyrolytic carbon and silicon carbide to provide a high-integrity containment for fission products, Fig. 3.4. These spheres are compacted in a graphite matrix to provide fuel compacts of the desired geometry. In the prismatic core, fuel rods are inserted in removable graphite moderator blocks, through which the coolant also passes; in an alternative design annular fuel compacts are inserted in graphite tubes or integrally bonded to the tubes, which are located in channels in the graphite moderator block and are directly cooled, Fig. 13.5. In the pebble-bed reactor the fuel compacts are of the same composition, i.e. coated particles dispersed in a graphite matrix, but are formed into spheres approximately 60 mm diameter. The spherical compacts form a moving-bed reactor core directly cooled by helium and allowing lower fuel temperatures for the same gas outlet temperature as the indirectly-cooled prismatic design.

A similar concept of large-diameter spheres, but with BeO as the matrix material, was developed in Australia, but has not been pursued.

3.4 Water reactor fuel elements

There is a big diversity of types of water reactor fuel element, with the pressurized (PWR) and boiling (BWR) light water-cooled and-moderated types predominating. Other commercially important reactor systems are the

Fig. 13.4 HTR coated particles (sections)

Fig. 13.5 HTR fuel components

pressure-tube type pressurized heavy water reactor (PHWR) typified by the CANDU/PHW developed in Canada, and the graphite-moderated, H_2O-cooled RBMK reactors being installed in quantity in the USSR. Additionally, a light water-cooled, steam-generating heavy water reactor, SGHWR, has been developed in the UK, but has now been abandoned, and a pressure-vessel type PHWR, the MZFR, has been developed in Germany.

Most water reactor fuel elements have features in common, including smooth-surface zirconium alloy cladding containing enriched UO_2 fuel in pellet form, full core height fuel pins, and square fuel assemblies (Fig. 13.3 and 13.6).

The following are some of the exceptions.
1. A few PWRs have stainless steel cladding.
2. The Canadian CANDU/PHW has short (495 mm) circular fuel assemblies fuelled with natural UO_2, which are shuffled axially in the horizontal pressure tube to achieve maximum utilization of the fuel.
3. The German MZFR has natural UO_2 fuel located in thin circular flow-control tubes (separating the heavy water moderator and coolant) within a pressure vessel. The Atucha MZFR type reactor has the longest fuel assembly of any power reactor at 5.3 m active length.
4. The Russian RBMX reactors have two circular assemblies in each vertical pressure tube, fixed on a tie-bar and with a combined length of 6 m.
5. The British SGHWR has full-length circular assemblies in vertical pressure tubes.

The CANDU/PHW fuel is good example of extreme neutron economy and the fuel assemblies are all Zircaloy 4, a Zr, Sn, Fe, Cr alloy, with cladding only 0.38 mm thick. Despite the need for neutron economy all other water reactor fuel elements contain some in-core components of stainless steel or nickel base alloys.

Some typical design parameters for a range of water reactor fuel elements are given in Table 13.1. In view of their predominance the design features of the PWR and BWR fuel elements only are described in more detail.

3.4.1 *PWR*
These reactors have square, full-core-height fuel elements, [4,5] comprising a basic support structure of unfuelled stainless steel or Zircaloy 4 tubes, attached to top and bottom end fittings, to which a number of spacer grids, generally in the range six to nine are attached (Fig. 13.3). The fuel pin arrays vary from 14×14 to 17×17, less the support tubes, the trend over recent years having increasingly been towards the latter type of element with smaller-diameter fuel pins (See Chapter 4, §5.22). A number of elements have rod control clusters attached at their top end. Slender control rods of Ag, In, Cd, alloy, clad in type 304 stainless steel, are located in support tubes (guide thimbles) and are withdrawn on reactor start-up. they are not normally used for power control or shaping. In a typical Westinghouse element, there are 24 guide thimbles and a central instrumentation tube for in-core neutron

flux and coolant exit temperature monitoring. The bottom end fitting is positively located on the core bottom grid plate, and the element is spring loaded against the hold down system either at the bottom or top end, to allow for differential thermal expansion and irradiation growth relative to the core structure.

The fuel pins in PWR elements are usually floating, i.e. not anchored at either end, and are centred to provide equal end clearances. This arrangement influences the amount of fuel rod bow that develops in service. The spacer grids grip the pins, each of which is held centrally in its grid cell by two integral springs which load it against four support dimples with a load in the range 4.5–7 kg. Spacer grid details vary widely and are considered to be proprietary by most suppliers. The material of construction is generally Inconel, but there is a trend towards the use of all Zircaloy structures for neutron economy reasons. If spring loading is lost, there is a risk of serious fretting, which can rapidly penetrate the fuel rod cladding; relaxation characteristics of the spring are therefore important. Most PWR spacer grids also incorporate flow deflectors to promote mixing of the coolant between sub-channels and improve heat transfer performance.

The heart of the fuel element is the fuel pin, and the design features of a typical fuel pin are shown schematically in Fig. 13.6. It is essentially a long straight Zircaloy 4 tube, containing close-fitting UO_2 pellets and having a short unfuelled length (the gas plenum) at one or both ends, usually at the top. The gas plenum in most pins contains a helical stainless steel spring, whose sole purpose is to hold the UO_2 pellet stack together during transport and handling operations. The fuel tube is sealed at each end by Zircaloy 4 end plugs.

Fuel pin design determines the performance characteristics under irradiation, the burn-up level that can reliably be achieved, and the response to operational manoeuvring of the reactor power. The following are important aspects of the fuel pin design.

1. *Choice of cladding thickness.* All PWRs have cladding which is elastically stable under the coolant pressure of about 150 bar. This, combined with a desire to minimize creepdown rate leads to thickness/diameter ratios of about 0.05 or greater. The 17×17 Westinghouse design has a clad outer diameter of 9.5 mm with a wall thickness of 0.57 mm.

2. *The UO_2–clad gap.* This has an important effect on the UO_2 temperatures, and hence on the release of fission products, which is highly temperature sensitive, as well as on the component of UO_2 swelling caused by bubbles of these fission gases on UO_2 grain boundaries. There is also a desire to avoid immediate contact at power between the UO_2 and the cladding tube, and in the PWR this is usually delayed to some time within the second cycle (second year of operation). A typical design clearance (cold) is about 0.15 mm diametral.

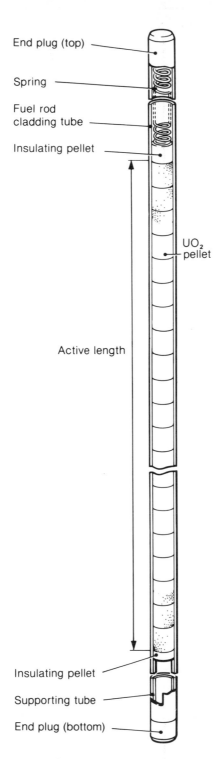

End plug (top)

Spring

Fuel rod
cladding tube

Insulating pellet

UO₂
pellet

Active length

Insulating pellet

Supporting tube

End plug (bottom)

Fig. 13.6 PWR fuel pin design (schematic)

3. *The condition and surface treatment of the cladding tube.* The metallurgical condition influences creep rate and response to fission-product-induced stress-corrosion cracking. There is no consensus opinion, but PWRs in general use the Zircaloy 4 tubing in a cold-worked condition, with a stress relief anneal at around 500–540 °C for 4 hours. Surface condition has been a contentious issue: most suppliers do not now employ either pickling or autoclaving as finishing processes, but prefer abrasive grit blasting for the bore and belt, grinding for the outside surface, which minimize the risk of contamination and consequent accelerated corrosion. There are, however, arguments for a very smooth bore surface (minimum gap thermal resistance after contact and reduced risk of stress-corrosion crack initiation), and alternative processes are being studied.

4. *Filling gas composition and pressure.* Helium is used universally as a fuel pin filling gas because of its good thermal conductivity. It also serves as a tracer gas when the pin is tested for leaks in a mass spectrometer. The original practice was to fill pins at atmospheric pressure, but two major disadvantages became apparent. Firstly, the clad rapidly crept down on to the UO_2 under the high coolant pressure and would then creep into any gaps in the UO_2 pellet stack, sometimes resulting in clad rupture. The mode of creepdown ovalizes the cladding, which leads to high local bending stresses on a power increase, again with the risk of clad rupture. Secondly, the gap thermal conductance reduced rapidly as fission gases, xenon and krypton, were released from the UO_2 causing a feedback effect on UO_2 temperature and fission-product release. For these reasons it has become the practice to pre-pressurize PWR fuel pins with helium to pressures in the range 20–30 bar (cold). The initial hot pressure is therefore in the range 60–90 bar, thus halving the initial clad compressive stress.

5. *The plenum volume.* There is some gas storage volume available within the UO_2 pellet stack, and the additional volume required to ensure that the internal gas pressure never exceeds the coolant pressure is small, of the order of 20 ml.

6. *Design of UO_2 pellets.* Most PWRs use solid UO_2 pellets; a few however, including the Russian PWRs, have hollow pellets, and there is an increasing interest in this type. In operation the UO_2 pellets are subject to large temperature-gradients, which result in centre-to-edge differential expansion and the development of major radial and sometimes transverse cracks. It is common design practice to avoid transmitting large axial strains to the cladding, by including a hemispherical or flat-bottomed dish of less than full pellet diameter in each end face. The controlling radius for determining pellet stack axial expansion is where the dish runs out, usually about 0.5–1.0 mm from the outside of the pellet. Another important function of

the dishing is to provide some volumetric allowance for UO_2 swelling. These dishes enhance the adoption of a 'wheatsheaf' shape in the hot pellet (ends larger diameter than centre), but this disadvantage, and transverse cracking at mid-pellet length, are both reduced by having a pellet length/diameter ratio not much greater than unity. The use of pellets chamfered on their edges has been shown to minimize inter-action with the cladding, but has not been widely adopted for large-scale manufacture.

7. *Pellet density*. It is conventional in PWRs to use pellets of about 95 per cent density, i.e. 10.4 g/cm³. Full theoretical density (10.95 g/cm³) is not achievable with normal production methods of pressing and sintering, but the density in use is deliberately chosen to provide further accommodation for solid fission products having a larger volume than that of the fissile atoms from which they are produced. The swelling rate due to solid fission-product accumulation is at most 0.4 per cent per 10 000 MW d/t U burn-up, and peak pellet burn-up in PWRs is currently about 55 000 MW d/t U, so that the built-in allowance is more than sufficient. No allowance is made for fission gas bubble swelling, because this is readily restrained by the cladding and coolant pressure.

The differences in PWR fuel pin design between the various suppliers are relatively small. There are some differences in cladding thickness, clad–fuel gap, pin diameter, gas plenum location, pre-pressurization and UO_2 pellet density and design, but all designs appear to have achieved satisfactory full-life reliability. No major changes are foreseen, but minor ones may arise as a result of present pressures to improve uranium utilization and to increase burn-up, so reducing the quantity of fuel to be stored or reprocessed.

3.4.2 *BWR*

The fuel elements in boiling-water reactor[6] are square assemblies of full-core-height fuel pins, but, unlike PWRs, several of the pins also act as structrual members, joining the top and bottom fittings. They have a smaller number of larger-diameter pins than PWRs, typically 7×7 or 8×8, with one or two positions occupied by unfuelled water-filled tubes (water rods), used to control flux peaking effects and to contain in-core instrumentation. BWR fuel elements also have a square external shroud-tube, which is attached to the top end fitting, and relative to which the remainder of the assembly can slide. The bottom end fitting sits in a circular hole in the core plate and can be gagged to a variable orifice size. There are no control rods within a BWR assembly, but cruciform shaped rods occupy the space between some groups of four adjacent elements. The shrouds assist easy insertion and withdrawl of these rods (which contain an array of stainless steel tubes filled with B_4C) and also permit control of the steam/water ratio (steam quality) within the individual elements.

Typical BWR elements have a *handle* attached to the upper end fitting on which the hold down bars rest to prevent levitation. There is spring loading of fixed and floating fuel pins around spigots attached to the top end caps, which engage in the top ending fitting. Floating pins are also spigoted into the bottom end fitting.

The fuel element shrouds are of Zircaloy 2 and of substantial thickness to resist dilation from differential pressure, and distortion. The aim in a BWR is to reuse these shrouds, and their corrosion resistance is of importance.

BWR fuel assemblies have spring spacer grids to maintain pin spacing. These are generally of a zirconium alloy with non-integral Inconel springs attached and bearing on two or four neighbouring pins. The spacer grids are attached to a non-fuelled *water rod*, in a central or near-central position.

Fuel pin design principles and construction are similar to those of a PWR, but the somewhat lower ratings, the two-phase coolant, and the lower coolant temperature and pressure result in some differences. The pins are larger in diameter, those of the GE 8 × 8 design being 12.5 mm and clad in 0.86 mm wall Zircaloy 2, which has lower H_2 pick-up than Zircaloy 4 in the more oxygenated coolant of a BWR. Because of the reduced coolant pressure, pins have not commonly been pre-pressurized, but there is a recent trend in this direction, using cold helium filling pressures of around 5 bar. This additional pressure results in lower fission gas release and avoids extreme situations in which a gas release can run away by a feedback mechanism resulting from deteriorating gap conductance. Pre-pressurization also helps to minimize UO_2 pellet–clad interactions (PCI), a more common cause of clad rupture in BWR pins because of the large local power changes brought about by the necessary periodic control rod movements. Because of this greater sensitivity to PCI and resultant clad rupture, which is almost certainly a fission-product stress-corrosion cracking phenomenon, there is a tendency to use more fully stress-relieved or even fully annealed cladding and UO_2 pellet designs, optimized to reduce the amount and effects of wheatsheafing.

3.5 Fast reactor fuel elements

All operating fast reactors are liquid sodium cooled and are able to be operated at near to atmospheric pressure because of the high boiling-point of sodium. Sodium is also an excellent heat transfer medium and high linear heat ratings can be used, thereby minimizing core size. Relatively small diameter fuel pins are used to minimize the plutonium inventory. The fuel material is generally a UO_2–PuO_2 mixture, although highly enriched UO_2 can be used. The main advantages of the FR are that it is an efficient burner of plutonium, and that the unmoderated neutrons are able to convert ^{238}U to ^{239}Pu more effectively than in a thermal (moderated) reactor, and so produce useful fissile plutonium from the non-fissile ^{238}U.

The basic design style of a FR fuel element,[7, 8] usually known as a sub-assembly, is not greatly different from that of a PWR, but there are important differences in detail which arise from the high linear heat rating, the high fast neutron flux environment, the high burn-up, and the low-pressure coolant.

All FR sub-assemblies have external shrouds, known as *wrappers* in the UK, of hexagonal form, and containing a hexagonal array of fuel pins (Fig. 13.7). The purpose of the wrapper is two-fold: firstly it isolates the coolant flow to the sub-assembly and enables it to be independently controlled by means of an inlet gag; secondly it helps to avoid snagging between neighbouring sub-assemblies when charging and discharging. There are disadvantages in wrappers, apart from cost, in that they tend to distort due to creep and void-induced swelling, and they may not be a permanent feature of future fast reactor sub-assemblies. A standard sub-assembly for the Dounreay PFR has 325 fuel pins in a wrapper which is 142 mm across hexagon faces. This reactor has a free-standing core, the aim being to avoid interaction between neighbouring sub-assemblies, these are arranged in groups of six around a leaning post in an empty lattice position, and in some of which control or shut-off rods operate.

The lower end of a sub-assembly carries a conical spike assembly, which fixes into the bottom core-plate (or diagrid) and through which the sodium enters from the lower plenum. The spike carries filters and flow control gags.

High fast neutron irradiation doses cause voids to develop in most metals and alloys by the agglomeration of the atomic-scale defects they produce (vacancy condensation); these materials are therefore subject to swelling and also to bowing if differential irradiation doses occur across a component.

The slender nature of FR fuel pins, together with the high linear heat rating and tendency to bow as a result of void swelling, means that a lot of spacer grids are required to maintain pin spacing and avoid pin-to-pin contact. These spacers are honeycomb type made with hexagonal dimpled unit cells, the dimples on three faces centering the pins and applying restraint by interference where necessary. Structural members, known as grid-legs in PFR, locate each spacer grid at the six corners and are rivetted to the wrappers although such attachment may be dispensed with to ease subsequent disassembly problems.

In PFR sub-assemblies the fuel pins are floating, i.e. not anchored at either end, with end stops and an axial growth allowance. It is likely that all pins will be anchored at one end in later designs.

As well as having a fertile blanket of breeder assemblies disposed radially around the core, FRs have upper and lower axial breeder zones, which form parts of the core sub-assemblies. The fertile material used is natural or depleted UO_2, and may be contained within the pins together with the fissile material, or in separate pins. In PFR the lower axial breeder is contained within the pins, while the upper axial breeder is a separate assembly of large diameter wire-wrapped pins known as the mixer–breeder. This region is designed to promote mixing of the hot sodium, to equalize the outlet temperature, and to provide a mixed sample for failed-element detection purposes.

The use of wire-wrapping, a wire wound helically around the pin and attached at both ends, is an important alternative design style to the use of spacer grids. The pins are stacked together within the wrapper and naturally adopt a close-packed hexagonal array, the desired minimum spacing being determined by the wire diameter. Such designs have certain advantages in terms of manufacturing cost and ease of disassembly after irradiation. They have, however, the disadvantages, that to avoid fretting the initial fit must be tight, and that the void swelling of cladding and wire-wrap leads to distortions of the hexagonal symmetry and in the limit to local pin-to-pin contact, which degrades heat transfer. Which of the design styles, wire-wrapped or with spacer grids, ultimately becomes the preferred choice for large power reactors will depend on many factors not yet established with the necessary confidence, especially the behaviour of the designs at the high damage doses required for economical commercial operation.

The fuel pins in FRs are typically of small diameter, 6–8 mm, and full-core length (0.9 m active length, 2.2 m overall in PFR). The cladding material is generally an austenitic stainless steel in the cold-worked condition, e.g. M.316, but other alloys, especially high-strength and void-swelling-resistant alloys such as Nimonic PE16, are of interest. The mixed-oxide fuel, UO_2-PuO_2 in the ratio of about 4:1 may be co-precipitated or dry blended. It can be used in the form of solid or hollow (annular) pellets or as granules or spheres fed into the can and consolidated by vibration (Vipak).

Because of the low coolant pressure, high burn-up, and high rating, leading to high fission-gas release, a larger gas plenum is needed than in most other reactor fuel pins and this is most effective if located at the lower, cooler, end of the pin. The lower axial breeder is inserted above the plenum, supported by a platform fixed by crimping of the can, and is separated from the fissile core fuel by a further spacer, sometimes of wire mesh. Details of these inner pin features vary from design to design. A typical fuel pin design is shown schematically in Fig. 13.8.

There is some interest in the use of a carbide fuel, UC–PuC, in FRs, because it has a higher density of heavy atoms, beneficial in maximizing the amount of breeding, and a higher thermal conductivity, which permits the use of larger-diamter pins and reduces manufacturing cost. The use of such a fuel material is quite feasile, because the carbides are compatible with molten sodium, but the manufacture and reprocessing of carbide fuel is more difficult, and at this stage of FR development only a limited effort is devoted to it in the UK. More active interest is being shown in other countries, notably the US.

The radial breeder elements in FRs are an important feature of the reactor, in that neutrons which would otherwise be wasted convert ^{238}U atoms into fissile ^{239}Pu in this region. They are relatively lowly rated, although they have the unique feature of increasing in rating over their

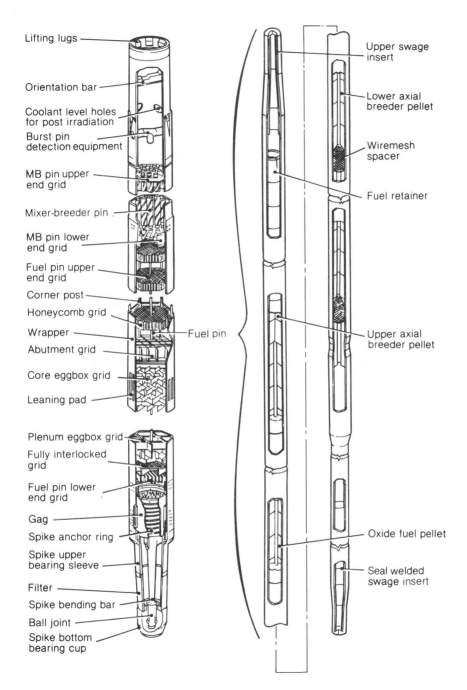

Lifting lugs

Orientation bar

Coolant level holes
for post irradiation

Burst pin
detection equipment

MB pin upper
end grid

Mixer-breeder pin

MB pin lower
end grid

Fuel pin upper
end grid

Corner post

Honeycomb grid

Wrapper — Fuel pin

Abutment grid

Core eggbox grid

Leaning pad

Plenum eggbox grid

Fully interlocked
grid

Fuel pin lower
end grid

Gag

Spike anchor ring

Spike upper
bearing sleeve

Filter

Spike bending bar

Ball joint

Spike bottom
bearing cup

Upper swage
insert

Lower axial
breeder pellet

Wiremesh
spacer

Fuel retainer

Upper axial
breeder pellet

Oxide fuel pellet

Seal welded
swage insert

Fig. 13.7 Cut-away PFR sub-assembly

Fig. 13.8 PFR fuel pin design features
(schematic)

lifetime in reactor because of the build-up of plutonium, some of which is burnt up by fission and contributes to the thermal output of the reactor. These elements are also contained in hexagonal wrappers, the same size as the core sub-assemblies, but they comprise bundles of large-diameter, wire-wrapped pins, consisting of solid UO_2 pellets clad in stainless steel.

All FR sub-assemblies, both core and breeder, have bearing pads in certain positions on the wrappers. These pads may be hard faced with a range of commercial alloys and maintain minimum wrapper-to-wrapper spacing.

Other types of FRs have been considered and continue to be assessed and evaluated as possible competitors to the liquid-metal-cooled reactor. The most likely alternative would be gas-cooling with either helium or carbon dioxide. Such a reactor would inevitably involve some changes to the fuel sub-assembly design because of the change to a high-pressure gas environment.

3.6 Miscellaneous fuel elements

The fuel element types already described cover all those of significance in large power reactors at present in use or likely to be used within the next 20 years. There is however another class of fuel elements in use in many small test reactors and also in naval vessels. These elements comprise an assembly of flat or curved plates, or more rarely, thin-walled concentric tubes of progressively increasing diameter. The plates are made by roll-bonding a sandwich with a metallic uranium alloy in the centre and either aluminium or zirconium on the outside. The uranium is usually of very high enrichment (93 per cent ^{235}U) and is partly in solution in the matrix metal, partly present as intermetallic compounds. In most materials-test reactors (MTRs) the matrix metal and cladding are an A1 alloy because they operate in contact with a low-temperature D_2O moderator, while in naval PWRs they are of a zirconium alloy.

A variant of the plate-type element has a cermet fuel comprising UO_2 particles dispersed in a metal matrix by powder metallurgy techniques and again roll-bonded to sheets of the matrix metal.

Figure 13.9 shows a typical example of the MTR type of fuel element. The central hole is to provide space for the insertion of test specimens.

Fuel elements for merchant ship propulsion, or for ice-breakers, have generally been of fairly conventional PWR rod–oxide type, the main differences being that they have a more robust structure to withstand the ship's motion and are designed to tolerate the frequent power-cycling involved in manoeuvring.

4 Fuel materials

The common fuel materials in use in power reactors are U-metal alloys and UO_2, which is usually enriched. In fast reactors (U, Pu)O_2 with around 20

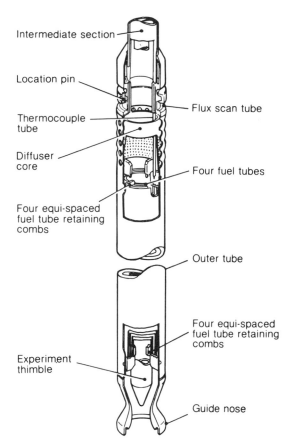

Fig. 13.9 MTR fuel element

per cent Pu is the most favoured material. This section concentrates on these materials, but some mention is made of other possible fuel materials, some of which have had limited in-reactor testing.

4.1 Chemical processes for UF₄ and UF₆ production

The starting point for the manufacture of uranium fuel materials at the fabricators' plant is a crude U_3O_8 concentrate, commonly known as *yellow-cake* because of its bright yellow colour. The yellowcake is produced at or close to the mining operation.

The uranium ores of commerce have only a low U content; typically, for instance, the US and Canadian ores are below 0.5 per cent U_3O_8. Concentration is therefore essential. The common physical methods of concentration are not particularly effective and are not widely used, but oscillating

(Wilfley) tables have been used successfully in the Congo to separate out the pitchblende from the gangue, and in South Africa some gold ores are treated by froth flotation, the uranium content being concentrated with the gold.

By far the most common methods of concentration are chemical ones, which involve leaching the whole ore body with mineral acids or with sodium carbonate.[9]

In the acid leaching process the crushed ore is ground with water and then leached with dilute sulphuric acid in the presence of an oxidizing agent, commonly manganese dioxide or sodium chlorate, the uranium entering the liquid phase. Dilute or strong acid processes can be used depending on the state of the ore, the stronger acid being more appropriate when the U-bearing ore particle sizes are larger. The carbonate leaching process is more appropriate when the ore body contains limestone or similar basic constituents which would consume uneconomic quantities of acid. It is more selective for uranium than is acid leaching, in that smaller amounts of non-uranic materials are dissolved. Fine grinding is required and leaching temperatures are at boiling-point or above, autoclaves being required in the latter case. Air or oxygen, and sometimes other oxidants, are used.

The uranium liquors are then separated from the reject solid material, filtration being normal in the carbonate leaching process, with the uranium being precipitated from the liquid fraction as sodium diuranate. In some cases the product of the acid leaching processes may be filtered, but more commonly thickening or settling are used to separate the impure U-bearing liquid. In this case ion exchange is normally used for separation of uranium from other metals in the liquor. Resins are avilable which are reasonably selective towards uranium sulphate, and a highly concentrated uranium-bearing solution can be obtained. Uranium may then be precipitated from the solution as sodium or ammonium diuranate, the latter being calcined directly to U_3O_8, in which concentrated form it is sent to the processing plant.

Though the processes described above are by far the most common, some variants are used in special cases. In some low-grade ores the uranium is contained along with iron pyrites and these can be oxidized by bacteria, the sulphide content being converted to sulphuric acid, which then acts in the same way as in an acid leaching process. It is possible to use bacterial methods directly on some ore bodies *in situ* without mining, grinding, and crushing.

The next stage of the process is the purification of the U_3O_8 concentrate (yellowcake) and conversion to a pure uranium tetrafluoride, from which uranium metal or uranium hexafluoride for enrichment is made. Though there are some variants, most processes, including the UK one, proceed by a method involving solvent extraction of uranyl nitrate.[10]

The U_3O_8 concentrate is dissolved in nitric acid and heated to produce a slurry which is aged and filtered. The crude uranyl nitrate is then purified by counter-current solvent extraction using tri-*normal*-butyl phosphate in odourless kerosene (TBP–OK). This has the important effect of separating out such elements as boron and cadmium, which have a high neutron capture cross-section, and metals such as molybdenum, which, like uranium, have volatile fluorides. On heating to about 60 °C the TBP–uranyl nitrate complex breaks down to give a purified uranyl nitrate solution, the solvent being returned to the process. The uranyl nitrate solution is then concentrated by evaporation and thermally denitrated at 300 °C to yield uranium trioxide. This is subsequently hydrogen reduced to uranium dioxide at about 500 °C and converted to uranium tetrafluoride by reaction with HF, both in fluidized beds. When enriched fuels are required, the volatile compound uranium hexafluoride is the feed material to both the gaseous diffusion and centrifuge processes. The pure UF_4 is converted to UF_6 by reaction with fluorine. This is done in a fluidized-bed reaction vessel containing an inert bed of calcium fluoride into which the UF_4 is fed, together with a mixture of nitrogen and fluorine. The reaction is very rapid and the resultant gas mixture is taken to condenser vessels cooled to -40 °C, where UF_6 condenses, leaving the N_2–F mixture to be recycled. The UF_6 is then melted and poured into transit containers. Schematic flow-sheets of these processes are shown in Figs 13.10 and 13.11.

4.2 Uranium metal

The process in common use for producing U metal is the reduction of UF_4 by magnesium, calcium, or a similar strongly electropositive metal. In the UK process, UF_4 and magnesium turnings are blended and compressed into large pellets which are charged to a reaction vessel where a thermite type of reaction is allowed to take place. The resultant billet is cleaned of slag, remelted in a vacuum induction furnace, and cast in graphite moulds into rods about 25 mm in diameter and 1 m long. Small amounts of iron and aluminium are added in order to reduce the tendency to swelling in the reactor. At this remelting stage any other required alloying elements can be added. The cast structure is textured and subject to strong irradiation growth effects in the reactor. This structure must be modified by heat treatment and, in the UK, the grain size is refined by rapid induction heating to the beta phase, about 660 °C, and quenching. In some cases, particularly for the longer elements, bars are subsequently annealed at 550 °C, a process which minimizes bowing tendency. The beta-quenching is a continuous process, the bar being traversed through the induction coil and quenching spray. The product is not texture free, but has a balance of positive and negative growth textures in the outer and inner regions which prevents any large net irradiation growth. The fine grain size in the outer region of the heat-treated rod is essential to avoid the development of a very rough surface by growth of individual grains at low irradiation temperatures.

It is possible to produce uranium bars by casting into cylindrical billets and extruding or rolling these to the diameter required in the gamma phase at

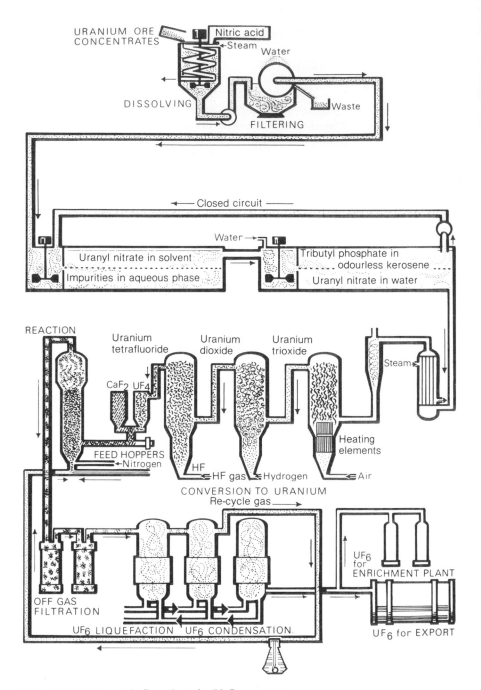

Fig. 13.10 Schematic flow sheet for $U_3O_8 \rightarrow UF_6$

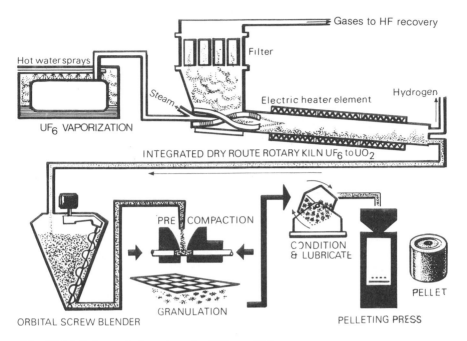

Fig. 13.11 Schematic flow sheet for $UF_6 \rightarrow UO_2$

about 800 °C. Thesé processes leave a highly developed texture in the material, which is difficult to remove by heat treatment and which promotes irradiation growth. For this reason these methods have not been favoured.

Finally the uranium rods are machined to size and grooved.

4.3 UO₂ Fuel

In the process of producing pellets of enriched UO_2 for use in metal–oxide fuel elements, the first step is the enrichment of uranium in the form of UF_6. This is taken to an enrichment plant where the fraction of the ^{235}U isotope is increased. Two basic processes are in large-scale operation, the classical diffusion method and the newer gas centrifuge. In a diffusion plant, advantage is taken of the fact that the flow of a gas through a membrane is, for a given pressure difference, inversely proportional to the square root of the molecular weight of the gas. Thus, using UF_6 gas passing through a membrane with suitable-sized holes, the gas at the lower-pressure side of the membrane is about 1.0043 times richer in ^{235}U than on the higher-pressure side. This of course is a very small advantage, and large plants on the cascade principle with many stages are necessary to get sufficient enrichment in ^{235}U to be commercially useful.

Gas centrifuge enrichment also works on the small difference in molecular weight. As the gas passes through the rotating cylinder, the heavier fraction will tend to pass along the outside, being taken there by the greater centrifugal force, and the lighter (^{235}U) fraction will tend towards the axis. In practice a counter-current principle is used, which is more efficient than a single stage, and the cascade principle is employed to get the necessary degree of enrichment. These techniques are discussed in greater detail in chapter 12.

An important aspect of enrichment processes is the extent to which the *tails* are depleted. It is common practice today to operate with depletion to a level of about 0.2 per cent ^{235}U.

The next stage is to convert the enriched UF_6 to UO_2. In the integral dryway route (IDR) used in the UK, UF_6 is converted directly to UO_2 by passing it, together with steam, through an inclined rotary kiln with a counter current of hydrogen and with appropriate zonal temperature control (Fig. 13.11). Ceramic-grade UO_2 powder is discharged from the lower end of the kiln into transport containers.

Other processes have been employed in other countries, and in some cases in the UK. Many of these were variants of the uranate precipitation process, and perhaps the most commercially exploited is the ammonium uranyl carbonate process (AUC).

The final stage in the production of the UO_2 fuel is to form the fine powder into pellets by pressing and high-temperature sintering.[11] Prior to this stage, the enrichment of the powder must be right. This may have been assured at the UF_6 enrichment stage, but for various reasons it is sometimes necessary to adjust the enrichment. In the old UF_6 conversion route this could be done in the wet stage prior to precipitation of ammonium di-uranate (ADU). In the new IDR process, enrichment adjustment is done by dry-blending of powders.

To achieve satisfactory pellet sintering the *activity* of the powder is important and the specific surface area has an important influence. It is usual therefore to grind the powder in a high-pressure air, fluid-energy mill, so that 80 per cent is of sub- μ m diameter corresponding to a specific surface area of about 3 m²/g.

There are two commonly used routes from powder to pellet, the so-called binder and binderless routes. In the former, which has been the standard UK route for many years, the powder is formed into the free-flowing granules necessary for consistent die-filling by slurrying with a proprietary binder and solvent, and spray drying. Alternative granulation processes are available. In the binderless route, the fine powder is pressed into briquettes, which are crushed and sieved to yield granules of suitable size.

The pellets are made in single or multi-punch hydraulic pelleting presses, the resulting 'green' pellets being de-bonded at 800 °C (binder route) and finally sintered in continuous hydrogen atmosphere furnaces at about 1650 °C. Density control within a specified range, commonly 10.4–10.7 g/cm³, is usually required (theoretical density 10.95 g/cm³), and if the powder as manufactured tends to yield a higher density, additives can be used to create additional closed porosity. These additives are compounds that decompose at low temperatures into harmless gases, and leave voids in the matrix which create the thermally stable closed porosity. It is bad practice to achieve density control by under-sintering, since the fuel will then densify in service with adverse effects on performance.

The use of additives is more convenient with the binderless pellet route, and there is therefore a trend to adopt this route more widely. The product made with additives is known as BNFL CONPOR fuel. (CONPOR is a registered trade name.)

UO_2 pellets are made in various sizes ranging from as small as 5 mm diameter up to 16 mm, in solid or annular form, the central hole being formed by the use of a central mandrel in the pelleting die. Pellets are in most cases finished to size by grinding with SiC or diamond wheels. Chipping can be important in affecting performance, especially in gas-cooled reactors, and very careful treatment in the green state and during grinding is necessary to meet the required quality standards.

An alternative to the use of UO_2 in pellet form is to fill the cans with particulate fuel, although this is not common in thermal reactors. The vibro-compaction process discussed earlier is used to ensure consolidation and hence the fuels have come to be known as Vipak. Both high- and low-energy processes have been used, the former with an irregular shaped UO_2 particle feed produced by crushing and sieving, the latter with spheres formed by rolling fine powder granules incorporating a binder, the resulting spheres being de-bonded and sintered. The density achievable by vibro-compaction is lower than that of sintered pellets, but there is no fuel–clad gap which partially off-sets this disadvantage. Multi-component feeds, i.e. mixtures of different size fractions, are used to improve the density of Vipak fuels, three components giving close to the maximum achievable density of about 85 per cent.

There is increasing interest in UO_2 produced by gel precipitation as a spherical feed suitable for vibro-compaction, or possibly for pelleting, but this interest is mainly for mixed UO_2–PuO_2 fuels referred to below.

4.4 UO_2–PuO_2 fuel

The chemical and ceramic processing techniques used to produce mixed UO_2–PuO_2 fuels, mainly used in fast reactors, are substantially the same as those used for UO_2, and the mixing is either done at the nitrate solution stage, followed by coprecipitation or by later dry-blending. Homogeneity is important for physics and thermal performance reasons as well as in relation to the solubility of the irradiated fuel during chemical reprocessing.

The main practical differences in the fabrication of mixed-oxide fuels are related to the added toxicity and beta/gamma radiation levels encountered. These mean that much of the fabrication has to be done remotely in sealed lines or glove boxes, and add to the problems of plant maintenance, repair, and replacement. The processes involved in powder preparation, pelleting, and sintering are similar to those for UO_2 fuels, but details of equipment and techniques differ to accommodate the remote-handling requirement. Whether such fuels are best made in free-standing glove-boxes, or in sealed-face production lines is still debated, and different practices are employed in different countries, e.g. the UK and France. The greater simplicity of the French glove-box technique is offset by higher radiation doses to the operators, and it may not be possible using this method to achieve the mean dose of 0.5 man-rem/y currently being set as a design target. One of the major problems as a source of radiation is fine powder settling in various parts of the plant. It is to minimize this problem that interest in the gel process for producing high-density spherical feed material directly from solution has been stimulated.

In the gel precipitation process, heavy-metal nitrate solution is blended with a solution of predetermined viscosity containing a gelling agent and a modifying agent. The resulting solution is formed into droplets prior to precipitation in concentrated ammonia solution. After ageing, the spheres are washed with demineralized water and dried. The spheres are then debonded in carbon dioxide, and subsequently sintered in argon containing 5 per cent hydrogen at a temperature in the region of 1500 °C. Two sizes of oxide fuel sphere (800 and 80 μm diameter) have been used in a vibro-packed system giving within-pin densities of approximately 81 per cent of theoretical. The use of three sizes of spheres is feasible and gives a higher density, but introduces added difficulty in requiring concurrent feeding.

4.5 Coated particle fuel

There are three main stages in the manufacture of coated-particle fuel for use in HTRs, namely fabrication of the spherical fuel kernels, application of the coatings, and consolidation of the coated particles in a graphite-based compact.[3]

Depending on the fuel management schemes adopted, kernels of various compositions and sizes are made. The range includes fissile kernels of highly enriched uranium oxide or carbide with sphere diameters of 200 μm, thoria (ThO_2) breeder spheres at approximately 500 μm diameter, or mixed fissile–breeder spheres, either low-enriched UO_2 at 800 μm diameter or UO_2–ThO_2 in the range 450-800 μm. Oxide kernels may be fabricated directly from ceramic-grade powders by a controlled agglomeration growth process followed by sintering in hydrogen as for standard fuel pellet production. Carbide kernels may be made in the same way by incorporating carbon in the spherical granule, to allow carbide formation by the carbo-thermic reduction of UO_2, followed by sintering, both processes being carried out in either vacuum or flowing argon. However, wet processes are generally preferred because of the need to refabricate highly active materials when the thoria cycle is employed. Processes such as sol-gel, gel-precipitation, and the so-called resin bead process have been developed. The gel routes have also been exploited for fast reactor fuel production as already described. In the resin bead process, fuel nitrate solutions are passed through a column of ion-exchange resin fabricated in the form of small spheres. After exchanging the uranyl ion, the resin bead is heat treated to decarbonize it, followed by sintering or, alternatively, some of the carbon is retained for the formation of a carbide sphere. Controlled porosity may be incorporated within the sphere by means of the carbon burn-out process or by the use of additives such as in BNFL CONPOR pellet fuel.

Pyrolytic carbon and silicon carbide coatings are applied in fluidized-bed reactors by the pyrolysis of appropriate hydrocarbons and silanes. In most designs of coated particle, a low-density buffer coat of carbon is first deposited, followed by an inner high-density isotropic pyrocarbon coating generally derived from propylene. The coatings are deposited at 1200–1300 °C with the particles fluidized by argon carrier gas. Silicon carbide coatings, close to theoretical density, are deposited from methyltrichlorosilane decomposion, using hydrogen or hydrogen–argon as a carrier gas at a temperature of about 1500 °C. Finally, an outer high-density isotropic carbon coating is deposited, again using propylene (Fig. 13.4). The use of the fluidized-bed technique facilitates excellent heat transfer from the indirect-heating elements, allows coatings to be applied sequentially by simply changing gas composition and temperature, and produces coated particles of good spherical shape. The role of the buffer coating, 50–100 μm thick, is to absorb fission fragment damage and provide additional porosity to reduce fission gas pressure. The inner pyrocarbon, 30–40 μm thick, seals the buffer coat and provides a barrier for fission-product gases, while the silicon carbide coating, 30–80 μm thick, provides a barrier for solid fission products. The outer pyrocarbon coat of approximately 30 μm thickness acts to chemically protect the silicon carbide and, because of irradiation shrinkage, places the underlying silicon carbide coat under comprehensive stress for most of the fuel particle life.

As the final stage, the coated particles are mixed with graphite powder and appropriate binders and pressed into fuel rods for the prismatic-core design or large spheres for the pebble-bed reactor. A variety of recipes and fabrication techniques have been employed, but to achieve the best heat transfer and irradiation performance, the compact needs to be of high density and contain the least amount of non-graphitizable binder carbon. To achieve uniform distribution of the coated particles it is preferable to 'over-coat' them with graphite powder containing a small amount of resin binder. The resinated graphite powder is fed continuously to a batch of coated particles as they are tumbled in a rotating drum. The overcoated particles are then warm pressed in steel dies to form fuel rods about 15 mm diameter and 60 mm long or isostatically pressed in rubber moulds to form spheres of 60 mm diameter. The compacts are then subjected to controlled heat treatment up to 800°C to carbonize the resin binder and then to 1800 °C to stabilize and remove hydrogen.

4.6 **Thorium-based fuels**

Thorium, a metal found more abundantly in nature than uranium, generally as monazite sands, can be used as a fertile material in power reactors. The ^{232}Th isotope, which forms the bulk of the naturally occurring material, is converted by neutron capture into the fissile ^{233}U isotope. There is therefore potentially available a new large source of fissile fuel material.

The manufacturing route for ThO_2 involves sulphuric acid leaching, NH_3 neutralization, which precipitates the Th as sulphates and phosphates, dissolution in HNO_3, and solvent extraction with tributyl phosphate, which removes most of the co-occurring rare earths. The oxide is finally produced by calcining the oxalate or nitrate, and it can be fabricated into pellet form by pressing and sintering.

The use of ThO_2 in HTRs has been referred to. It can be used as a breeder in FRs, or as a fertile material in thermal reactors, mixed initially with ^{235}U or ^{239}Pu. In some reactors, especially the HWRs studied in Canada, a self-sustaining cycle can be maintained once a sufficient inventory of ^{233}U has been built up.

There are severe problems in utilizing a Th fuel cycle because of the need to chemically reprocess the irradiated fuel, to extract the fission products and recycle the ThO_2 and ^{233}U. The activities of decay products of ^{232}Th and ^{233}U are such that re-fabrication into fuel elements requires heavy shielding, unless it can be done very quickly before the activities build up to harmful levels.

There is sufficient irradiation experience with ThO_2–UO_2 fuels to establish that performance is good, and to show that no major design changes, as compared with UO_2-based fuel elements, would be needed.

4.7 **Other fuel materials**

Uranium compounds other than the oxide have been used in fuel elements, notably the carbide UC, which has the twin advantages of higher uranium density and better thermal conductivity. Its main disadvantage is a lack of compatibility with the commonly used coolants CO_2 and H_2O. At operating temperatures in LWRs and CAGRs, the UC would react rapidly to form UO_2 with a substantial volume change. This means that, if a defect did form in the fuel cladding it would soon enlarge and result in a major release of active fuel material to the coolant circuit. UC is, however, quite stable in contact with liquid sodium, and it is a most promising material for use in sodium-cooled fast breeder reactors. Its irradiation behaviour has been investigated and found to be satisfactory in terms of fission-product release and dimensional stability. The use of UC in FR fuel elements would result in larger-diameter fuel pins for a given rating, thus reducing costs. The higher uranium atom density would result in a higher breeding gain, reducing the *doubling time* the time needed to double the original plutonium inventory.

UC is made by carbo-thermic reduction of UO_2. It is a more difficult operation than UO_2 manufacture, and even on a large commercial scale would cost more. As with UO_2, the fuel may be fabricated into pellets by pressing and sintering or used as suitable size powder or granules for vibro-compaction.

The major reason holding back the wider use of UC–PuC as a fuel for FRs is that it requires different and more complex manufacturing and chemical reprocessing techniques.

Uranium nitride has attractive properties as a fuel material; it has a higher U density than the oxide and also a better thermal conductivity. Unfortunately nitrogen also has a higher thermal neutron capture cross-section (1.88 barns) than oxygen (0.0002 barns), and in thermal reactor would require more enrichment of the fuel. The compatibility of UN with CO_2 and H_2 coolants is better than that of UC, but may not be good enough in practice. If the low neutron-absorbing isotope ^{15}N were freely available, a nitride fuel made with it could be an ideal fuel in FRs, where the high neutron absorption of natural N is a drawback.

UN is made by high-temperature reaction of N_2 or NH_3 with U metal powder or UH_3. Again, as with UC, the chemical reprocessing route is more complex than for UO_2.

A uranium silicide, U_3Si, was made and extensively tested in Canada as a possible fuel for their natural uranium CANDU–PHW reactors. As with the other alternatives to UO_2, the attraction lies in the high uranium density and thermal conductivity. U_3Si did show undesirably large swelling, caused by a phase change, but could be stabilized by additives. Work on this alloy has not been continued, the major snag being the rather poor compatibility of U_3Si with water. U_3Si was made by melting the constituents together and casting. The Si present introduces reprocessing problems.

Uranium (or U–Pu) metal or alloys have from time to time been reappraised as possible fuel materials for LWRs and FRs. In the LWR application the compatibility with H_2O is the main problem, but dimensional stability could also be a difficulty in going to high burn-up. Metal fuels have been used with some success in FRs, and their future use is by no means discounted in some quarters.

The penalty in terms of fuel material choice imposed by the need to consider how the material will behave in the event of a leak developing in the cladding is a very severe one, and if freedom from leaks could be guaranteed it is more than likely that some alternative to UO_2 would have gained wider acceptance. With the very low defect rates now being achieved in CAGRs and LWRs (approaching 1 in 10^4 pins), and bearing in mind the contribution to these leaks made by the unfavourable properties of UO_2, it is far from certain that some alternative will not eventually come along, especially for those reactor types in which a leaking fuel element can readily be replaced. Table 13.2 shows the relevant properties of the fuel materials discussed.

Table 13.2 Fuel material properties

	Density	Mass of U per volume of fuel	Melting-point	Coeff. of linear thermal expansion	Thermal conductivity
	(g/cm^3)		(K)	$(K^{-1} \times 10^{-6})$	(W/mK)
Uranium: pure	19.05	19.05	1403	15.2 (300 K)	32.0 (775 K)
cast rods	18.75	19.05	1403	18.5 (875 K)	
Uranium dioxide: pellets	10.3–10.7	9.1–9.4	3075 approx.	11.0 (295–1275 K)	5.0 (775 K) 2.5 (2375 K)
Uranium carbide	13.62	12.97	2675	11.0 (295–1275 K)	29.9 (775 K) 24.5 (1275 K)
Uranium nitride	14.32	13.52	3125	10.0 (295–1875 K)	18.6 (775 K) 25.1 (1275 K)
Uranium silicide	15.6	15.0	1938	17.5 (295–1025)	14.9 (375 K)

5 Fuel cladding manufacture

The only materials of commercial importance as fuel cladding are magnesium and zirconium alloys, stainless steels, and some Ni-base alloys. Some of the important properties of typical alloys used in power reactors are shown in Table 13.3. The selection of the preferred cladding alloy depends ultimately on three main factors: temperature of operation, corrosion resistance, and neutron absorption, although weldability is also of importance. Mechanical properties, such as creep strength and ductility also play a part, particularly in FBRs where the coolant pressure is low and the cladding is stressed in tension, but this has not been a dominant characteristic. In FRs the swelling caused by irradiation is also a factor, possibly a key one.

In all LWRs and FRs the fuel pin cladding has no extended heat transfer surface and takes the form of a plain thin-walled round tube, while for Magnox and AGRs an extended heat transfer surface is required and is an important part of the manufacturing operation. The metallurgical condition most suited to optimum irradiation performance is also of importance in relation to the choice of manufacturing method. In most cases a fine stable grain size and freedom from preferred crystallographic orientation is desirable, while the degree of cold work and residual stress level has to be controlled within limits.

5.1 Magnox

The fuel cladding in Magnox reactors is unique in having a large extended heat transfer surface, and demands a more complex manufacturing operation than any other. In the UK reactors the Magnox A180 cans (Table 13.3), have a wall thickness of about 2.0 mm, and a fin height of 9 to 10 mm. A fine

Table 13.3 Cladding material properties

		Main alloying elements	Melting range	Cross-section thermal nuetrons	Coeff. of linear thermal expansion
		(% by mass)	(K)	(barns)	$(K^{-1} \times 10^{-6})$
Magnox Al 80		Al 0.80 Be 0.005	920	0.065	24.5 (295–375 K) 26.2 (295–475 K)
Stainless steel	Type 304	Cr 19.0 Ni 10.0	1675–1725	2.87	16.5 (295–475 K) 18.4 (285–875 K)
	Type 316	Cr 17.00 Mo. 2.5 Ni 12.00	1645–1675	2.90	16.4 (295–475 K) 18.7 (295–875 K)
	20/25	Cr 20.00 Ni 25.0	1665–1725	3.21	16.7 (295–475 K) 17.9 (295–875 K)
Zirconium alloys	Zircaloy-2	Sn 1.5 Cr 0.1 Fe 0.14 Ni 0.06		0.193	
	Zircaloy-4	Sn 1.5 Cr 0.1 Fe 0.21	2090 approx.	0.194	6.15 (305–575 K)

grain size is needed to avoid creep-induced grain boundary cavitation, leading to leakage in elements to be used near the bottom of the reactor channels, where the cladding temperature is relatively low (200–350 °C). Since cans produced with a fine grain size tend to undergo rapid grain growth to very large grain size in channel positions where they operate at higher temperatures (350–420 °C), the very fine grain size has to be sacrificed in the upper elements. It is, however, less important in these positions because the irradiation growth responsible for the creep-induced cavitation in the cladding is less at higher temperatures. These special requirements have led to two different types of Magnox fuel elements depending on their service temperature, the high-temperature elements having a stable coarse grain size with several grains across the wall.

The helical design of Magnox can is manufactured from an extruded hollow billet by an extrusion process followed by a warm twisting operation, in which the straight finned can is twisted on a mandrel to preserve the correct bore diameter. The twisting conditions are controlled to leave the right amount of cold work in the can to recrystallize to a fine grain size, when those destined for low-temperature service are subsequently pressurized on to the uranium rods. The cans are finished by four axial slots machined at 90° intervals for insertion of gas flow splitters.

The herringbone design of can used in the later Magnox reactors is manufactured entirely by machining from a thick-walled, hollow extrusion with four thick ribs at 90° apart. The ribs are partially machined away to generate the channel support lugs and the fins machined by fly cutter.

5.2 AGR

Because of the high temperature of operation and CO_2 environment, a type of austenitic stainless steel, 20/25Nb, is used for AGR fuel cladding. This has 20 per cent Cr, 25 per cent Ni, and 0.5 per cent Nb as its principal alloying ingredients. Ideally a higher-strength material would be desirable, and a version strengthened by formation of a dispersion of TiN, by nitriding, is under development. The steel is melted in vacuum induction furnaces and remelted by the consumable electrode vacuum arc-melting process, to yield a high-purity ingot. The ingot is fabricated to tube by forging followed by extrusion to thick-walled tube hollows, and a combination of tube reduction, a cold-pilgering type operation, and single- or multi-stage cold drawing with interstage stress-relief annealing. These processes naturally yield the fine grain size required to ensure that there are of the order of ten or more grains across the tube wall. The final product in the partially cold-worked condition is straightened and cleaned prior to machining on the external heat transfer surface which is a single spiral fin 0.2 mm high. Multistart helical finning which has heat transfer advantages is currently under investigation.

5.3 **LWR**

Zirconium alloys are now almost exclusively used for cladding in both BWRs and PWRs, combining as they do low thermal neutron capture cross-sections (0.19 barns), good corrosion resistance, and quite adequate mechanical properties. In reasonable thicknesses, 0.5–0.7 mm, such cladding is elastically stable, at the diameters used, under LWR coolant pressures.

Two essentially similar alloys, Zircaloy 2 and Zircaloy 4 (Table 13.3) have been most widely used, the former in boiling-water environments, the latter in pressurized water. The main factors in this selection are corrosion and hydrogen pick-up rates.

Zirconium occurs in nature mainly as $ZrSiO_4$ and is combined with a significant amount (up to 3 per cent) of hafnium, a strong neutron absorber, which has to be separated by fractional crystallization of phosphates. The metal is then produced as a sponge by Mg reduction of $ZrCl_4$. An alternative method involving thermal decomposition of ZrI_4 (Van Arkel process) produces a high-purity crystal bar Zr, but is more expensive, especially for large sections. This grade of Zr is still used in the USSR for the production of cladding alloys. The Zr sponge is double vaccum-arc melted, at which stage the alloying elements are introduced, forged to bar of about 125 mm diameter, machined and bored, copper clad, and extruded to a tube hollow of about 50 mm outer diameter and 8 mm wall thickness. This hollow is then tube-reduced to final size with the necessary interstage annealing in vacuum furnaces to avoid oxygen and nitrogen pick-up. The latter is particularly detrimental to corrosion resistance, the former to ductility. Detailed control of reduction parameters is needed in the tube-reducing process to control the orientation of hydrides that form in fuel cladding as H_2 is absorbed during the corrosion process. This control can also influence the degree of preferred orientation, which is related to both irradiation growth and creep. The tube so produced has a very fine grain size, and so there is no problem in ensuring that there are 50 or so grains across the wall.

Zirconium alloys were notorious in the early years of their use as nuclear fuel cladding for erratic and sometimes catastropic corrosion behaviour. These occurrences were eventually traced to N_2 pick-up (which is neutralized by an appropriate Sn addition), fluorine contamination of the surfaces by pickling liquors (HNO_3–HF), and the size and distribution of intermetallic compounds. Most fuel cladding for LWRs was at one time pre-autoclaved in steam, giving a glossy black surface layer of oxide to check for freedom from contaminants, but this process is rarely used now, and surface finishing is by abrasive treatment, which removes the surface layer.

The tube after final heat treatment, which is determined by the required properties and differs for BWRs and PWRs, is straightened out and cut to length when it becomes the fuel can, no heat transfer surface being required.[4]

5.4 **FR**

Cladding material for FR fuel pins requires good creep strength, because it operates under tension in service, and the best possible resistance to neutron-induced void swelling. The most commonly used material has been the austenitic stainless steel M316 or variants of it, containing 17 per cent Cr, 14 per cent Ni, 2.5 per cent Mo, and 1.7 per cent Mn. Other higher nickel content alloys (up to 50 per cent) have been made and used experimentally. As with the AGR steel, these alloys are made in tube form from vacuum-melted ingot by forging, extrusion to hollows, and a combination of tube reduction and cold-drawing with appropriate interstage anneals, followed by a final anneal to control the structure, degree of cold work, and grain size.

5.5 **HTR**

In HTRs the fuelled pellets may be separately clad in graphite tubes, or inserted in massive graphite blocks. In both cases the graphite components are made by machining from extruded bar stock.

5.6 **Other reactors**

In MTRs and some naval reactors, the fuel elements are in plate form. The cladding in these cases is basically thin sheet produced from ingots by conventional forging and rolling techniques.

6 Fuel assembly

Fuel assembly operations involve the sealing of the fuel material into the cladding, and the putting together of the various assembly components by a variety of engineering-type operations.

One of the most critical operations is sealing the cladding, and the high standard of integrity demanded for end seals has necessitated considerable development over the years. The difficulties of obtaining a comprehensive assessment of weld quality purely by non-destructive examination led workers to concentrate on weld process control. Automatic welding has been the subject of continuous development commencing with relatively simple cam-operated or clock-controlled machines, but more recently using mini-computers in the control systems. The computerized systems have been further refined and developed, providing very sophisticated units capable of operating in open- or closed-loop mode. Programing in normal engineering terms provides considerable flexibility in operation, and the monitoring facilities permit detailed checks on the operations. The result is that very few weld failures have occurred in fuel elements in service.

Some of the operations comprising the assembly process will now be described by fuel element type.

6.1 **Magnox**

In the Magnox case the assembly process is relatively simple. One end of the can is sealed with a cupped end cap, screwed in and tungsten–inert gas (TIG) edge welded. The uranium rod is then inserted with an Al_2O_3 insulating disc at each end, the free space filled with helium, and the second end sealed in a similar fashion.

The elements are next pressurized, either hydraulically at 250 °C, 105 MPα, to give a fine grain size for low channel positions, or in CO_2 at 515 °C, 7 MPα, for the high channel positions. This operation deforms the can into the anti-ratcheting grooves machined into the uranium rod at intervals along its length and in close-packed end arrays. Subsequently the elements are cleaned by a citric acid pickle, which also decontaminates them with respect to uranic surface contamination. After drying, the canned rods are ready for inspection and final assembly, consisting of making the necessary attachments to the canned rod to provide its support features.

The ancillary components are the upper and lower end fittings, which suport the elements one on another, and the splitter cage, fabricated from strip and forged braces, used only on helical elements. The end fittings are forged or cast in a strong Mg alloy, such as MN80 (0.8 per cent Mn), and the splitters and splitter braces are made from either Al80 or heat-treated Zr55 (0.55 per cent Zr). Spring stabilizers, used on all helical elements and some of the herringbone designs, are integral with the upper end fittings and have a hinged link arm actuated by a Nimonic 80 spring, which bears against the channel. Sticking of one element to another is avoided by use of a thin stainless steel shim attached to the male and fitting. The end fittings, which are screwed into the end caps, are locked in position by drilling and pegging.

6.2 **AGR**

The length of finished tubing constituting the AGR can is sealed at one end by a welded-in end cap, the UO_2 pellet stack and insulation pellets are loaded, the free space filled with helium, the second end welded, and the sealed pin is then pressurized to deform the cladding into the anti-stacking grooves in the pellets. In AGR reactors a mass of absorber within-core is undesirable, so that the end caps used are thin-walled components welded to the can by both a TIG edge weld and a resistance seam weld, the lower end weld carrying an extension piece for subsequent locking. This double welding operation is used because of the thin sections of the can and end cap.

The fuel element cage is assembled by fitting into the outer graphite sleeve the bottom grid, the lower inner sleeve, the intermediate brace, the upper inner sleeve, the top brace, and screwing in the retaining ring. The pins are then fed in through the braces and locked into the bottom grid by rolling over the extension pieces. Subsequently, at the reactor site, eight such fuel elements are assembled onto a tie bar, with appropriate end components, for loading into the reactor.

6.3 **LWR**

The sequence of operations in making LWR fuel pins is the same as for the AGR, except that pins are not pressurized after sealing and plenum spacings or bottom support stools are introduced. Insulating pellets, or natural UO_2 pellets, may be used at the end of the pellet stack, but are not universal. In most LWRs end caps are solid, machined from rod, and attached to the cladding tube by TIG welding (usually a spigot butt weld), or in some cases by resistance butt-welding.

In LWRs the use of Zircaloy cladding leads to tight restrictions on total H_2 content within the sealed pin, since even milligram quantities of hydrogen, which comes mainly from moisture trapped in the UO_2 pellet, can cause local hydriding on the bore surface and rapid clad failure. The high coolant pressure in PWRs, 150 bar, can cause significant clad creepdown and ovalization, and in extreme cases has resulted in failures where the clad has crept down into inter-pellet gaps resulting from UO_2 densification. This higher clad creep rate also makes PWR fuel pins more susceptible to pellet–clad interaction on reactor power increases after a time at reduced power (the power-ramp or PCI phenomenon), leading to high clad stresses and the danger of fission-product-induced stess-corrosion cracking. For these reasons, PWR fuel pins have for some years been helium filled to above atmospheric pressure, in the range 20–30 bar. This increased helium filling pressure has a further advantage in reducing UO_2 temperature and hence fission-gas release, although being more troublesome in causing rapid clad swelling and rupture under some accident conditions. There is a recent trend to pre-pressurize BWR fuel pins to a lower pressure, 5–7 bar, to mitigate the continuing power ramp problem in these reactors.

Also in the context of power ramp defects, a number of potential remedies are being widely tested which could result in the clad bore surfaces being coated or clad with lubricants or metal barriers. Alternatively, means of minimizing fission-product release, such as the use of large grain size UO_2 fuels are under development.

In the assembly of LWR fuel elements it is normal practice to build up a sub-assembly of intermediate spacer grids and any full-length structural members, and to attach it to either the top or bottom support plates, before inserting the fuel pins. Depending on the particular design and reactor, the intermediate spacer grids may be made of stainless steel, Inconel 718, or a zirconium alloy, may be of *egg-box* or ferrule construction, and may be brazed or welded. The springs to apply the stabilizing force to the fuel pins are integral in some designs and separately attached in others. In one design the springs are in a strong zirconium alloy, in all others they are in a strong Fe or Ni base alloy and are hardened by standard heat-treatment processes.

Top and bottom support plates are fairly massive components of stainless steel or iron, made by casting and machining or machining from wrought bar stock.

6.4 CANDU

The fuel pins for the CANDU/PHW reactor are designed for a low burn-up, not more than 9000 MW d/t U, and maximum neutron economy. The pins are 495 mm long and have no gas plenum; the cladding at 0.38 mm wall is the thinnest of any water reactor, and there are no intermediate grids, so the pin spacing and support on the pressure tube is determined by appendages brazed or welded on the cladding at mid-length and close to each end. The end caps are thin discs, resistance welded to the cladding and to thin end-support plates.

6.5 FR

In principle, the manufacturing processes for FR fuel pins are similar to those described for LWRs, although UO_2 pellet loading and second-end welding must be done remotely or in sealed glove boxes, because of the toxicity of PuO_2. Radiation levels from traces of fission products, actinides, and spontaneous fission are high enough to cause problems, although heavily shielded facilities do not need to be employed.

In pellet-type pins, the lower (plenum) end is sealed before the can goes into the active line. The lower support platform is fixed in position by external crimping, and the lower axial breeder (UO_2) is inserted. In the active line, core fuel (UO_2–PuO_2) and upper axial breeder, when integral, are inserted either automatically or manually through glove ports, with the necessary intermediate spacers and upper-stack support device, prior to second-end sealing. Welds are commonly straight-butt or spigot-butt TIG welds.

Vipak pins are again pre-prepared with lower end cap, support platform, and lower axial breeder, before inserting vertically into the active line and attaching to the vibrator. The prepared UO_2–PuO_2 spherical feed material, generally two-size components, is loaded in two stages. Firstly, the can is filled with the larger size fraction, after which the smaller is infiltrated. An upper stack support feature is then inserted, followed by the upper axial breeder where integral, and second-end welding, usually in the vertical orientation.

If the pins are for assembly into wire-wrapped bundles, the wire wrap is then wound on to the pin at the required tension and attached by welding to the end caps. Before leaving the active line the pins are decontaminated and checked for surface activity.

Radial breeder elements for FRs have fairly standard metal–oxide type fuel pins which are made by conventional methods. In the case of PFR these pins, and those of the upper axial breeder, are wire wrapped, the thick wire being attached to the end caps by welding after the pin has been fuelled and sealed.

Spacer grids in FRs generally apply forces to the pin by interference, the whole structure of a unit cell acting as a spring. Detailed design can be varied to minimize large variations in the forces applied over the range of tolerances. These grids are usually made of stainless steel using an egg-box or ferrule construction, and are either spot-welded or brazed.

The wrapper is a key component of a FR element, which is attached either mechnically or by welding to the upper- and lower-end support units. The hexagonal wrappers have to be made to tight tolerances and can be made from seamless or welded round tube by a drawing operation, or by welding together two pre-forms, followed by drawing to size.

Insertion loads can cause problems with the very slender pins in FRs, hence the need for sophisticated grid designs. The grids are spaced and held in position by axial support members on the outside, which may in turn be riveted or welded to the wrapper.

The alternative wire-wrapped design of FR element requires a different assembly technique in which the bundle of pins is pre-assembled into a hexagon array and loaded simultaneously into the wrapper.

7 Inspection and quality control

It has already been emphasized that reliability in operation is an important requirement of fuel elements, because of the heavy outage penalty incurred by an enforced shutdown, and also the loss of investment represented by a fuel element that has to be prematurely discharged.

Good reliability requires freedom from design faults, together with careful control of material quality and manufacturing standards.

Quality control, as practised in the fuel element manufacturing industry, requires the identification and control of key features which have a bearing on performance. An integral part of the development process is to identify such items and to establish suitable limits. The important aspects fall into the following areas:

(1) material composition, properties and defects,
(2) dimensional tolerances, and
(3) manufacturing standards.

Some examples may help to clarify features falling into each category. A material supplied with a wrong composition may corrode excessively; develop extreme brittleness in service; be incompatible with other materials in contact with it; have excessive neutron absorption. The ^{235}U content in UO_2 is obviously vital to the nuclear performance; the N_2 content of Zircaloy is vital to its corrosion behaviour; excessive B in stainless steel enhances irradiation embrittlement; these are all examples of material composition demanding rigorous control.

Material properties are determined not only by composition, but also by the thermal and mechanical processing history in arriving at the finished form. Heat treatment and cold work both affect such important properties as creep strength, ductility, and corrosion resistance, by virtue of their effects on the constitution and structure of the material. They also affect grain size and preferred orientation, both important in determining performance under irradiation.

Defects in materials can result in leak paths or in premature rupture under service conditions. For example, a central rolling defect in Zircaloy rod used for end-cap manufacture resulted in many leaking fuel pins in early LWR fuel. While no materials are completely free from defects, a tight control on what is acceptable is needed.

Dimensional tolerances may or may not be important, but regardless of their importance they must be given to instruct the manufacturer. Tolerances can control such important aspects as leakage flow through a joint, fretting of one component against another, ability to assemble components, fuel element pressure drop, heat transfer performance, and fission-product release. This later is a good example, where a 0.025 mm increase in fuel–clad gap can result in an unacceptable increase in fission gas release. On the other hand a 25 mm error in axial position of an intermediate brace or spacer grid may be quite unimportant. In deciding on non-critical tolerances, what the manufacturer may be reasonably expected to consistently achieve is a useful guide. Too wide a tolerance range may be confusing and not helpful; too tight a tolerance range may cause the manufacturer serious problems and result in excessive costs.

Manufacturing standards are those features of the finished product not determined primarily by the materials or components going into manufacture, but by the operations carried out on them. Weld quality, as determined by depth of penetration, freedom from defects, alignment, and impurity pick-up, is an important aspect that has to be closely controlled, especially in what are classed as critical welds. Similarly, with brazed joints or hard-facing deposits the bonded area and freedom from defects must be to known acceptable standards. The performance of Magnox elements and AGR fuel pins depends on the cladding being correctly pressurized in to the anti-ratcheting grooves, so that the degree of penetration is an important manufacturing issue.

The way in which the design intent with respect to all these areas is shown to be achieved, i.e. that the product is of acceptable quality, is the province of quality control (QC).

The design intent is expressed in the form of a set of detailed, dimensioned, drawings, which include materials specifications and surface finish requirements. The hierarchy of documents underlying the drawings consists of:

(1) materials specifications,
(2) product specifications,
(3) process specifications, and
(4) quality-control plans.

It is neither possible, necessary, nor indeed desirable to specify fully any material or component. The material specification covers composition and properties, within limits, of the material in the form to be supplied. It may be that this is effectively a product, as in the case, say, of PWR cladding tube, so the material and product specifications in this case are the same, except in respect of dimensions and possibly surface finish, which are covered by drawing references. Product specifications cover all components which, when assembled together, comprise the fuel element. There may also be separate product specifications for sub-assemblies, e.g. fuel pins, and for the completed element. These cover all operations carried out in converting the material to the product, and their effects, for example, welding, machining, heat treatment, mechanical treatment, and chemical treatment.

In general, reliance is placed on product specifications, and the use of process specifications by the customer is infrequent, although the manufacturer will require them. This is partly because they involve commercially sensitive know-how which the manufacturer is not prepared to divulge to the customer. There may, however, be intangible effects of a particular process which are not covered by the product specification, such that a process change can result in the product behaving differently, either better or worse. This situation may be covered by specifying that for a material or product which has been evaluated and found to be satisfactory, the manufacturer should not vary the route in any significant respect without the approval of the customer. This is effectively saying that the supplier of a material or product made by a particular series of processes has been approved, and that repeat orders can be accepted with confidence. This is a formal process used by some fuel manufacturers to cover the intangibles involved in manufacturing operations.

It is necessary to check that materials and products made to specifications meet those specifications; even though any prudent manufacturer will have carried out his own tests. The checking is the inspection procedure, which covers such methods as chemical and physical analysis, mechanical testing, non-destructive testing (especially for defects), destructive testing, and measurement. Inspection may be done on a 100 per cent basis or on a sampling basis; the nature of the test, however, frequently dictates a sampling basis, otherwise all the product would be consumed. Such is the case for chemical analysis and mechanical testing, while weld radiography and fuel pin leak testing, as examples, can be and usually are done on a 100 per cent basis.

The importance of the feature being inspected will determine the confidence required that all components are within specification and hence the frequency of sampling and inspection. Even though 100 per cent inspection may be possible, it may not be justified, bearing in mind that inspection can be expensive. It is not unusual for example for inspection to account for 25 per cent of the cost of a component.

The inspection requirements for a material, component or sub-assembly are presented in the form of a quality control plan, which details what inspection is to be carried out, what methods are to be used, what sampling frequency is required, and how the results are to be reported, and if necessary, stored. A set of quality control plans, together with the knowledge that the inspection called for is being carried out and interpreted correctly, represents the ultimate assurance that a fuel element is fit for service. However, mistakes can be made, equipment can go wrong, and unremitting vigilance is required, including frequent checking of equipment and instruments against standards, to ensure that the quality control plan is working effectively.

Inspection can be done in a variety of ways. It may be wholly or partly delegated to a manufacturer (subject to checks), or to an independent inspection authority, or it can be done by the fuel manufacturers' own inspectors, who should be independent of the production department.

This chapter has dealt with many of the design features and materials properties which determine whether a fuel element will behave well in service. It has been shown that a large number of criteria need to be taken into account in optimizing the design of fuel for any particular reactor. There is, however, one criterion in particular which needs always to be borne in mind. The primary barrier to the escape of active fission products consists of the cladding, end caps, and welds that seal in the fuel material. It is an overriding aim in design, manufacture, and inspection to ensure that this barrier is of the highest quality, and will not be breached in the aggressive and demanding environment of the reactor core.

Acknowledgements

The author would like to thank those many colleagues in SNL, RNE and BNFL Springfields who have given helpful suggestions, advice, and information, which have greatly assisted in the writing of this chapter. In particular, thanks are due to Mr. D. H. Willey for provision of much of the background material.

References

1. ELDRED, V.W., HARRIES, J.E. and SHAW, R.A. The performance of uranium/ magnox fuel elements in the UK reactors. In BNES. *Proc. Int. Conf. on Nuclear Fuel Performance, London, October 1973*, paper 8. British Nuclear Energy Society, London (1975).
2. WADDINGTON, J.S., RAVEN, L.F., and THORPE, G. Development of fuel elements for the AGR. *Nucl. Energy*, **18** (4), 283–287 (1979).
3. ALLEN, P.L. FORD, L.H. and SHENNAN, J.V. Nuclear fuel coated particle development in the Reactor Fuel Element Laboratories of the UKAEA. *Reactor Technol.*, **35**, no. 2 (1977).
4. FRIEDLANDER, G.D. Optimising PWR fuel and design. *Electl. Wld.*, 50–53 (May 15, 1979).
5. ARNSBERGER, L.P. and NAKASATO, S. The PWR 17 × 17 assembly–a design incorporating improved safety margins. In *Proc. Crest Specialist Meeting on the Safety of Water Reactor Fuel Elements, Saclay, France, October 1973*, paper SNI 1/7. Nuclear Energy Agency OECD, Paris (1973).
6. KLEPFER, H.H. and TROCKI, T. Fuel for the BWR 6. In *Proc. Int. Conf. on Physical Metallurgy of Reactor Fuel Elements, Berkeley Nuclear Laboratories, England*, 1973, paper 39. Metals Society, London (1975).
7. HOLMES, J.A.G. *Design of oxide fuel for fast reactors. In* IAEA. *Proc. Symp. on Fuel and Fuel Elements for Fast Reactors, Brussels, July 1973*. International Atomic Energy Agency, Vienna (1974).
8. BISHOP, J.F.W. Performance development of the PFR fuel and the use of PFR as a fuel development facility. In BNES. *Proc. Int. Conf. on Fast Reactor Power Stations. London, March 1974*. BNES, London (1974).
9. KENNEDY, R.H. Ore to Concentrates. In *Facilities and technology needed for nuclear fuel manufacture*, pp. 1-32. *IAEA*-158 (1972).
10. ROGAN, H. Production scale processes and plants in the UK. In *The Conversion of uranium ore concentrates to nuclear grade uranium hexafluoride and to enriched uranium dioxide*, pp. 53–62. In *IAEA*-158 (1972).
11. DORAN, J. Manufacture of fuel elements from enriched UO_2 powder in the UK –a review. In *IAEA*-158, pp. 269–276 (1972).

14

Nuclear fuel reprocessing in the UK

R. H. ALLARDICE, D. W. HARRIS, A. L. MIILS

Nuclear fuel reprocessing has been carried out on an industrial scale in the United Kingdom since 1952.

Two large reprocessing plants have been constructed and operated at Windscale, Cumbria and two smaller specialized plants have been constructed and operated at Dounreay, Northern Scotland.

At the present time, the second of the two Windscale plants is operating, and Government permission has been given for a third reprocessing plant to be built on that site. At Dounreay, one of the plants is operating in its original form, whilst the second is now operating in a modified form, reprocessing fuel from the prototype fast reactor.

This chapter describes the development of nuclear fuel reprocessing in the UK, commencing with the research carried out in Canada immediately after the Second World War. A general explanation of the techniques of nuclear fuel reprocessing and of the equipment used is given. This is followed by a detailed description of the plants and processes installed and operated in the UK.

Contents

1 Introduction

Nuclear fuel reprocessing is the term used to describe the separation of fissile and fertile nuclear materials from fission products and other impurities in fuel which has been irradiated in and discharged from a nuclear reactor. The products of reprocessing operations are normally pure compounds of the fissile and fertile materials, which are then subject to further processes to convert them into material (metal or oxide) suitable for reuse as fuel in nuclear reactors. The impurities or fission products are separated in re-processing in the form of solid, liquid, or gaseous wastes, which have to be stored or disposed of in a suitable manner.

In the United Kingdom the development of the relevant processes and industrial equipment arose initially from a Government decision to develop nuclear weapons based on plutonium. Reprocessing facilities have since been extended in order to handle the fuel arisings from the United Kingdom's civil nuclear power programme, and spare reprocessing capacity is available on commercial terms to overseas customers.

The first industrial-scale irradiated-fuel-reprocessing operations in the United Kingdom were carried out at the Windscale Works, Cumbria, in the B204 plant, which commenced operation in 1952. It was designed to handle fuels from the early gas-cooled reactors. This was superseded by a larger plant (B205) built and operated at Windscale, which reprocessed all the fuel arisings from the United Kingdom's nuclear power programme. Modifications to the Windscale plants in 1969 enabled fuel from the Windscale advanced gas-cooled reactor (WAGR) and from light water reactors (LWR) to be processed as well.

A new fuel-reprocessing plant. THORP (thermal oxide reprocessing plant), is currently being designed and built by British Nuclear Fuels Limited for installation at Windscale with an operating date in the late 1980s.

Two smaller reprocessing plants have been constructed at Dounreay in

Northern Scotland: the materials-testing reactor (MTR) fuel reprocessing plant, which commenced operation in 1958 and is still operating, and the Dounreay fast reactor (DFR) fuel reprocessing plant, which operated from 1960 to 1975. This latter plant was closed down when the Dounreay fast reactor was taken out of service. It was then decontaminated and rebuilt to enable it to reprocess the fuel arisings from the prototype fast reactor (PFR) also situated at Dounreay. This plant commenced PFR fuel reprocessing in 1980. (See Table 14.1)

At the present time, large-scale nuclear fuel manufacturing and reprocessing operations in the United Kingdom are carried out on a commercial basis by British Nuclear Fuels Limited (BNFL). Small-scale fuel fabrication, reprocessing operations, and research are carried out by the UKAEA because of their direct links with the main-line UKAEA reactor development programme, especially the liquid-metal-cooled fast reactor project.

1.1 Reprocessing objectives

The basic physical processes which occur in a nuclear reactor are fully described in chapters 1 and 2. The main nuclear reactions are:

^{235}U + slow neutrons → fission products + fast neutrons + gamma radiation,

fast neutrons + moderator → slow neutrons,

^{238}U + slow neutrons → ^{239}U + gamma radiation,

and

$$^{239}U \xrightarrow[\text{Beta emission}]{} {}^{239}Np \xrightarrow[\text{Beta emission}]{} {}^{239}Pu.$$

The plutonium isotopes, ^{238}Pu, ^{240}Pu, ^{241}Pu, ^{242}Pu, and the higher actinides, americium and curium, are also formed.

Fuel discharged from a nuclear reactor still contains important quantities of fissile nuclides, mainly ^{235}U and ^{239}Pu, which can be reused. It also contains ^{238}U which is of value as a fast reactor fuel. The primary purpose of reprocessing is to recover these materials in a chemically purified form, and in particular to free them from fission products, some of which are deleterious to the neutron economy of the reactor. For example, certain rare-earth fission products (^{149}Sm, ^{151}Sm, ^{154}Eu, ^{157}Gd) are very strong absorbers of slow neutrons. In certain fuels moreover, fission products have adverse metallurgical effects. There is a limit to the amount of fission product accumulation that can be tolerated and a limit to the damage that the fuel containment and support structure can be permitted to suffer, for example, due to irradiation swelling, before the fuel has to be removed from the reactor. The reprocessing of irradiated fuels serves to restore the nuclear reactivity of the fuel by removing fission products and replacing fissile material. Mechanical integrity is restored by incorporating the purified fuel together with new canning and support materials into replacement fuel elements.

As will be seen, the chemistry of the separation process is not basically

Table 14.1 UK reprocessing plants

	Dates	Fuel treated	Approximate capacity t/d	Reactors served etc.	Products
Windscale plants					
B204†	1952–64	U metal		Windscale Early Magnox	U (depleted), Pu
B205	1964–	U metal	7	Magnox (incl. overseas) Product of head-end plant	U (usually depleted) Pu
Modified plants	1969–73	UO₂	1.3	WAGR, LWR, etc.	U/Pu feed to B205
THORP	late 1980s	UO₂	5	AGR, LWRs (incl. overseas)	U, Pu
Dounreay plants					
MTR reprocessing (D1204)	1958–	U–Al (highly enriched U)	0.0006	MTRs (UK and overseas)	U (highly enriched)
DFR reprocessing (D1206)	1960–75‡	U–Mo (highly enriched U)	0.014	DFR, (MTR)§	U (highly enriched), Pu
PFR reprocessing (D1206)	1980–	UO₂–PuO₂	c. 0.03	PFR	U (depleted natural), Pu

† The B204 plant is being refurbished as a development facility in aid of THORP. The B205 plant is still fully operational
‡ not fully active until 1962
§ MTR fuel was processed in this plant from 1966 until 1975 in addition to DFR fuel

difficult, and there are several techniques available. The new features arise from the intense radioactivity of the fission products and from the need for criticality control.

The fuel cycles for uranium-fuelled thermal reactor and for fast reactor systems are illustrated in Fig. 14.1, which summarizes the stages by which fissile-material reactivity and fuel-canning materials are replaced.[1] Table 14.2 shows decontamination factors for a modern reprocessing plant together with the major fission-product arisings to be handled. So far as nuclear metallurgical and chemical requirements go, a rather small degree of de-contamination from fission products would suffice. However, very much more stringent demands must be met to obtain a uranium product pure enough to handle with no greater precautions than are necessary for natural uranium, and a plutonium product pure enough to handle in a glove box. The decontamination necessary to achieve this varies according to the circumstances, but those specified for the second reprocessing plant at Windscale in the UK may be quoted as typical.[2]

High recoveries of uranium and plutonium are required in reprocessing for a variety of reasons: to retrieve the valuable fissile material in spent fuel, to avoid the hazards of uranium and plutonium in waste streams, and to avoid the danger of an accumulation of fissile material in the plant which might become *critical*. In currently operating large plants, uranium recoveries of better than 99.8 per cent and plutonium recoveries of better than 99.5 per cent are the rule. Considerable effort is expended by plant operators to ensure both good recoveries and accountancy of uranium and plutonium to International Safeguards requirements.[3]

The chemical form of the uranium and plutonium products must be such that they can readily be stored for subsequent reuse. Suitable end-products from most processes are nitrate salts in nitric acid solution, or oxide. Further chemical conversions may be regarded as the initial stages of fuel fabrication rather than part of reprocessing, though the dividing line is to some extent arbitrary.

Discharge of radioactive materials from a reprocessing plant to the environment must be strictly controlled. Suitable precautions are needed to avoid biological hazard to plant operators or the general public from the radiations emanating from the plutonium and fission products. These precautions exert a major influence on the choice of separation processes.

The task is to design and operate a plant which may process many thousands of curies of fission-product activity daily, while keeping the radiation exposure of operating staff to very low levels in accordance with internationally agreed recommendations, (see Chapter 18).

Plutonium is primarily an alpha emitter but the fission products, which comprise tritium as well as some forty elements in the middle of the periodic table, are generally beta and gamma emitters and vary widely in chemical character. They include the noble gases krypton and xenon, the rare earths and radioactive iodine. The half-lives vary from seconds to years and *daughter*

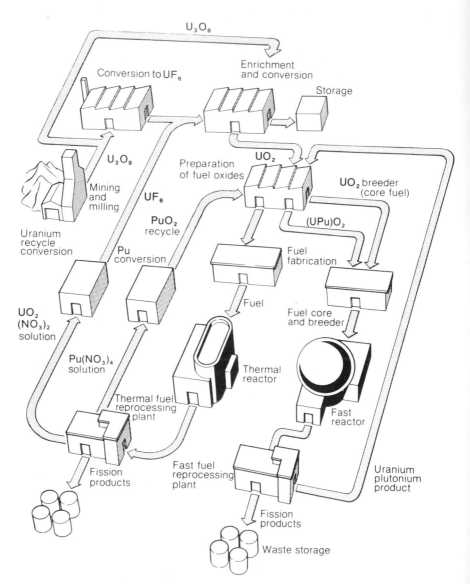

Fig. 14.1 Fuel cycles for thermal and fast reactor systems

Table 14.2 (a) Decontamination factors†

	Decontamination factor for: Fission products	U	Pu	Separation plant recovery (%)
U product	10^7	—	3×10^5	99.97
Pu product	3×10^8	10^7	—	99.8

† Magnox fuel, rated at 2.7 MW/t, 3000 MW d/t burn-up, cooled for 130 days

(b) Important fission-product arisings in spent fuel

Fuel	Burn-up (MW d/t)	Quantity of fission product (g/kg fuel) Zr	Ru	I
Magnox	3 600	0.4	0.24	0.02
LWR (3% enriched)	36 000	3.6	2.2	0.2
Fast reactor (15% PuO_2–85% UO_2)	50 000			
(core section only)		6.4	8.0	0.4
(core plus integral breeder)		3.2	4.0	0.2

products are formed by radioactive decay, so affecting the separation chemistry.

The effective range of alpha and beta radiation is very small and personnel protection can be provided by taking precautions to prevent ingestion or inhalation of active particles together with some light shielding for beta-radiation protection. The greater penetrating power of gamma radiation requires the introduction of massive shielding materials such as concrete, iron shot concrete, or lead to separate the operators from the equipment, which is then operated by remote control.

2 Evolution of reprocessing technology in the United Kingdom

Industrial-scale reprocessing of nuclear fuel has been carried out in the United Kingdom for nearly 30 years. Over this time the process requirements have changed from a demand for plutonium for nuclear weapons to a requirement to recover valuable uranium and plutonium as an essential component of the United Kingdom's electricity supply industry. At the same time a nuclear fuel reprocessing service has been offered on a commercial basis to countries not having reprocessing plants.

Over this period of time nuclear reactor design has advanced from the old air-cooled *piles* to compact, highly complex liquid-metal-cooled fast reactors. Improvements in reactor design have led to an increase in the radioactivity and mechanical complexity of the fuel to be processed, together with an increase in the proportions of fissile material in the fuel.

Developments in the field of nuclear fuel reprocessing have been noteworthy for the high safety standards which were adopted at the outset and have been maintained and improved upon throughout. Control of radioactive emissions to the environment and radiation exposure of operators have evolved with the development of reprocessing plants. The overall development of these plants and processes must be seen as a continuously evolving process commencing over 25 years ago.

The United Kingdom's involvement in reprocessing research commenced in Canada as part of a combined UK–Canadian–French effort towards the end of the Second World War using a small quantity of plutonium supplied by the USA. The passing of the MacMahon Act in the USA in 1946 prevented further access to American atomic energy information, and led to the setting up of an independent research effort initially in Montreal and later at the Canadian Nuclear Research Centre at Chalk River, Ottawa, where laboratories suitable for handling radioactive materials were built.

The Canadian effort was mainly concentrated on the design of a batch separation plant to be built at Chalk River. In 1946 the build-up of the British team of chemists and chemical engineers began. Its remit was to develop a process for the separation of plutonium from irradiated uranium and to supply data to design teams being formed in Britain.

The combined efforts of design teams, initially at Risley (Lancashire) and ICI, Widnes, and then later at Harwell, resulted in the first Windscale separation plant which commenced operation in 1952. This has been recorded in detail,[5] and the historical context has been described by Professor Gowing.[6]

The wartime American plant at Hanford used a precipitation process for the extraction of plutonium from irradiated uranium. Ion exchange, volatilization, absorption, and solvent extraction processes had also been studied.[7] The American objective backed by wartime urgency had been to extract plutonium only. The uranium was rejected, together with the fission products to the waste streams and stored.

By the time the British research efforts had commenced, there were anxieties about the adequacy of uranium supplies, and thus the added objective of recovery of uranium became part of the process requirements. Uranium recovery was to prove an important factor in the context of fast reactor development in the UK.

From the onset of the work it was clear that there were several criteria which had to be satisfied if a process was to be acceptable. The process had to

(1) be able to produce the required products from a variety of feedstocks;

(2) be chemically and physically suited to operation in high radiation fields;

(3) provide the degree of uranium/plutonium/fission product separation specified;

(4) be suited, if possible, to continuous rather than batch operation;

(5) be capable of engineering design, construction, and operation by techniques currently in use in the chemical engineering industry;

(6) give waste streams that could be handled by available methods; and

(7) allow a high operational plant availability with a minimum of direct maintenance.

In general terms these criteria still govern plant and process design today.

Precipitation processes were rejected because of the loss of uranium and attention was focused on solvent extraction. The early work on this process screened over 200 possible extractants. The Canadians had chosen triglycoldichloride (*trigly*) for their batch solvent extraction process. The British team short-listed about six extractants for further investigation, mainly ethers, but excluding substances such as diethyl ether which, while they had excellent extractive properties for uranium and plutonium, were rejected on account of their flammable and explosive nature. Of the two extractants finally selected, dibutylcarbitol (*Butex*) was the first choice. In addition to favourable extraction properties and good radiation stability, it offered the possibility of concentrating the highly radioactive liquid wastes from the process, although non-evaporable substances are added at the U–Pu separation stage. The second choice extractant, methyl isobutylketone (*Hexone*), is less favourable in this respect since non-evaporable substances have to be added to the process at the initial extraction stage in order to improve its efficiency. Later, as is noted at the end of this chapter, an extractant with superior chemical and physical properties to Butex or Hexone was discovered in the USA. In due course, this extractant, tri-*n*-butylphosphate, supplanted all other extractants in the commercial reprocessing of nuclear fuels.

Thus the early British development effort concentrated on the Butex process, using at first small laboratory-scale extraction columns and later a pilot plant again using column extractors. Simultaneously with this chemical engineering work, studies of the chemistry of plutonium and fission products were also being carried out at Chalk River by British scientists. Part of this work involved the development of analytical methods that would be ultimately required for process control. About 60 milligrams of plutonium (two-thousandths of an ounce) were available for all the basic chemical and chemical engineering studies required. Shortage of material meant that after completion of a set of experiments all the products and waste arisings had to be processed to recover the original plutonium before proceeding to the next series of experiments or tests. As facilities for radioactive work became available at Harwell the work was transferred to the UK, where in addition to the separation process methods for the production of metallic uranium and plutonium, procedures for waste-handling and storage received attention. In the summer of 1948 the first British-made plutonium—4 milligrams—was recovered at Harwell from a uranium fuel slug that had been irradiated in Harwell's GLEEP reactor. Figure 14.2 shows a general view of the small laboratory-scale reprocessing facility used in the work.

The UK project both in Canada and in the United Kingdom reported to Christopher (now Lord) Hinton, who subsequently built the first British

Fig. 14.2 The original laboratory apparatus for the separation of plutonium from irradiated uranium, Harwell 1948

reprocessing plant at Windscale based on chemical and engineering data obtained in the two countries. During the early work in Canada, there were exchanges of information on reprocessing topics between British and American scientists and engineers. Following President Eisenhower's 'Atoms for Peace' policy in 1953, detailed exchanges of technology were made by many countries at the Geneva Conferences on the Peaceful Uses of Atomic Energy.

In addition to solvent extraction, other separation processes were extensively investigated. ICI, who were involved in the atomic energy project from an early stage, providing industrial-scale design information on the Butex process, also explored the possibility of the *fluoride volatility* process. It had been established that uranium metal was soluble in bromine (or chlorine) trifluoride, and that the uranium could then be recovered by distillation as uranium hexafluoride, which is a suitable feed material to a uranium diffusion plant. Fission products, with the principal exception of tellurium, are not volatile, nor do they have volatile fluorides. Thus the scheme offered the basis of an economical separation process, provided plutonium could also be recovered. It later transpired that a major programme along similar lines had been carried out in the USA. However, the ICI process could not be established in time to influence the choice of process and plant at Windscale, and in fact plutonium has never been satisfactorily recovered by this process.

Pyrometallurgical processes involving slagging and volatilization were also studied. In these processes, the decontamination of uranium and plutonium from fission products and from each other is orders of magnitude less than that obtained by solvent extraction, thus necessitating a remotely controlled fuel refabricaton process.

The success of solvent extraction led to the abandonment of these processes at an early stage in the UK although interest in pyrometallurgical processes was rekindled to some extent in the mid 1950s, both in the UK and the USA, and is discussed later.

From the early studies of radioactivity, the risks to workers handling radioactive materials were appreciated, and as a result, a new breed of scientist, the health physicist became an important presence in all nuclear laboratories and factories. The responsibility of the health physicist is to monitor, check, and advise on all matters concerning the interrelation between the health of the worker and the work in which he is involved. This includes the measurement of radioactivity in the environment at and around the workplace, and the measurement and control of the radiation to which the worker might be exposed. General and personal radiation-measuring instruments were provided, and workers were subjected to regular stringent medical examinations, including frequent blood tests and urinalysis; these measures continue today.

Health physics control has played an important and ever-increasing role in work with radioactivity and will continue to do so in the future. Then, as now, the health physicist was totally independent from the development scientists and engineers, so he was able to be involved in the work in an impartial manner. Today, the health physics interest forms a considerable part of any total project team. The nuclear industry is probably the first major industry to have recognized the need for health physicists and health and safety experts from its inception, and to have fulfilled that need from the beginning.

Based on the microscale chemical investigations of three decades earlier, solvent extraction processes and plant have evolved to a high degree of sophistication. Not only has there been a continuing process of technical and scientific improvement, but also there has been the establishment and maintenance of high industrial health standards within the industry itself.

3 Reprocessing technology

3.1 Features of reprocessing schemes

All methods of reprocessing nuclear fuel have a number of common features. The plant and process can be subdivided into the following general sections.

1. *Fuel receipt and storage*. Irradiated nuclear fuel from the reactors is accepted and stored until the reprocessing plant is able to process it;

2. *Fuel breakdown and preparation for reprocessing*. Depending on the type of reactor, the complexity of the fuel element or sub-assembly varies. In this section of the plant the fuel element or sub-assembly is stripped of as much of its non-fuel components as possible, and the fuel component itself is then usually broken down into smaller pieces. If the reprocessing scheme is an aqueous processs, these small pieces are then dissolved, say in nitric acid, for reprocessing. In a non-aqueous process, fuel is prepared as appropriate, e.g. by fluorination in the case of the fluoride volatility process.

3. *Separation*. The various types of process have already been noted. Since this chapter is primarily concerned with reprocessing in the United Kingdom, it will concentrate on the solvent extraction process. The process consists of the primary separation of the fissile and fertile materials (plutonium and uranium) from fission products and the purification of the plutonium and uranium.

4. *Waste handling*. Gaseous, liquid, and solid wastes are generated as a result of reprocessing activities. These wastes might be merely contaminated with radioactive materials, or they might be radioactive in their own right. Treatment and, where appropriate, disposal schemes are required.

5. *Solvent recycling*. The organic solvent used in solvent extraction is recycled for reuse in the main stream processes.

Before describing the reprocessing plants and processes in the UK, it is perhaps useful to enlarge upon the concepts noted above and to include in the description some definitions of technical terms used in the work.

3.1.1 *Fuel receipt and storage*

Nuclear fuel is transported from the reactor to the reprocessing plant in large containers so constructed as to prevent the radiation from the fuel penetrating to the outside of the container. These containers are known as *flasks*. At the reprocessing plant these flasks are unloaded into storage *ponds*, which are essentially large tanks filled with water. The water provides both shielding to absorb neutron and gamma radiation from the fuel and also removes any heat generated by the fuel. In addition, water, being transparent, enables the operator to see his work. Fuel is stored in these ponds. The current trend is for fuel to be placed first in sealed water-filled containers or *bottles*, which are then stored in the pond water, thus preventing direct contact between the fuel itself and the water in the pond.

3.1.2 *Fuel breakdown*

At the appropriate time, fuel is removed from the pond. This is usually when the short-lived radioactive fission products, in particular iodine-131, have decayed, and when the decay heat output from the fuel has fallen to a level which makes it convenient to handle in the reprocessing plant itself.

Fuel element design varies with the type of reactor in which it is used (Fig. 14.3). In general terms, however, fuel, which is usually a metal or an oxide, is contained in a can made of material such as Magnox (a magnesium alloy), Zircaloy (a zirconium alloy), or stainless steel to form a fuel element or fuel pin. This containment is called the cladding. One or more fuel elements may be built into an array usually surrounded by a *wrapper*, which has lifting lugs and other necessary mechanical items attached, to form a *sub-assembly*. It can be seen from Fig. 14.3 that the complexity and physical dimensions of the fuel element assemblies vary widely with reactor type. Chapter 13 describes fuel element construction in detail.

With Magnox fuels, the Magnox cladding is stripped off the fuel, uncovering the metallic uranium fuel rod. Stainless-steel-clad AGR sub-assemblies are dismantled, packed into pin containers, loaded into a stainless steel basket, and placed in the dissolver. The stainless steel items remain undissolved in the basket and are removed as solid waste. The basket is reusable. Similar techniques are used for LWR fuels which might be clad in stainless steel or Zircaloy, but the assemblies are sheared into small pieces directly without dismantling.

The more complex fast reactor sub-assemblies are dismantled, and the individual fuel pins only are chopped into short lengths into a stainless steel basket which is placed in the dissolver. The fuel is then dissolved out of the cladding, which remains as a solid in the basket. As with the AGR and LWR fuels above, this is retrieved at the end of the dissolution cycle and the stainless steel is dealt with as solid waste.

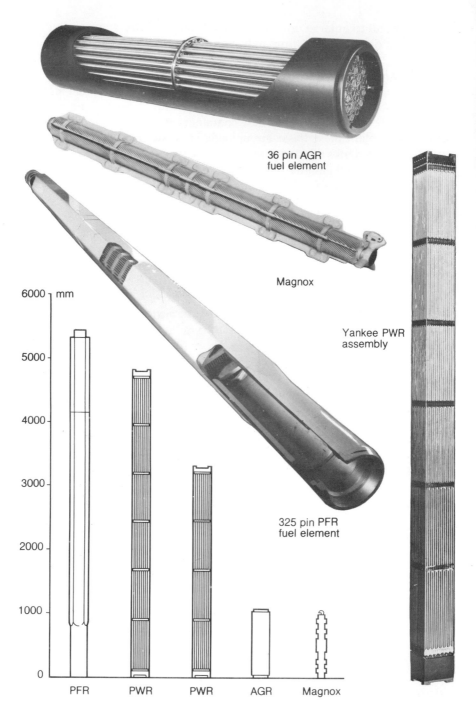

36 pin AGR
fuel element

Magnox

Yankee PWR
assembly

325 pin PFR
fuel element

Fig. 14.3 Nuclear fuel sub-assemblies

3.1.3 *Fuel dissolution*

Irradiated metallic or oxide fuels containing uranium, plutonium, and fission products can be dissolved in strong, boiling nitric acid to provide the feed to the solvent-extraction process for separating uranium and plutonium from fission products.

The solution from fuel dissolution contains uranium, plutonium, and fission products as nitrates. The uranium is present as uranyl nitrate ($UO_2(NO_3)_2$), but the plutonium content of the fuel exists as a mixture of plutonium tetranitrate ($Pu(NO_3)_4$) and plutonyl nitrate ($PuO_2(NO_3)_2$). The proportion of each species depends upon the dissolution conditions.

Dissolution reactions in nitric acid are complex and their discussion is beyond the scope of this chapter. The following equation is a simplified representation of nuclear fuel dissolution for metal fuels where M represents uranium.

$$M + 2HNO_3 + \frac{3}{2} O_2 \rightarrow [MO_2(NO_3)_2] + H_2O$$

The reaction can produce a uranyl–plutonium nitrate solution which is about 3M in HNO_3, an appropriate acidity level for the subsequent solvent-extraction process, and one which minimizes corrosion of the stainless steel dissolver at working temperatures (about 100 °C). Most of the fission products formed in the fuel during its period in the reactor will also dissolve in nitric acid. The more important fission products for various reactor fuel types from a reprocessing point of view, are listed in Table 14.2.

3.1.4 *Separation*

As has already been noted, solvent extraction is used exclusively in nuclear fuel reprocessing in the United Kingdom. The following is a simplified explanation of both the principles and practice of solvent extraction; details of specific processes are given in subsequent sections.

The process of transferring a substance in solution in one liquid into solution in a second liquid which is wholly or partly immiscible with the first is known as *solvent extraction* or *liquid–liquid extraction*.

Solvent extraction is in essence a very simple technique. It may be illustrated by assuming that two substances A and B are both soluble in water as shown in Fig. 14.4. This solution is then mixed with a suitable liquid which is not soluble in water, say benzene, and after mixing the two liquids are allowed to separate under gravity. The less dense benzene phase will float on top of the water. If B is more soluble in the organic solvent than in water, the benzene phase will contain B but not A, which is assumed to be insoluble in benzene. A and B have now been separated. By mixing the organic phase alone with, for example, caustic soda solution, B is *stripped* from it back into an aqueous phase, from which it can be recovered for use, and the organic phase can be recycled back to the beginning of the process.

All nuclear fuel reprocessing in the United Kingdom has been carried out

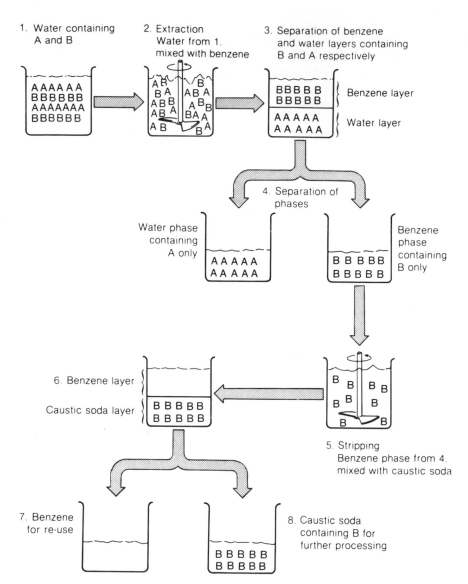

Fig. 14.4 Solvent extraction—an explanation

by variations of this simplified version of the solvent extraction process described above but, of course, with the appropriate chemicals. The aqueous phase is nitric acid; the immiscible phase is an organic liquid which preferentially extracts uranium and plutonium rather than fission products.

The chemical nature of the substances involved often influences the extent of the separation that can be achieved in a solvent extraction process. It is possible, for example, by adding a substance known as a *salting-out* agent,[9] to enhance the extraction of one component into the solvent while not affecting the extraction of the other components. The salting-out agent itself might or might not be extracted.

Some substances such as plutonium nitrate can exist in nitric acid solution in different chemical forms; the most common forms of plutonium nitrate are as follows:

(1) plutonium trinitrate, $Pu(NO_3)_3$;

(2) plutonium tetranitrate, $Pu(NO_3)_4$; and

(3) plutonyl nitrate, $PuO_2(NO_3)_2$.

Plutonium tetranitrate and plutonyl nitrate are both soluble in a number of organic liquids, whereas plutonium trinitrate is not generally soluble in such substances. Thus, by converting plutonium nitrate from one form to another, the plutonium fraction in the process can be made soluble or insoluble in the organic phase as required.

The initial solution whose components are to be separated is known as the *feed*, which is usually an aqueous solution. The organic liquid mixed with the feed for the purposes of extraction is called the *solvent* or the *solvent phase*. This is immiscible with the feed. A solvent consisting of a solution of two or more substances is a *mixed solvent*. A liquid waste stream from a solvent extraction plant is termed a *raffinate*, and this may be either an aqueous or an organic stream.

The transfer of a substance from the aqueous phase to the organic liquid phase is termed *extraction* and the chemical contained in the organic solvent for this purpose is known as the *extractant*. Some extractants are too powerful to be used in an undiluted form, while others have viscosities and densities or other physical properties which make them inconvenient for practical use. These extractants are usually diluted with an inert organic liquid called the *diluent* which itself plays no chemical part in the solvent-extraction process. Hence in some processes the solvent phase is a mixture of extractant and diluent. The solvent-rich solution containing the extracted solute(s) is the *extract*. The reverse process, in other words the transfer of a substance from the organic phase back to the aqueous phase, is known as *stripping* (see Fig. 14.4).

A third process, *scrubbing*, occurs after the extraction section. In this process contaminating substances in the extract are removed by treatment by a fresh aqueous phase containing only the acid component of the feed.

The essential step in solvent extraction is the intimate contact of two immiscible liquids for the purpose of mass transfer of constituents from one phase to the other followed by the physical separation of the two immiscible liquids. Any device or combination of devices which accomplishes this once is a *stage*. If the liquids leaving a stage are in equilibrium, so that no further change in concentration would have occurred within them after a longer contact time, the stage is termed *theoretical* or *ideal*. A multistage *cascade* is a group of stages arranged for counter-current flow of liquids from stage to stage for purposes of enhancing the extent of separation. In counter-current flow the liquids are frequently contacted without repeated physical separation and recontacting in discrete stages. Such methods are known as *differential* or *continuous-contacting* methods (Fig. 14.5). One sequence of extraction, scrub, and stripping operations is called a *cycle*.

Plutonium nitrate exists in two forms in the dissolver solution. Plutonium tetranitrate is more extractable than the plutonyl form and it is usual therefore to *condition* the dissolver liquor and convert the plutonyl form to the more extractable tetranitrate. This is readily done by passing oxides of nitrogen through the liquor. Uranium in the dissolver solution exists entirely as uranyl nitrate $UO_2(NO_3)_2$, which is readily extractable.

The conditioned dissolver solution is the *feed* to the solvent extraction process. In the first extraction cascade the plutonium and uranium are separated from the bulk of the fission products. Some 99.9 per cent of the uranium and 99.98 per cent of the plutonium is extracted into the solvent phase, leaving 99.5 per cent of the fission products in the aqueous raffinate. In a subsequent cycle, the uranium is extracted into the solvent phase, leaving the plutonium in the aqueous phase. This is achieved in most flowsheets by converting the plutonium to the trinitrate form. Typically only about 0.01 per cent of the uranium follows the plutonium into the aqueous phase, and about 0.02 per cent of the plutonium is extracted with the uranium. Since the uranium and plutonium streams retain small amounts of fission products, they are further purified by separate solvent extraction cycles.

Reduction of plutonium tetranitrate to plutonium trinitrate is accomplished by a number of reagents, ferrous sulphamate typically being used in the UK. The ferrous ions react according to the reaction:

$$Fe(II) + Pu(IV) \rightleftharpoons Pu(III) + Fe(III).$$

Nitrous acid is almost invariably present in nitric acid solutions. Sulphamate reacts with nitrous acid, converting it to water and nitrogen, and hence prevents it acting as an oxidizing agent, which would otherwise reoxidize the Pu(III) back to Pu(IV).

A number of organic extractants have been examined for use in nuclear fuel reprocessing schemes as listed in Table 14.3. In the United Kingdom Butex was used in the first Windscale separation plant (B204) and it is also used in the so-called 'oxide head-end plant'. Tri-*n*-butyl phosphate has been used in the second separation plant at Windscale (B205) and also in all the

Countercurrent continuous flow in a mixer settler contactor

M = mixing compartment
S = settling compartment, discrete settling in each settler

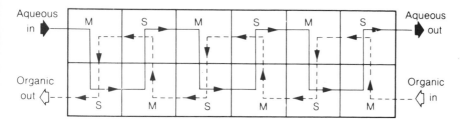

Flow patterns in contactors

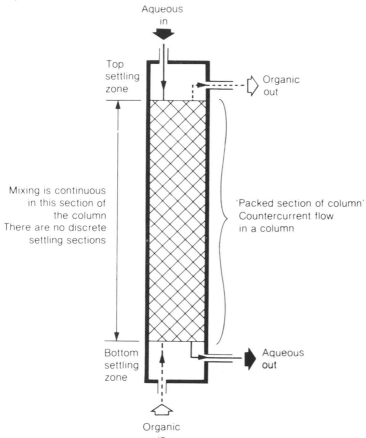

Fig. 14.5 Countercurrent solvent extraction

Table 14.3 Extractants and diluents for solvent extraction processes in the nuclear industry

Solvent	Alternate name	Diluent	Used to separate	Salting-out agent†
Ethyl ether	Diethyl ether	None	U from ores	$Ca(NO_3)_2$, $Mg(NO_3)_2$
Hexone	Methyl*iso*butyl ketone	None	U, Pu from fission products	$Al(NO_3)_3$
TBP	Tri-*n*-butyl phosphate	Kerosene	U, Pu, and Th from fission products and U and Th from ores	HNO_3
Butex	Dibutyl ether of diethyleneglycol	None	U, Pu from fission products	HNO_3
Penta-ether	Dibutyl ether or tetraethyleneglycol	None or butyl ether	U, Pu from fission products	HNO_3
Trigly	Triglycol dichloride	None	U, Pu from fission products	$Al(NO_3)_3$
TTA	Thenoyl trifluoroacetone	Benzene	Pu from fission products	None
DIPC	Di-*iso*propyl carbinol	None	U from fission products	None
DOPPA	Dioctyl pyrophosphoric acid	Kerosene	U recovery from ores	None
Cello-solve	Ethyleneglycol monobutyl ether	None	U from fission products	$Al(NO_3)_3$
DIPE	Di-*iso*propyl ether	None	U from fission products	$Al(NO_3)_3$

† A salting-out agent is used to improve the extraction of a substance in a liquid-liquid extraction system. It may or may not be itself extracted in the process. See reference 9.

processes used at Dounreay. Thenoyl trifluoroacetone has had a limited specialized application at Windscale in the plutonium purification plant.

Processes based on tributyl phosphate are called *Purex* processes.

It has already been noted that the effectiveness of a solvent extraction process depends upon the quality of extraction into the organic phase. The ratio of equilibrium concentrations of a given species in the organic and aqueous phases is known as the *distribution coefficient*, which is a measure of the effectiveness of extraction. Distribution coefficients for U, Pu, and some fission products in both typical Butex and Purex systems are given in Table 14.4. It will be appreciated from this table that distribution coefficients play a significant part in the actual choice of extractant.

3.1.5 *Waste handling*
Nuclear fuel reprocessing generates solid, liquid, and gaseous wastes. Some of these wastes are chemical wastes and are treated as such, but the bulk of the waste arisings are radioactive, or are radioactively contaminated, and require special treatment or storage.

At the present time the general principles of waste management are firstly, to reduce the quantity of waste arisings at source as much as possible, and secondly, to reduce the radioactive content of the bulk of the waste to as

Table 14.4 Approximate distribution coefficients of U, Pu, and some fission products between Butex and aqueous phase, and 20% vol. TBP and aqueous phase

	Butex	Butex	TBP
	3N nitric acid in aqueous phase	8M ammonium nitrate and 0.2M nitric acid in aqueous phase	3M nitric acid in aqueous phase
U(VI)	1.5	3.0	20
Pu(VI)	1.8	2.5	3
Pu(IV)	7	7	7
Pu(III)	0.01	0.002	0.01
Ru†	0.1		0.1 to 10^{-3}
Cs	10^{-4}		10^{-4}
Zr	2.5×10^{-4}		3×10^{-3}
Nb	2.5×10^{-3}		0.01

† The chemistry of ruthenium in solvent extraction is very complex. The distribution coefficient range given covers most of the chemical species of ruthenium present. An average distribution coefficient is typically about 2×10^{-3}

low as possible, using operational control or possibly a decontamination step. It should be noted that waste highly contaminated with α-, β- and γ-radiation emitters is more expensive to store than highly β-, γ-contaminated waste, etc. The order of cost per unit volume of radioactively contaminated wastes to be stored is:

decreasing
cost

high α, β, γ
low α—high β, γ
high α—low β, γ
low α—low β, γ
low β, γ,

there being a difference of an order of magnitude in cost across this range. The terms 'high' and 'low' as used here are purely relative.

An important category of waste arises from the fuel breakdown procedures and consist largely of non-fuel sub-assembly items. Because they have been in a nuclear reactor these components have become radioactive in their own right and require to be stored until their induced radioactivity has decayed. They emit mainly beta and gamma radiation. Any alpha contamination from, say, plutonium is removed before storage. Other 'solid' items from plant operations such as ventilation filters may contain radioactive particles and have to be stored.

Decontamination of solids invariably gives rise to liquid wastes. Liquid wastes can be reduced in volume by evaporation, with the concentrate being stored, as is done with the highly active aqueous raffinates from solvent extraction (Fig. 14.6), and these liquid concentrates can themselves be converted into a solid glass by a vitrification process.[10] The non-radioactive distillate from the evaporation process can usually be discharged to sea quite safely.[11,12] Intermediate-level radioactive liquors can usually be deconta-

Fig. 14.6 Highly active liquid storage tanks at BNFL Windscale

minated by a *floc* process, which is a fairly standard water treatment process, or alternatively by ion exchange. In this process, the radioactive component is incorporated into or on to a solid, which is then stored, leaving a decontaminated liquid for discharge.

Organic raffinates are cleaned chemically and are recycled to the plant. The arisings from the chemical cleaning are dealt with according to thier radioactivity content.

Some of the gaseous radioactive isotopes such as iodine-131 and xenon isotopes have conveninently short half-lives and decay in the fuel during storage, before reprocessing. Other radioactive gases such as tritium, carbon-14 as carbon dioxide, krypton-85, iodine-129, have longer half-lives (12.3 y, 5.73×10^3 y, 10.76 y, and 1.7×10^7 y, respectively). The extent to which it is decided to remove these isotopes from gaseous wastes will be based on assessments of the very small but widespread radiation doses resulting from their world-wide discharge to the atmosphere. Chemical and physical techniques are available in principle for some degree of their removal from wastes. Thus iodine can be removed from gaseous wastes by washing the gas stream, either with mercuric nitrate solution or with another appropriate chemical that forms an iodine compound. In large installations absorbers such as charcoal or silver zeolite can be used.

Krypton can be removed by washing the gas stream with halogenated hydrocarbons such as chemicals of the *Freon* series, or alternatively by cryogenic means. It should be noted that these gas removal processes do create a secondary waste for disposal.

The need to comply with the legal requirements concerning the storage and disposal of radioactive materials has been noted.[10] A further principle is now superimposed on waste discharge procedures, that of *ALARA* (as low as is reasonably achievable).[14] There is a great deal of research in progress throughout the world on waste procedures and disposal and decontamination techniques, which means that less and less radioactive material is being discharged to the environment for dilution and dispersion,[15] the radioactive fraction of the wastes being concentrated and retained for long-term storage in a similar manner to the high-level wastes.

3.1.6 *Solvent recycling*

The organic extractant and the diluent used in the solvent extraction process are valuable chemical reagents, and, after stripping to remove uranium and plutonium, they are themselves processed by washing with caustic soda and sodium carbonate solution, followed by nitric acid washing before being recycled back to the main solvent extraction process for reuse.

The purpose of the solvent-washing procedure (which is usually carried out in mixer–settlers) is to remove traces of fission products, uranium, and plutonium, which are retained by the solvent phase and are not removed by the conventional stripping procedure, and solvent degradation products.

During its passage through the main solvent extraction process, the organic extractant and the diluent undergo slight decomposition due to the α, β-, and γ-radiation present to form so-called *degradation products*. These degradation products (e.g. monobutyl phosphate, dibutyl phosphate, etc.) are chemically related to the extractant and diluent, but even when present in very small quantities they interefere with the efficiency of the solvent extraction process. (They form very strong chemical complexes with plutonium and zirconium and prevent their efficient processing. Other degradation products affect the physical separation of solvent and aqueous phases in the solvent extraction process by stabilizing emulsions.)

The aqueous wash liquors thus contain the degradation products and become waste streams, leaving the purified extractant and diluent available for reuse.

Work is in hand to modify solvent wash procedures so that the quantity of aqueous waste is minimized. Alternative techniques using solid absorbers for the removal of degradation products are also being investigated.

4 Criticality control

In general terms, nuclear fuel reprocessing is part of the process chemical industry. Owing to the radioactive nature of the materials being processed, plant and equipment have to be installed behind massive concrete shielding. This complicates plant and process design to some extent, but the processes and equipment used are familiar to the non-nuclear chemical engineer and chemist.

An important difference between traditional and nuclear chemical engineering is the need in the latter for *criticality* avoidance.

Before the nuclear fuel reprocessing techniques and equipment used in the United Kingdom are examined, it is appropriate first to look at the topic of criticality and methods for its avoidance.

Whenever amounts of fissile isotopes (plutonium, uranium-235, or uranium-233) exceeding a few hundred grams are involved, attention must be paid in reprocessing plant design and operation to the possibility of producing a self sustaining chain reaction—the condition known as criticality. The effects of such a reaction would depend greatly on the way in which the condition had been achieved, but under no circumstances could a supercritical accumulation of plutonium be held together under sufficient pressure for a major explosion to be possible. Local overheating of solutions and possibly local disruption of plant and equipment is the worst that might be expected. Local radiation levels could be very high, but the normal biological shielding provided in a reprocessing plant would reduce external radiation levels substantially.

However, safe operation of a reprocessing plant processing fissile isotopes can be assured in a number of ways. In all cases the question is really one of plant design and process control. A higher assurance of safety from accidental criticality can be given when control can be exercised independently of the plant operator. The five methods of criticality control used in reprocessing plants are given in §§4.1–4.5.

4.1 Mass control

If the total mass of fissile material in a specific vessel or section of the plant is limited to the minimum safe mass then no matter how the material is disposed, criticality is impossible. In a reprocessing plant, mass limitation depends on an adequate fissile-material accounting system and strict control of operating procedures.

4.2 Volume control

If a vessel or piece of plant has a total volume less than the critical volume for a chemical compound containing fissile isotopes (allowing for possible blockages or overflows), then criticality cannot credibly occur, even under operational or accident conditions leading to abnormally high levels of neutron moderation or neutron reflection.

4.3 Concentration control

This technique is particularly suited to large-scale continuous processing at low fissile concentrations, as in thermal reactor fuel reprocessing. The limiting critical concentration is very simply stated, but the integrity of criticality safety lies in controlling the many process variables including non-fissile feeds to the plant. In addition in-line instrumentation is necessary to monitor the fissile concentration and to detect accumulations of fissile-bearing solid particles. This system has been used successfully in the first and second Windscale reprocessing plants.

4.4 Geometric control

The term covers a variety of techniques. It is essentially the limitation of the physical dimensions of individual pieces of process equipment, so that criticality is impossible for all credible variations in the concentration, disposition, or chemical form of the fissile material and its local environment. The limited dimensions used are usually based on infinite cylinder or slab geometry, where the surface-area-to-volume ratio is high, thus facilitating the escape of neutrons from the equipment. This method of plant design is particularly suited to the processing of low-volume, high-fissile-content fuel, such as fast reactor fuel, and has been employed successfully in the design of the Dounreay fast reactor fuel reprocessing plant.

4.5 Neutron poisons

Criticality control methods such as those listed above can be made less restrictive by the introduction of neutron absorbers such as boron or cadmium into the equipment design (e.g. as part of the structural materials or as packing) or into the solutions being processed (e.g. by including cadmium, nitrate, or gadolinium nitrate in process liquors). These *neutron poison* techniques have not been used much in practice in reprocessing plants in the UK because of the necessity of ensuring that the absorbing materials are present in the required form and concentration at all times, and that they have not been dispersed by plant corrosion or separated from the plant solutions by chemical reactions. Neutron poisons therefore reintroduce a new element of managerial control, and require additional checks by neutron-monitoring or chemical-analysis procedures. The technique has been used in some plants in the form of fixed absorbers external to the process equipment to isolate or shield one plant unit from another, thereby preventing inter-action leading to a chain reaction. Because of the greater absorption effi-ciencies of the absorbers for thermal neutrons external absorbers are frequently used in conjunction with neutron-moderating materials, and often consist of sandwiches of cadmium sheet and a hydrogenous material.

5 Reprocessing equipment

Because the equipment in which nuclear fuel reprocessing is carried out is operated behind massive concrete walls in order to protect the operator, reprocessing equipment has to be carefully designed with a high degree of robustness and reliability if a remote maintenance approach is not being followed.

The early plants were designed on the assumption that once they became radioactively contaminated, it would not be possible to decontaminate them sufficiently ever to allow man to enter them. It is now realized that this is not so. Indeed, reprocessing plants have been entered, and in one instance, at Dounreay, the plant was completely decontaminated, stripped, and re-furbished for reuse.[16]

It should be emphasized that although the ability to decontaminate and enter plants has been demonstrated, it does not lessen the requirement for excellence in plant design. Further details of this work are given in chapter 16. What it does indicate is that the restoration of a site when the plant is no longer required is now a possibility (although it has to be emphasized that these decontamination exercises do themselves generate a considerable quantity of radioactively contaminated waste material).

The variation in complexity of the fuel sub-assemblies to be reprocessed has already been noted. Various techniques are used in order to prepare fuel for dissolution, depending upon the fuel to be processed. These will be described here briefly, together with other techniques that are, or have been, in use in various plants, although not necessarily in the United Kingdom.

5.1 **Mechanical decladding**

Magnox fuel was originally declad mechanically by passing it through a die which stripped the cladding off the fuel and retained it, allowing the fuel itself to go forward for dissolution. An improvement on this technique was employed at Dounreay for the niobium-clad uranium alloy fuel used in the Dounreay fast reactor (DFR). Here the cladding was peeled off the fuel rather like a banana skin as the fuel travelled forward through cutting wheels. This technique allowed distorted fuel pins to be handled. The uranium alloy fuel was then cracked off its central vanadium tube and sent for dissolution.

Subsequent modifications to this fuel, which included replacement of the niobium cladding by stainless steel, enabled fuel cropping (see §5.2) to be used.

The successful 'banana-peeling' technique used at Dounreay led to its application in Magnox fuel decladding at Windscale, where this method is still used in place of the original technique.

Figure 14.7 illustrates various decladding techniques.

5.2 **Shearing**

Some types of fuel cladding, for example, stainless steel or Zircaloy, are not amenable to the 'banana-peeling' technique. Here, shearing or cropping are possible alternatives.

In the course of modifications to the Dounreay fast reactor fuel reprocessing plant already noted in §5.1, a cropping technique was introduced. Here the fuel, as single pins complete with cladding, was fed into a hydraulic cropping device to obtain short (25 mm) lengths of fuel inside its clad. The fuel and clad is then fed to the dissolver where the fuel is leached from the clad, the clad remaining essentially unaffected by the leaching action in the dissolver. Advanced gas-cooled reactor and water reactor (LWR) fuels are similarly broken down for leaching; however, instead of cropping individual fuel pins a number are sheared together in a massive shear. In the AGR case the pins are close-packed into a container for this operation, but LWR sub-assemblies are sheared directly.

In the prototype fast reactor fuel reprocessing plant the outer *wrapper* of the sub-assembly is partly removed by cutting it with a laser beam, and the individual fuel pins (325 per sub-assembly, each one stainless steel or Nimonic clad) are removed one at a time for cropping prior to leaching. In this case neither the wrapper nor other constructional matter other than the immediate fuel cladding enters the dissolver. The cladding has to be removed from the dissolver at the end of the fuel-leaching period.

A major problem with the above technique of fuel comminution is that if the ends of the sheared or cropped sections are significantly distorted, access of leaching acid to the fuel might not be satisfactory, and a loss of fuel might result due to incomplete dissolution. Fuel not dissolved, or trapped in the distorted cans might require a further recovery procedure.

(a)

Fig. 14.7 (a) Magnox fuel clad stripping, Windscale

(b)

Fig. 14.7 (b) Chopped AGR fuel, Windscale

Fig. 14.7 (c) Laser cutting of PFR sub-assembly wrapper, Dounreay

(d)

Fig. 14.7 (d) PFR fuel dismantling machine, Dounreay

5.3 **Total dissolution**

Total dissolution of fuel and cladding is unusual and in general is reserved for special one-off fuels. It has mainly been used in the USA. This technique is however used in the reprocessing of materials-testing reactor fuels (MTR's such as DIDO and PLUTO at Harwell) in the MTR fuel-reprocessing plant at Dounreay. MTR fuel consists of an enriched uranium–aluminium alloy sandwiched between layers of aluminium. Mechanical separation of the uranium-aluminium from the sandwich is not posssible, and at Dounreay the whole sub-assembly, with the exception of lifting lugs and location pieces, is sheared and dissolved in nitric acid–mercuric nitrate to give an aluminium nitrate–uranyl nitrate–fission-product nitrate solution for reprocessing. Mercuric ion is used as a catalyst to aid the dissolution of the aluminium.

An alternative technique used at Eurochemic (Mol, Belgium) is to dissolve the aluminium preferentially in caustic soda to produce sodium aluminate, which is removed, and then to dissolve the uranium in nitric acid.

5.4 **Chemical decladding**

Fuel cladding materials such as stainless steel, Zircaloy, aluminium, and Magnox are soluble in a variety of acids and alkalis, and in metals such as molten zinc. These techniques are known as chemical decladding and have been used both in the USA and Europe (Eurochemic), but not in the United Kingdom. Fuel cladding as such, even after leaching, represents a convenient, compact waste form, whereas wastes from chemical decladding are voluminous and usually dilute. Further, since they contain chemicals capable of dissolving, say, stainless steel, a problem arises concerning the materials of construction for storage or processing of such wastes.

An associated technique is that of selective chemical attack on the fuel cladding by a relatively small quantity of reagent, or a reagent offering good waste-handling properties. The objective here is to change the physical structure of the cladding, making it more brittle for example, and hence more amenable to mechanical rather than chemical removal. These techniques are still at the research and development stage.

5.5 **Fuel dissolvers**

The vessels in which the dissolution or leaching is carried out can be conveniently divided into two categories—continuous dissolvers and batch dissolvers. (For convenience the term 'leach' will not be used further, the term 'dissolve' will be taken to include leaching unless noted otherwise.) Different types of dissolvers are shown in Fig. 14.8.

As the name suggests, in a continuous dissolver (Fig. 14.8(a)) the prepared fuel is fed to the dissolver continuously, together with acid and any other reagents required for dissolution. The product, a solution of fuel and fission products, usually as nitrates in nitric acid, is obtained continuously from the

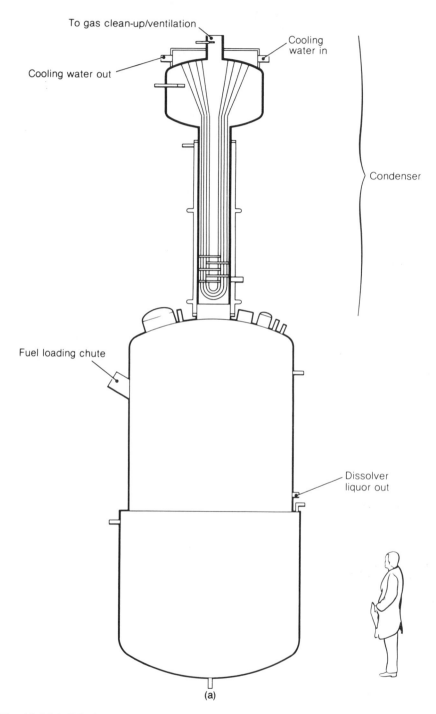

To gas clean-up/ventilation

Cooling water in

Cooling water out

Condenser

Fuel loading chute

Dissolver liquor out

(a)

Fig. 14.8 (a) Windscale B205 continuous dissolver (schematic)

dissolver and fed to the reprocessing plant. In batch dissolvers (Fig. 17(b), the fuel is fed in discrete quantities to the dissolver, together with the necessary reagents. Dissolution is carried out, and the dissolver is then emptied before the next fuel batch is loaded for dissolution.

Declad metal fuel, e.g. Magnox fuel, is dissolved at Windscale in a continuous manner. DFR fuel, MTR fuel, and oxide fuels (AGR, LWR, and PFR fuels) are dissolved in batch dissolvers, since they are not separated from their cladding prior to dissolution. Work is in hand in the United Kingdom and in many other countries to develop a continuous dissolver for sheared and cropped fuel. The difficulties associated with ensuring a smooth flow of the insoluble fuel cladding through the dissolver are considerable (Fig. 14.8(c)).

Sheared MTR fuel is placed directly into the dissolver for total batch dissolution. For the other fuels listed, the cropped or sheared material, together with sub-assembly structural items, is placed into a perforated drum or *basket*. This is then placed in the dissolver, which contains the appropriate chemical reagents. At the end of the dissolution period, the dissolver product liquor (fuel nitrate, fission-product nitrates, and nitric acid) is removed, and the basket containing the now fuel-free cladding is also removed and emptied and is then available for the next batch.

In general, dissolvers are made of stainless steel of sufficient thickness to allow for any corrosion likely to occur, but the DFR fuel dissolver was made of titanium because the reagents required to dissolve the uranium–molybdenum fuel (nitric acid–ferric nitrate) are highly corrosive to stainless steels at 100 °C. It is of interest to note that a special *graded* joint is used at the transition from the titanium dissolver to stainless steel plant in order to avoid electrolytic corrosion between the two dissimilar metals.

Criticality considerations impose constraints on dissolver design and operation. In the Magnox dissolver, at Windscale, low levels of fissile material (plutonium) are present, and no precautions are necessary other than ensuring that a build-up of plutonium in the dissolver does not occur.

MTR fuel, which contains highly enriched uranium, is dissolved under batch conditions, with a mass limitation applied to each batch. The dissolver is emptied after each dissolution.

Low-enriched reactor fuels, such as AGR or water reactor fuels, are dissolved in large dissolvers, where soluble poisons are used for criticality control. Cadmium has been used for AGR fuel reprocessing, and gadolinium nitrate is proposed for the thermal oxide fuel reprocessing plant (THORP).

Fast reactor fuels contain higher concentrations of fissile material. These fuels are dissolved in limited-geometry dissolvers, the physical size and geometric layout of the dissolver being carefully controlled by design. Hence criticality cannot occur for the fuel type specific to the dissolver design and process. A consequence of geometric limitation is that the size and volumetric capacity of such dissolvers is small (say 150–500 litres per dissolver). It might therefore be necessary for larger fast reactor fuel re-

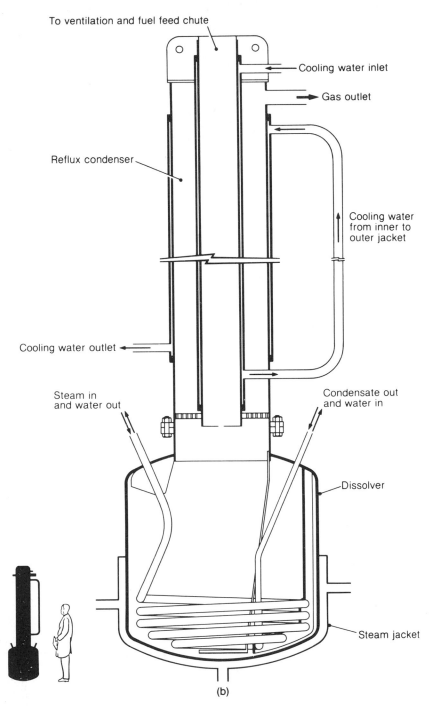

To ventilation and fuel feed chute

Cooling water inlet

Gas outlet

Reflux condenser

Cooling water from inner to outer jacket

Cooling water outlet

Steam in and water out

Condensate out and water in

Dissolver

Steam jacket

(b)

Fig. 14.8 (b) Irradiated fuel dissolvers (batch, mass limited), eg. MTR fuel dissolver (schematic)

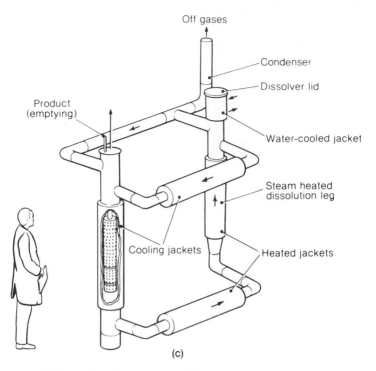

Fig. 14.8 (c) PFR fuel dissolver (schematic)

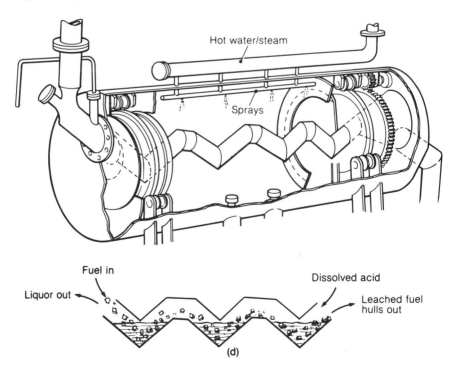

Fig. 14.8 (d) Irradiated fuel dissolver (geometrically limited) eg. CFR fuel dissolver (hypothetical)

processing plants to have more than one dissolver providing a feed to the solvent extraction plant.

5.6 Dissolver product clarification and chemical adjustment

Irrespective of the efficiency of the fuel dissolution process, some of the fission products formed during reactor irradiation are not soluble in the dissolution reagent, usually nitric acid. As the burn-up of the fuel increases from Magnox to AGR/LWR to Fast Reactor fuels, so the quantity of insoluble material increases.

These insoluble fission products can block pipes and interfere with the operation of the solvent extraction plant, and it is now becoming customary to remove the insolubles from the process liquors immediately after fuel dissolution. A centrifuge is used for this purpose, but this does give rise to problems associated with its operation under radioactive plant and criticality safety conditions. An additional problem arises in that these insolubles are, in part, radioactive and hence thermally hot. (They consist of an 'alloy' of Mo, Ru, Rh, Tc, Pd, together with a second phase of the type Pd_3U and small quantities of Zr, Nb, Sb, etc; their exact composition depends upon the actual fuel being processed.) There is therefore a heat rejection problem superimposed on to the other design problems. Figure 14.9 shows a large-capacity centrifuge of the type likely to be used in the THORP plant, and the much smaller limited-geometry machine used for fast reactor fuels.

The use of magnetic fields as an alternative means of filtering the liquors is under development.

In addition to the problem of solids, the dissolver product might require some degree of chemical adjustment, especially from a batch dissolver, owing to fuel variations. Two adjustments are most usual:

(1) acidity, which is accomplished by addition of concentrated or very dilute acid, and

(2) plutonium valence.

The solvent extractability of the various chemical forms of plutonium nitrate has been noted. $Pu(NO_3)_4$, the preferred extractable form of plutonium, has a tendency to oxidize to $PuO_2(NO_3)_2$ in hot nitric acid, which is precisely the dissolver condition. If the concentration of $PuO_2(NO_3)_2$ in the dissolver product is unacceptably high for solvent extraction, it is decreased at this point in the process by adding a reducing agent, say, sodium nitrite, or by passing oxides of nitrogen into the liquor and reducing the $PuO_2(NO_3)_2$ to $Pu(NO_3)_4$. These chemical adjustments are known as *conditioning*, and the vessel in which it is carried out is called a *conditioner*. Again, such a vessel must comply with the criticality constraints imposed upon the whole plant (Fig. 14.10).

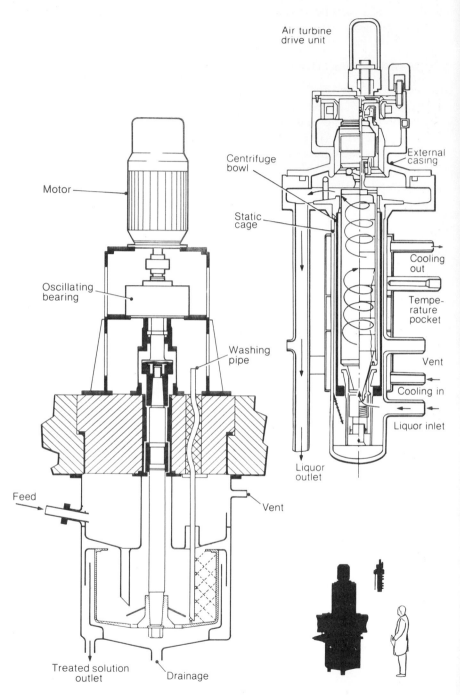

Air turbine
drive unit

Centrifuge
bowl

Static
cage

External
casing

Cooling
out

Tempe-
rature
pocket

Vent

Cooling in

Liquor inlet

Liquor
outlet

Motor

Oscillating
bearing

Washing
pipe

Feed

Vent

Treated solution
outlet

Drainage

Fig. 14.9 Centrifuges for liquor clarification

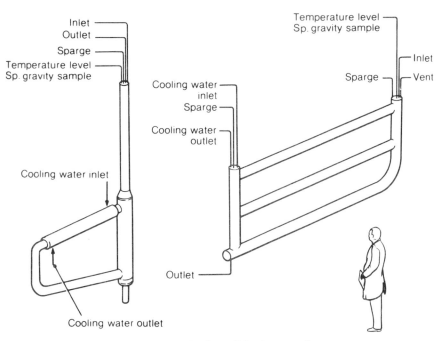

Fig. 14.10 (a, b) Geometrically limited conditioning vessels

5.7 Solvent extraction equipment

The equipment used to bring the organic and aqueous liquids into contact with each other and to separate the two liquids is called a *contactor*.

The mixing and separating stages can be carried out in a simple stirred vessel, but this method of operation is time-consuming and can result in inefficient separation of the two phases. The more usual equipment design employs continuous counter-current solvent extraction, in which the two phases flow in opposite directions and the contact time is adjusted by variable flow rates, and, in some instances, by variation in the energy applied to the mixing device.

The simplest design of equipment is a packed column consisting of a vertical tube filled with metal or ceramic rings or shapes (Fig. 14.11), which break up the liquid phases and force them into tortuous paths through the column. Phase separation is achieved solely by the difference in density of the two phases, the lighter organic solvent flowing up through the column and the heavier aqueous phase down. In such a device mixing is not vigorous and flowrates are comparatively low. As a result the column must be very tall to achieve good extraction of the desired solute into the solvent. The tall columns add to the height and complexity of a radioactive reprocessing plant because of the need for heavy shielding and for liquid transfer systems which do not use mechanical pumps (this is to obviate the need for direct maintenance of radioactive equipment).

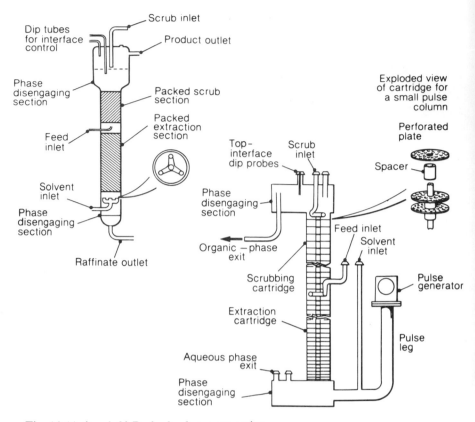

Fig. 14.11 (top left) Packed column extraction
Fig. 14.12 (bottom right) Pulsed column extractors

The efficiency of the extraction column can be greatly increased and the required height reduced by *pulsing* the feed, so that the phases are drawn back and forth (reciprocated) through the packing, or alternatively through perforated plates (Fig. 14.12). The pulse can be applied by means of an external piston or diaphragm, or by applying air pressure by fluidic control. The equipment can be so arranged that the amplitude and frequency of the pulse can be varied within pre-calculated design limits. Pulsed columns have been operated successfully in irradiated-fuel reprocessing plants in the UK (Windscale second plutonium purification plant), in Europe (Eurochemic and Marcoule plants) and in the US (Hanford and Idaho Falls). Pulsing reduces the height of column required compared with the height of a packed unpulsed column carrying out the same process by a factor of up to five.

The principal alternative to the column for the counter-current solvent extraction process is a device known as the mixer–settler. In this equipment the organic and aqueous phases are repeatedly mixed and separated in banks of horizontal stages, each stage consisting of a mixing chamber at one end and a longer settling chamber at the other. The sizes of the mixing and settling chambers depend on the flowrates of the two phases, the desired contact time, and the physical properties of the solvent and aqueous solutions (particularly differences in density, viscosity, and coalescence). The method of agitation used in the mixer to mix the solutions also provides the propulsion energy to move the liquids from stage to stage. Many designs of mixer–settler have been used in reprocessing plants ranging from the geometrically limited units used at Dounreay (designed to be safe from accidental criticality), to the large units used in the second Windscale reprocessing plant and in France (Cap de la Hague and Marcoule) (Fig. 14.13). In general the US design of pump-mix mixer–settler (Fig. 14.13), in which the agitator is used to lift the heavier phase into the mixer has not been used in the UK, where all the designs have concentrated on positioning such items as cannot be guaranteed maintenance-free outside the biological shielding.

One of the possible disadvantages of the mixer–settler as a contactor for irradiated-fuel reprocessing plant is that the contact time and the liquid hold-up in the equipment are comparatively large, so that, in the highly active parts of the plant, radioactive solids particles can be deposited in the settlers and at the interface between the two phases. This can lead to the organic solvent being exposed to high levels of radioactivity with consequent chemical breakdown (known as radiolytic solvent degradation). The effect is more pronounced in large-scale plant such as the second Windscale reprocessing plant than in the small geometrically limited equipment at Dounreay. The large liquor hold-ups also increase the fissile material inventory of the plant. This means that the volume of liquid and the time required to run the plant up or down are both high.

A development of the mixer–settler concept designed to minimize liquor hold-up and contact time, is the centrifugal contactor (Fig. 14.14). This contactor is well known in the oil and pharmaceutical industries, and versions suitable for nuclear fuel reprocessing are being developed and introduced in the USA and in France. In this design the settling section in the mixer–settler, which is chiefly responsible for the large hold-up of liquids, uranium, and plutonium, is replaced by a small centrifugal separator mounted on the same shaft as the mixing vanes. The centrifugal units are more efficient than conventional mixer–settlers, contain only 2 per cent of the volume, and need only a small fraction of the time to reach equilibrium operating conditions or to be flushed out. However, as can be seen from Fig. 14.14, centrifugal contactors are mechanically complex and consequently expensive to manufacture. The advantages listed above have to be set against the high degree of operational reliability and low maintenance requirements demonstrated by the more conventional mixer–settler designs. Centrifugal contactors have not yet been adopted in the UK reprocessing plants.

Some of the principal advantages and disadvantages of the main types of solvent extraction contactors used in irradiated-fuel reprocessing plants in the UK and elsewhere are listed in Table 14.5.

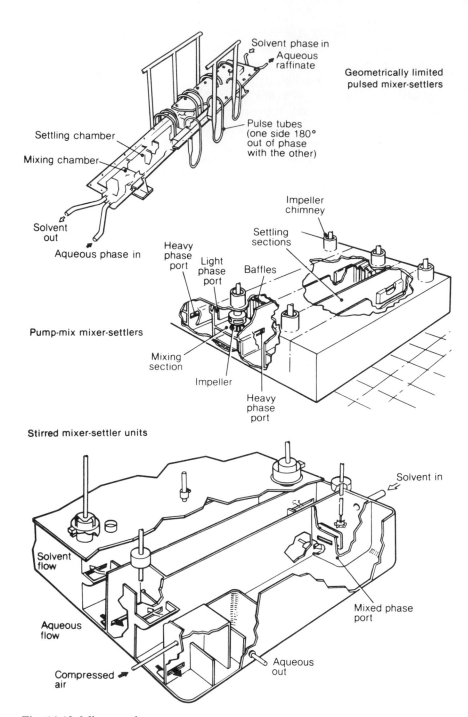

Fig. 14.13 Mixer-settler extractors

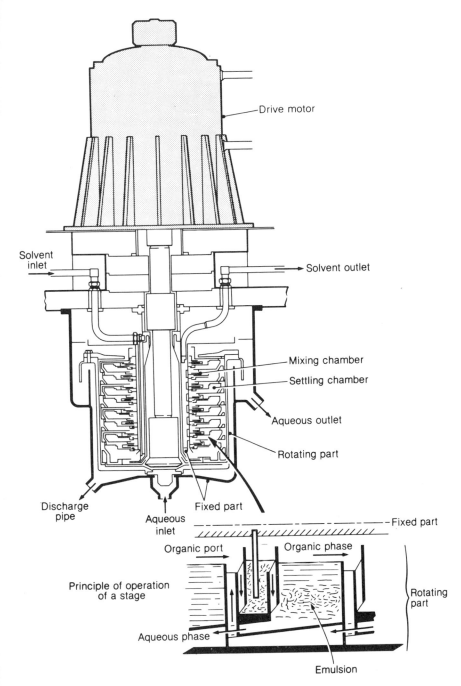

Fig. 14.14 Centrifugal contactor (schematic) (Courtesy of Robatel Ltd.)

Table 14.5 Equipment for solvent extraction reprocessing of irradiated fuels

Contacting equipment	Relative capacity for processing natural uranium fuels per volume of equipment	Relative capacity for processing enriched fuels in critically safe design	Amount of shielding per unit capacity	Flexibility	Reliability in plant service
Mixer-settlers					
Pump mix	medium	medium[1]	medium[5]	excellent	excellent
Centrifugal[4]	large	large[2]	small	good	good
Air ejector or air pulsed	medium–small	medium–small	medium	good	excellent
Columns					
Pulsed sieve plate	medium	large[3]	medium	good	excellent
Pulsed packed	medium	large[3]	medium	good	good
packed	small	medium[3]	large	fair	good

Notes
1. Height of mixer–settler limited to 75 mm. Not amenable to efficient *poisoning*
2. Built of stainless steel containing a *poison* such as gadolinium or boron which assists criticality control of the unit
3. Packing constructed of *poisoned* stainless steel thus allowing larger-diameter columns.
4. Still under development
5. With variable-speed drive and replaceable impellers

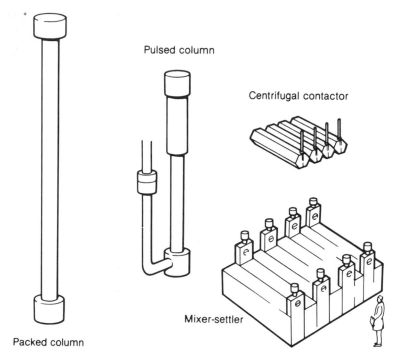

Pulsed column

Centrifugal contactor

Mixer-settler

Packed column

Fig. 14.15 A comparison of the volume of space occupied by various solvent extraction units

The choice of contactor is dictated by the design criteria adopted for the reprocessing plant. It is noticeable that the mechanical complexity of the contactors used has increased as experience of irradiated fuel reprocessing has expanded, and parameters other than plant simplicity and low maintenance requirements have become more important. It is conceivable that the low plant inventory and ease of flushing of the centrifugal contactor will ultimately be regarded as outweighing its cost and mechanical complexity in future reprocessing plant design. Fig. 14.15 shows in schematic form the relevant heights and floor areas occupied by the contactors discussed above.

5.8 Waste-processing facilities

At the present time most of the solid and the highly active liquid wastes generated in the UK are stored. Figure 14.16 shows a waste engineered store at Dounreay in Scotland. Placement and movement of the drums within the store is carried out remotely aided by television cameras. Less-radioactive solid waste is buried in specially licensed pits on reprocessing sites.

Highly active aqueous liquid waste is stored with or without prior evaporation, depending upon its fission-product content, in large tanks which are fitted with internal cooling coils (Fig. 14.6). Some tank designs

Fig. 14.16 The α, β, γ waste engineered store at Dounreay during commissioning

have less complex coils, but the external surface of the tank is jacketed for increased cooling. A considerable amount of work is under way on the conversion of the highly active liquid wastes to glass.[10, 11] Medium-active aqueous wastes are usually treated via a floc or similar process prior to their discharge. The solids are stored in tanks at the reprocessing site. These processes follow conventional chemical engineering practice.

The chemical treatment of gases has been described in §3.1.5. The chemical engineering processes of scrubbing, absorption, and cryogenic separation are well known.[17] A possible sequel to the cryogenic process for the separation and collection of krypton from off-gases is to implant the atoms of krypton into a nickel matrix by ion implantation[18] such that they are *fixed*. BNFL's THORP plans provide for removal of ^{129}I and ^{14}C, but ^{85}Kr will be discharged since cost-benefit assessment shows that removal is not justified. Unwanted organic liquids can be disposed of by incineration. Normally, however, the organic solvent is purified and recycled to the plant.

6 The United Kingom's irradiated-fuel reprocessing plants and processes

Irradiated-fuel reprocessing plants in the UK have been constructed and operated at Windscale in Cumbria and at Dounreay in Northern Scotland.

The Windscale plants process fuel from thermal reactors. To date this has included fuel from gas-cooled reactors, ranging from the aluminium-clad fuels of the early reactors through the Magnox clad fuels to the stainless-steel-clad fuel of the AGR, also Zircaloy and stainless-steel-clad fuels from overseas water reactors. Fast reactor fuel (stainless-steel-clad) and materials-testing reactor fuel (aluminium-clad) are processed at Dounreay.

The first UK reprocessing plant (B204) was constructed at Windscale and operated between 1952 and 1964. This was followed (chronologically) by the Dounreay materials-testing reactor fuel reprocessing plant which commenced operation in 1958 and is still operating.

The Dounreay fast reactor fuel reprocessing plant operated from 1960 until 1975, when it was closed down for major refurbishing and modification to enable it to reprocess the fuel from the prototype fast reactor, situated at Dounreay.

The second separation plant at Windscale (B205) commenced active operation in 1964 and is still operating. Part of the original separation plant at Windscale was modified and operated as a *head-end* section to the second reprocessing plant for oxide fuels from 1969 to 1973. Following a small accidental release of radioactivity into a working area, after the successful reprocessing of about 100 t of fuel, it was decided to modify the plant as a development facility.

The refurbished Dounreay fast reactor fuel reprocessing plant restarted in the autumn of 1980, reprocessing irradiated fuel from the Prototype Fast Reactor.

The new thermal oxide fuel reprocessing plant (THORP) is to be constructed at Windscale and is scheduled to operate in the late 1980s. It will process AGR and LWR fuels from UK and overseas reactors.

Table 14.1 shows the various UK plants and lists the relevant data. Because of the differences in types of fuel processed both on the Windscale and Dounreay sites it is convenient to consider each site separately, rather than to discuss the plants in the strict chronological order noted above.

6.1 The first Windscale reprocessing plant[19] (B204)

Following the decision to use solvent extraction to recover plutonium from the first Windscale nuclear 'piles', the first Windscale reprocessing plant was constructed, based on designs initially developed in Canada and subsequently and more fully in the United Kingdom. This plant, which used Butex as the extraction agent, was operated from 1952 until 1964. The major features of the plant and process are given in §§6.1.1.–6.1.9.

6.1.1 *Decladding*
The irradiated fuel elements comprised uranium metal rods approximately 300 mm × 25 mm dia. clad in aluminium. They had been irradiated in the air-cooled reactors at Windscale to a suitable low burn-up. After a period of

120 days' storage under water to allow short-lived fission products to decay, the cladding was removed from the fuel rod by extrusion through a die. The pieces of cladding were transferred in a shielded flask to a storage silo for long-term storage and the fuel rods were transferred by a second shielded flask to the main primary separation process via the dissolver.

6.1.2 Dissolution
Uranium rods, 6M nitric acid and oxygen, were fed continuously to the dissolver to produce a uranyl nitrate, plutonium nitrate, fission-product nitrate solution in 3M nitric acid. At an acidity level of 3M HNO_3 the following conditions existed.
1. The distribution coefficients for uranium and plutonium are high enough in a Butex system for efficient solvent extraction in a single column without the addition of a salting-out agent, which allowed the first extraction-cycle raffinate to be evaporated to a small bulk.
2. The corrosion rate of the dissolver at boiling temperature was low enough to be acceptable.
3. The metal (and therefore the radioactivity) hold-up in the dissolver was not excessive for continuous operation.

6.1.3 Conditioning and cooling
The solution was then conditioned as required after analysis to the correct acidity, plutonium valency, and uranium concentration, and cooled to 25–30 °C before being fed to the first solvent extraction column. The uranium concentration in the dissolver product was maintained at 300 g/1—a higher concentration could lead to crystallization—in order to minimize the volume of liquor in the first extraction column.

6.1.4 Separation of plutonium and uranium from fission products
The primary separation took place in a long packed column using Butex as the extractant for uranium and plutonium.

Extraction of the uranium and plutonium was greatly enhanced by the concentration of nitrate ions in the system so that it was possible, within the length of the extraction columns, to extract uranium and plutonium in significant quantities into the solvent phase at high nitrate concentrations and subsequently to remove them again by stripping into an aqueous phase at low nitrate concentrations. The success of the extraction process is largely dependent upon the efficiency and ease with which these transfers from solvent to aqueous and back can be accomplished.

6.1.5 Separation of plutonium from uranium
The dependence of the distribution coefficient of plutonium nitrate on the valence of plutonium has been noted above. It is also much lower than that for uranyl nitrate at all nitrate concentrations and this provides a basis for the separation of uranium from plutonium. Approximate value of the dis-

tribution coefficients are given in Table 14.4. From these it is clear that, by converting the plutonium to the Pu(III) form, a good separation of uranium from plutonium can readily be achieved.

6.1.6 *Primary separation plant*

Application of the principles above to the feed to the first extraction column of the primary separation plant (see Fig. 14.17) resulted in 99.9 per cent of the uranium and 99.98 per cent of the plutonium being extracted into the solvent, and 99.5 per cent of the fission-product activity being retained in the aqueous raffinate.

The residual fission-product activity, which was extracted into the solvent with the uranium and the plutonium, was mainly due to ruthenium with small amounts of zirconium, cerium, and niobium. The first-cycle decontamination factors for these nuclides were 10 in the case of ruthenium, 4000 for zirconium, 10 000 for caesium, and 400 for niobium. The relatively low ruthenium decontamination resulted from the ruthenium in the feed solution existing in the form of a number of species, some of which were extractable into the solvent. A nitric acid scrub was used to wash entrained fission products from the solvent but it was ineffective in scrubbing out ruthenium which had distributed into the solvent.

The acidity of the solvent extract containing uranium and tetravalent plutonium from the first extraction column was adjusted to $0.5M$ HNO_3 with ammonia. Ferrous sulphamate solution was added to reduce the plutonium to Pu(III), which rendered it inextractable in the solvent. The mixture of solvent and aqueous phases was fed to a second extraction column, the low acid—high salt system here serving to extract uranium (owing to the high salt level), plutonium being returned in the aqueous phase (owing to the low acid concentration). About 0.01 per cent of the uranium and one-third of the residual fission-product content followed the aqueous acid plutonium stream, and about 0.02 per cent of the plutonium and two-thirds of the residual fission-product content remained with the uranium in the Butex-extracted phase. The two streams were then fed to separate uranium and plutonium purification processes.

The Butex containing the uranium was heated, and the uranium was then stripped from the Butex by dilute nitric acid in a third column. The acid product was evaporated to reduce the volume to be handled in the subsequent cycles which used mixer–settlers and not columns. Any Pu(III) present was oxidized to Pu(IV) during the evaporation.

The plutonium contained in the nitric acid solution from the uranium–plutonium separation column was purified in two cycles. In the first cycle within the primary separation plant, the Pu(III) was re-oxidized with sodium dichromate and extracted into Butex in a simple extraction column, and stripped into dilute nitric acid in another column. This aqueous product was concentrated for feeding to a separate purification plant to purify further the plutonium stream from fission products and uranium.

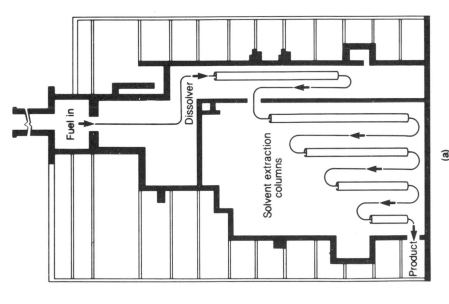

(b)

Fuel in

Dissolver

Solvent extraction
columns

Product

(a)

Fig. 14.17 The first Windscale reprocessing plant

6.1.7 *Uranium purification plant*

The concentrated uranium solution was stored for six months to allow ruthenium activity to reduce by radioactive decay. It was subsequently conditioned to 0.5M nitric acid, and heated with a reducing agent (hydrazine and ferrous sulphamate) to reduce any plutonium present to Pu(III) (which also served to modify the ruthenium species, so as to minimize their extraction into Butex). Ammonium nitrate was added to assist efficient uranium extraction into Butex in a mixer–settler system and the uranium was then stripped from the Butex into dilute acid. The decontamination factors achieved were 200 for Ru and 200 for Zr–Nb; the final product was pure enough to allow this liquor to be transferred to the Springfields factory and handled in the same way as natural uranium for conversion to uranium metal.

6.1.8 *Plutonium-purification plant*

The plutonium was completely converted from a mixture of valency states to the Pu(IV) state by first adjusting the acidity and then conditioning by use of reducing and oxidizing agents. Instead of Butex, tri-*n*-butyl phosphate in odourless kerosene (TBP–OK) was used as the solvent in this cycle, the change of solvent resulting in improved decontamination factors. The plutonium and uranium were extracted into the solvent phase, which was scrubbed with 3M nitric acid. Decontamination factors of 2000 and 10 were obtained from ruthenium and zirconium. U(VI) and PU(IV) were then separated in a further column using the difference in their distribution coefficients at low HNO_3 acidities. The efficiency of separation, and the level of Pu loss (to the uranium stream), were dependent upon the acidity in the stripping section, the acidity in the scrub section (higher portion of the column), and the uranium concentration in the feed. The plutonium-containing aqueous stream was concentrated by evaporation, prior to conversion to plutonium metal in the *finishing plant*.

6.1.9 *Chemical-engineering principles of the first UK reprocessing complex*

The basic differences in design philosophy between the UK and the US (totally remote, nil-maintenance plants versus the *canyon* concept of remotely maintainable plants) have been noted earlier. The design of the first radioactive reprocessing plant in the UK was based on the principles of using simple robust equipment with a minimum of moving parts and requiring minimal maintenance. In this context it was decided that the primary separation plant, which would handle the greatest mass of fissile and fertile nuclear materials and radioactivity, should be constructed of all-welded high-quality stainless steel with adequate thickness allowance for corrosion, all welds in the most radioactive sections of the plant being subject to visual and radiographic inspection and certification. In order to reduce the number of moving parts and maintainable units, all liquor transfers were made by gravity, vacuum lift, or steam ejector, and no pumps were used; also valves were not used on radioactive liquor lines, and valveless maintenance-free liquor *divertors* and *distributors* were designed and installed.

Packed columns depend upon the difference in specific gravity between the solvent and aqueous solutions to achieve flow and phase separtion, and they consequently needed to be very long. The number of stages available in each column was small, because the height equivalent to a theoretical stage of solvent extraction was large. A very high building—45 m—was required to house the columns. Aqueous flow was by gravity from the fuel dissolver situated in a cell fed at the top floor of the building. The flow of non-radioactive liquids was measured by *rotameters* and radioactive feed rates were controlled by a *constant-volume feeder* (a modern version of the Egyptian water-wheel) with the drive motor situated outside the biological shield.

To support the philosophy of eliminating direct active maintenance requirements, the installed primary separation plant had four separate cells containing dissolvers and first-cycle extraction columns, and two separate cells containing the other solvent extraction cycles and ancillary equipment. These cells were interconnected, and in the event of a major failure, operations could be transferred to a spare cell.

Experience from Chalk River suggested that the uranium and plutonium purification plants, which would handle lower levels of radioactivity, could use different contactors from the primary separation plant (where no maintenance would be possible), permitting a saving on the cost of the plant. Thus the uranium purification plant using Butex as solvent was equipped with mixer–settlers, which significantly reduced the building height. The original plutonium purification plant was a batch process using tributyl phosphate in odourless kerosene in stainless steel single-stage mixer–settlers. This was followed by another batch solvent extraction process using a chelating solvent (thenoyl trifluoroacetone—TTA—in benzene as diluent). TTA, which bonds tightly with plutonium, allowed for efficient plutonium purification. This rather complicated process with many secondary recovery processes was quickly replaced by a second plutonium purification plant which was built adjacent to the primary separation plant and used the TBP–OK solvent extraction process of the first purification plant. Operations were carried out in pulsed packed columns which reduced the height of the containment building required to 15 m (cf. 45 m height of the primary separation plant). Additionally, the low fission-product content of the feed to this plant permitted the use of some conventional valves and pumps.

Control of the total process was achieved by instrumentation backed by a comprehensive system of sampling and analysis. The primary separation process, which can be divided conveniently into four major sections— dissolution, fission-product separation, separation of uranium from plutonium, and primary purification of plutonium—was carried out in a single building. Uranium purification, plutonium purification, and fission-product evaporation were each housed in separate buildings which were added, in that order, at various intervals after the main plant had been commissioned and operated. For this reason there were, between each of the units, facilities for intermediate storage between main items of equipment.

The main sections of the process were operated from a control station which housed the instrument gauges and recorders required to indicate the performance of the section, and also controls through which the plant and performance could be adjusted as required. Items such as thermocouple leads were completely encased in all-welded stainless steel tubes, which prevented contamination and facilitated repair or replacement from outside the biological shield. Liquid levels and specific gravity measurements on the active plant were measured by means of air pressure feeding back to clock dials at the control station. Differential air pressures were also used to control the liquid interface position in the vertical separation columns.

A problem facing the plant designer was to establish points allowing operators to approach the plant safely for control and sampling purposes. The inaccessibility of the plant made essential the routine obtaining of samples by remote means from all appropriate points, since *ad hoc* methods of sampling would be difficult or impossible in a radioactive facility; thus sampling becomes a major factor in plant design and equipment layout.

Each of the main streams of the primary separation plant was led through the main biological shield into a shallow trough housed in a secondary biological shield. The trough in effect formed an opening in the pipe, and liquors, after flowing through the trough, passed back through the main biological shield to the next stage of the process. The various special instruments for the primary separation plant were mounted on this trough.

The plant described above operated in support of the military programme for about eleven years, and the operation was extended to take fuel from the Calder Hall and Chapelcross dual-purpose reactors. Figure 14.17 shows some of the plant items referred to above and Fig. 14.23 shows an outline flowsheet of the process.

In 1955 the Government decided to embark upon the first nuclear power installation programme, the power stations to be of the Magnox type and based on the Calder Hall design. The white paper announcing this decision envisaged a 1500 to 2000 MW(e) programme installed over 10 years. This programme would require 660 tonnes of natural uranium Magnox fuel to be reprocessed each year to enable the uranium to be utilised after enrichment, and the plutonium to be separated and stockpiled for use in the fast breeder reactor system which was being developed by the UKAEA at Dounreay. The required throughput exceeded the capacity of the first reprocessing plant complex at Windscale. Storage of the spent Magnox fuel was not a practicable alternative to reprocessing because the fuel element cladding gradually corroded in the water-filled ponds used for storage and would release radioactivity into the water.

Improved solvent extraction technology was then available as a result of

(1) UKAEA experience in flowsheet development and plant design for the higher burn-up fuels used in the fast reactor and materials-testing reactors (described later), and (2) world development of solvent extraction. In the course of this experience, tributyl phosphate (TBP) had become validated as the preferred extractant for use in the process.

It was intended that the second plant for reprocessing spent metal uranium Magnox fuel should be a centralized unit designed for long, continuous operation and requiring minimum maintenance, analytical, and process control effort. Thus the process was designed

  (1) to eliminate batch operations,
  (2) to avoid recycle of waste or product streams, and
  (3) to be suitable for centralized instrumental control and surveillance.

The plant was required to process up to 1500 tonnes per year of Magnox uranium fuel irradiated to 3000 MW d/t and stored for 130 days after discharge from the reactor and before reprocessing, and to produce uranium oxide and plutonium nitrate solutions with a fission product content low enough to enable them to be further processed in separate and lightly shielded facilities. The uranium and plutonium recoveries, degree of separation from each other, and fissile product decontamination factors specified for the plant are listed in Table 14.2.

Plant design would include an integrated storage, treatment and disposal system for liquid effluents, based on the evaporation of waste streams to minimum bulk and the recovery of nitric acid for reuse. The volume of highly active liquid effluent produced by reprocessing for permanent storage was not to exceed 100 litres/tonne (l/t) U processed, and the salt-bearing concentrate from the medium-active waste was to be stored for two years and treated by a flocculation process, the decontaminated liquor being discharged to sea, and the floc, containing the radioactivity, stored on site.

Experience gained on the first plant led to the acceptance of direct-maintenance techniques for the medium-and low-reactive plant and the control of nuclear safety by concentration limitation of plutonium throughout the plant, except in the final stages of concentration and storage of plutonium, which were carried out in geometrically limited euipment.

6.2 The second Windscale reprocessing plant[2, 20] (B205)

This plant was commissioned in 1964 and is still operating. During the design stages of the plant, process data and experience were available from the first Windscale plant described in §5, together with data from the two plants at Dounreay and from the USA. Diagram and flowsheets for various plants are shown together for comparison in Figures 14.18–14.23. With the exception of the Windscale Butex process all these other data were based on tributyl-phosphate (TBP) as extractant. A variety of hcad-end processes were used in the various plants owing to the different types of fuel being processed. These processes were assessed against the required specification and it was

Fig. 14.18 The second Windscale reprocessing plant (B205)—operating floor

decided that the plant would use mechanical decanning, continuous dissolution, and a three-cycle 20 per cent TBP–OK solvent extraction process
with no recycling of raffinates. The irradiated fuel to be reprocessed would
be Magnox reactor fuel as in the first Windscale plant.

The reasons for significant departures from the techniques and chemistry
used in the first plant will now be summarized.

The quantity of Magnox cladding associated with the required throughput
of fuel was 170 t/y and this could be removed by either chemical or mechanical
means. Chemical decanning was at first sight attractive, in that fuel elements,
rendered fragile by long irradiation and varying designs , could easily be
processed. The induced radioactivity in the cans was estimated to be about
10^4 curies per tonne (Ci/t) of Magnox, and since the solutions arising from
chemical decanning could not easily be decontaminated to a level suitable
for discharge, it was concluded that long-term storage of this activity would
be unavoidable. It was estimated that it would be cheaper to do this in metal
form rather than as a concentrated solution. Since considerable experience
was available at Windscale on mechanical decanning, this was the technique
to be selected.

The continuous dissolver in the first separation plant had operated well
apart from some intermittent surging in the liquid outlet system. A modified
design was therefore adopted for the new plant.

The Butex–TBP process had operated well in the first Windscale plant, giving products of a satisfactory quality and a manageable effluent. It had however the disadvantage of a low ruthenium decontamination factor in the first cycle. The additional throughput and the higher irradiation of fuel from the power programme could give rise to effluent discharge problems if Butex was used in the new plant. The material costs of the Butex process were much higher than those of a comparable TBP process. It was therefore decided to adopt an all-TBP process because of reduced operating costs and simpler effluent disposal.

Examination of results obtained in TBP processes suggested that a two-cycle solvent process, even with tail-end absorption processes, would be inadequate to meet the product specification with an adequate margin for operational variations and future demands. A three-cycle solvent extraction process was therefore chosen.

A more difficult decision was whether to separate the uranium from the plutonium in the first or the second cycle. Assessment showed that there was a still difference between these possibilities so far as volumes of effluents arising, wastes to store, purities of products, or usage of materials were concerned.

Separation in the second cycle ('late' separation) was chosen as it gave the following advantages.

1. The aqueous raffinate from the second cycle, containing 75 per cent of the radioactivity passing the first cycle, was free of salts and could be evaporated and directed to highly active liquor storage without penalty, an advantage of even greater value in event of a breakthrough of radio-activity from the first extractor.
2 It required one cycle less of purification of the plutonium stream.

These advantages were judged sufficient to outweigh the disadvantage of having two additional large contactors which could not be designed to be geometrically limited for criticality purposes.

Three alternative processes were considered for the plutonium purification step:

    (1) Evaporation followed by a single line of TBP–OK solvent extraction equipment, in which the nuclear safety would be controlled by concentration limitation guaranteed by instrumentation;

    (2) Evaporation of the separated plutonium stream followed by TBP–OK cycles, all in geometrically limited equipment; and

    (3) Batch or continuous ion exchange in either geometrically limited or mass-limited equipment.

Experience at Windscale suggested that, whilst ion exchange was well suited to recovery of plutonium from dilute streams, its use as a purification stage at the full throughput of the new plant was unattractive. It was decided therefore to select a TBP–OK process with its own solvent recovery system, making possible the introduction of alternative solvents if desirable. It was shown that satisfactory flowsheets and equipment were possible for either (1) or (2). A study of the relative merits of geometrically safe and concentration-controlled plants showed the latter to be simple to operate and control with an adequate margin of nuclear safety. A single line of concentration-limited contactors was therefore installed.

After two cycles of co-decontamination with uranium in the primary separation contactors and one cycle of plutonium purification, the plutonium product would be within specification. It was decided nevertheless to install an additional solvent extraction cycle to provide extra margins of decontamination from both fission products and uranium against future changes in feed to process, or against future changes in product specification.

Detailed development of a chemical process and its equipment is expensive. In order to minimize the development costs and, at the same time, carry out all the necessary research and development, the chosen process flowsheet was tested in a miniature pilot plant. Here, all the solids and liquids were of identical chemical compositon and radioactivity content to those in the full-scale plant, but the volumes and quantities handled were significantly smaller (c. 1/5000). This technique is now in general use for the design and proving of processes for radioactive plants.

Whilst most of the chemistry of the process can be confirmed in the manner described above, it will not give information concerning the operation of plant-scale equipment, and certain aspects of the chemistry, e.g. radiolytic degradation of the organic solvent, cannot be studied in the miniature system, since this is very dependent upon the scale of the final operation.

Full-scale equipment development was carried out using uranium metal and uranyl nitrate with no fission products or plutonium present.

Tentative flowsheets are drawn up on previous knowledge of distribution data for uranium, plutonium, and fission products. These data are supplemented as necessary with specially designed laboratory measurements. It is then possible to calculate the number of, say, extraction or strip stages required for a given part of the process, enabling the overall detailed flowsheet to be calculated.

Large-scale, non-radioactive trials were carried out to determine the hydrodynamics and hydraulics of the solvent extraction equipment—in this case mixer–settlers—which enabled the sizes of the contactors and other associated equipment to be estimated.

Supporting studies were also undertaken on the effects of solvent degradation products on the process, the behaviour of radioiodine and other fission products, and the methods of effluent disposal and treatment.

Before the final decisions on the flowsheet were taken and prior to pilot-plant trials, several studies were necessary to evaluate competing variants. Particular attention was given to the problem of solvent degradation. Workers in the US had clearly shown the importance of solvent degradation due to radiolytic and hydrolytic decomposition of the TBP to form mono- and di-*n*-butylphosphate (MBP and DBP), and of the petroleum-based diluents to give trace amounts of impurities. These impurities form complexes with uranium, plutonium, and some fission products, notably zirconium and niobium, which tend to inhibit the separation of fission products from uranium and plutonium and the stripping of the latter from the solvent.

Studies of TBP decomposition were carried out at Windscale and Harwell, and it was shown that, in the absence of recycle of aqueous process streams and intercycle evaporation, MBP and DBP formation did not occur to any important extent to the new process.

Equipment and flowsheet development had proceeded in parallel and an early start was made on the decanner, mixer–settlers, evaporators, and instruments. As the flowsheet and some of the design details were worked out, it become clear that development of other equipment was necessary. For example, a supporting programme of corrosion trials was instituted using non-radioactive liquors under process conditions, and in some instances fully radioactive liquors from the pilot plant were also used. After long irradiation the fuel rods were expected to be brittle and bowed, and it was doubtful whether the high loads imposed on the element by the punch-and-die technique of mechanical decanning, used for low-irradiation Calder Hall fuel, would be acceptable for decladding the new fuel.

Earlier work had suggested the use of a hydraulically loaded three-wheel cutter to cut the cladding and to decan the fuel elements. Development showed that the load to push the rod through the cutting wheels was considerably lower than that required for the punch and die. An automatic unit allowing removal of each component for maintenance was designed and extensively tested, firstly with inactive elements. A fully active, full-scale prototype was installed in a spare decanning pond at Windscale and approximated 130 tonnes of irradiated fuel were satisfactorily decanned, proving the system.

The choice between mixer–settlers and pulsed columns was a marginal one. Capital costs slightly favoured the mixer–settlers, and they also had the advantages of greater flexibility in process operation and less carry-over of solvent phase with the outgoing aqueous streams and vice-versa. They are also technically easier to instrument and have lower development costs because of the ease of direct scale-up of fission-product behaviour. On the other hand, pulsed columns have a smaller liquor hold-up and fewer moving parts. Mixer–settlers were eventually chosen and the chemical engineering work to establish design data for the range of contactors was successfully done on a single-stage model. Figure 14.13 illustrates the Windscale mixer–settler unit.

As equipment was designed, prototype units were constructed and tested under simulated process and maintenance conditions, and were modified as necessary to meet the stringent requirements imposed.

Additionally, it was necessary to develop a number of special items of equipment, such as low-lift centrifugal pumps and force lift ejectors for carrying out liquor transfers in the plant, together with flow controllers, metering pumps, pipe couplings, etc.

The success of the research and development programme was such that the plant is still operating at a throughput of 7 tonnes of irradiated fuel per day. An outline flowsheet for the process is given in Fig. 14.23.

Improved performance of Magnox fuel, with much increased fuel life, had reduced the load on the second reprocessing complex, with the result that its available spare capacity for alternative fuels was greater than that originally anticipated. It was decided therfore to utilize this spare capacity for the reprocessing of uranium oxide fuel from, for example, AGRs and LWRs.

6.3 The head-end plant

AGR and LWR fuels are stainless steel or Zircaloy clad and are built into *fuel sub-assemblies* (Fig. 14.3) rather than the single-fuel-rod type from Magnox reactors. It was decided to approach oxide fuel reprocessing in a way which would make maximum use of existing resources. To this end, utilization of the original Butex processing building (B204) was investigated as a site for a suitable head-end facility for the new complex. A scheme was drawn up which involved some extension of one half of the building, which would house a shear machine and a leach dissolver, and then lead into a single cycle of Butex solvent extraction (using the original static packed columns and their associated equipment). The product from the single-cycle Butex extraction would be collected batchwise and fed on a campaign basis into the main TBP–OK reprocessing unit, utilizing the considerable experience gained in the 1950s on the use of the two-solvent system. This proposal reduced capital expenditure to a minimum and also reduced considerably the development effort required. (Fig. 14.19 shows some plant items.)

A hydraulic shear of 300 tones maximum force was designed to shear the oxide fuel pins into short lengths (approximately 25 mm) suitable for the dissolver.

Fuel sub-assemblies were transported from the storage pond to the plant in a shielded container and lifted to the tenth floor of the old separation building by crane. The flask was lowered on to a platform, so that the door of the flask connected to a door in the shear cell. The sub-assembly was pushed hydraulically into the shear cell, being received in a container known as a *fuel envelope*. On completion of this operation the fuel envelope, one end of which was pivoted at the mouth of the shear, was displaced, bringing the fuel in a direct line with the shear pack and an incremental feed ram which was electrically interlocked with the hydraulics of the shear system. In this way, sections of the required length were sheared as required, and the system was able to accommodate fuel up to nearly 5 m in length and 0.3 m square section. The fuel passed from the shear system into a divertor and thence into a perforated removal stainless steel basket contained in one of a number of fuel dissolvers. These dissolvers were designed to cope with the criticality

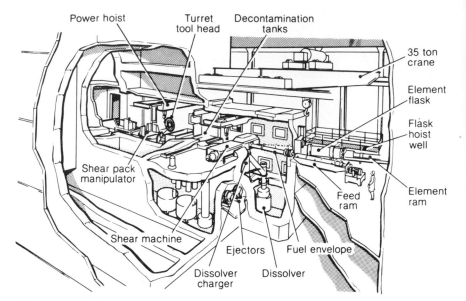

Fig. 14.19 Fuel reprocessing plant at Windscale (B204 head-end plant)

problems from the ^{235}U levels expected in all types of thermal oxide fuel. Higher enrichments could be accommodated by the use of suitable soluble poisons.

After an appropriate cycle the active dissolver liquor was transferred to a system of accounting tanks. The residual cladding or hulls were washed in nitric acid wash, which served as the dissolution feed for the next batch of fuel. The basket was removed to a monitoring station where the fuel content of the hulls was assessed by radioactivity measurements. Finally stainless steel hulls were tipped into a silo located immediately below the shear cell. The silo formerly contained medium-active equipment associated with the original chemical plant and was adequately shielded for this storage purpose. Zircaloy hulls were sent in a flask to B29 pond for storage.

The dissolver liquor was conditioned and fed to the single cycle of Butex extraction. This consisted of a single static packed column for the foward extraction and scrub, and a second packed column for the stripping of the constituents into aqueous solution. The latter was concentrated by continuous evaporation. The product was collected into one of a number of intermediate storage tanks having capacities varying from 20 to 200 tonnes of uranium, depending upon the uranium-235 content of the liquor. Liquors were then fed on a campaign basis into the main reprocessing plant (see Fig. 14.23).

The commissioning of this plant was completed in August 1969 and successful operating experience was gained on stainless-steel-clad fuel from

(a)

(b)

Fig. 14.20 The materials testing reactor fuel reprocessing plant, Dounreay:
  (a) Control face materials
  (b) Fuel stripping pond

the UK prototype reactors and Zircaloy-clad water reactor fuel from overseas. Optimum procedures for decontaminating and changing shear packs were applied and the versatility of the plant confirmed.

The plant was shut down after a release of airborne radioactivity on 26 September 1973 which led to a comprehensive review of the reprocessing of highly irradiated oxide fuels. An investigation into the incident[21] found that ^{106}Ru-contaminated gases had been released from within the biological shielding. The cause of the release was heat-generating reactions from an accumulation of solid fission products, which led to the possible ignition of zirconium fines (from fuel cladding) and the decomposition of the organic solvent (Butex). The gases produced, together with a volatile compound of ruthenium (RuO_4), escaped into the manned area of the plant. Modifications have been proceeding at low priority to adapt the plant to operate on long-cooled fuel without the solvent extraction cycle, thus providing a development facility in aid of THORP.

6.4 The thermal-oxide reprocessing plant (THORP)[22]

The continued construction of AGRs in the UK and LWRs elsewhere in the world, led BNFL to proceed with the design of a new large-scale plant to reprocess irradiated fuel from domestic AGRs and from overseas LWRs under commercial contract. Government approval in principle was given in 1976 for the construction of such a plant at Windscale and for the reprocessing of irradiated fuel for overseas customers, subject to the negotiation of satisfactory terms, including an option to return radioactive waste to the country of origin. On 1 June 1976 BNFL made application for outline planning permission for the expansion of Windscale Works including the construction of THORP with a designed reprocessing capacity of 600 tonnes of oxide fuel per annum (1200 t nominal design). As a result of objections from a number of corporate bodies and individuals, the Secretary of State for the Environment decided to examine the proposal to construct THORP, and set up an Inquiry under the Hon. Mr. Justice Parker for the purpose of hearing the cases for and against the proposed development. The BNFL case to the enquiry was based on the following main technical submissions.

1. Magnox fuel had been reprocessed successfully for 25 years; there was also experience of storing plutonium and of reprocessing oxide fuel. Therefore, BNFL had the necessary technical experience to develop and operate the proposed plant.
2. Reprocessing was desirable as an energy conservation measure.
3. Reprocessing of spent fuel from UK AGRs in operation or already under construction was essential on waste management grounds.
4. Existing plant in the UK was inadequate to deal with anticipated AGR spent fuel arisings.
5. A plant large enough to reprocess foreign spent fuel in addition to UK arisings would permit economies of scale and would bring a balance of payments advantage to the UK.

6. Foreign business existed which would justify construction of a plant of 600 tonnes annual capacity (1200 t nominal design).
7. Whilst the proposed development was not dependent on a decision to go ahead with the fast reactor system in the UK, it was essential if that option were to be kept open.
8. The reprocessing technology proposed was not novel, and the Nuclear Installations Inspectorate were confident that the proposed plant could be designed, built, and operated to high standards of safety.
9. The predicted emissions of radioactivity from the plant during routine operation gave no grounds for supposing that employees or the public would face any significant risk.

Following an Inquiry which lasted 100 days, and during which evidence was taken under oath from 146 witnesses and 161 written representations were considered, the Inspector concluded that oxide fuel from UK and overseas reactors should be reprocessed at Windscale, and that outline planning permission for THORP should be granted without delay. The Inspector's report[22] to the Secretary of State for the Environment was published on 26 January 1978, and it was subsequently debated in the House of Commons, following which authorization was given to proceed with the plant.

The design of the THORP plant is now (1981) well advanced and large-scale test rigs have been constructed at Windscale to examine various aspects of the plant. The plant will process irradiated UO_2 from AGRs and LWRs, including overseas reactors; thus it will have to handle stainless steel and Zircaloy-clad fuels which are enriched in ^{235}U.

The fuel will be received at the Windscale complex in shielded containers and stored pending reprocessing in steel *bottles*. This is more convenient if long-term storage of irradiated fuel is envisaged and reduces the possibility of contaminating the storage pond water.

As UO_2 fuel tends to disintegrate under irradiation, it will be sheared, together with its cladding, into, ca.5 cm lengths. The sheared, clad fuel will then be leached to dissolve the fuel and leave the cladding for disposal.

Fuel dissolution will take place under batch conditions: several dissolvers will probably be used and criticality control will be achieved by addition of a neutron absorber, gadolinium nitrate, with the nitric feed to the dissolver. There is some increase both in operating costs and in the volume of highly active raffinate owing to the addition of the neutron absorber, but this is outweighed by the ability to use volumetrically large items of plant.

As with the PFR fuel process (see §6.5.3) a centrifuge will be used to clarify the dissolver liquid (Fig. 14.9). Application of a magnetic field for solid–liquid separation is also being investigated for use in THORP.

Tri-*n*-butylphosphate has been selected for the extractant and kerosene as the organic diluent. The main solvent extraction process will be 'salt'-free; that is, no non-evaporable species will be used in the main line process. Thus uranium tetranitrate, $U(NO_3)_4$, will be used to reduce plutonium to Pu(III) in order to effect the uranium–plutonium separation; the uranium tetranitrate is oxidized to uranyl nitrate which then becomes part of the uranium product. The uranium for the manufacture of the tetranitrate is taken from the uranium product in the first instance, so there is no overall increase in the uranium throughput in the process.

Pulsed columns are being examined as possible contactors for the highly active cycle in THORP, owing to the decreased solvent degradation that occurs in these devices compared with mixer-settlers. Also, they have a greater volumetric throughput under conditions of limited geometry for criticality control.

Separation of uranium from plutonium will take place in the first cycle of the THORP process, allowing the use of large mixer-settlers as contactors, since criticality is no problem once the plutonium has been removed. The plutonium purification cycles will continue to use geometrically limited pulsed columns. It is proposed to use 30 per cent TBP–OK in the highly active cycles and the plutonium purification cycles, and 20 per cent TBP in the uranium purification cycles.

In addition to a 1/5000–scale fully active pilot plant, full-scale pulsed columns for use with uranium have been constructed at Windscale for experimental purposes.

It is intended to have the THORP plant operational in the late 1980s. A schematic flowsheet is given in Fig. 14.23.

6.5 Nuclear fuel reprocessing at Dounreay

6.5.1 *The materials-testing reactor (MTR) fuel reprocessing plant*[23]
This was the first plant to be built and commissioned at Dounreay. It commenced operation in 1958 and it is still operating.

As already noted, MTR fuel is basically an alloy of highly enriched uranium and aluminium, clad in aluminium rather like the filling in a sandwich. The burn-up or utilization of MTR fuel is very high compared to other reactor types: 50 times that of the Magnox reactors, 5 times that of LWRs, and twice that of fast reactors. Thus although the quantities of uranium to be processed per day are small, about 8×10^{-5} times the throughput of the second Windscale separation plant, the nature of the irradiated fuel requires full-scale plant operations for its processing and handling.

The Dounreay plant reprocesses MTR fuel from both UK and overseas reactors. The enriched uranium thus recovered is *blended-up* with ^{235}U to the required level and new fuel is fabricated for domestic use and export.

Upon receipt at the reprocessing plant MTR fuel is stored under water in *cruets*. These are trays designed to hold the fuel in a geometrical array, thus preventing criticality in the highly enriched fuel.

Immediately before reprocessing, as much non-fuel aluminium as possible is removed from the fuel element by underwater machine tools. The fuelled section of the sub-assembly (aluminium plus uranium–aluminium alloy) is

Fig. 14.21 Reprocessing plant: Dounreay fast reactor fuel solvent extraction cell

sheared into convenient sections for dissolution. Dissolution is batchwise, both the aluminium and the uranium being dissolved to give a solution of uranyl nitrate and fission-product nitrates together with *anion-deficient aluminium nitrate* in nitric acid. Aluminium nitrate has the chemical formula $Al(NO_3)_3$, but under certain conditions aluminium can be made to dissolve in aluminium nitrate without precipitation. This is done for MTR fuels where the dissolver product has the typical composition 67.5 g/1 A1; 3.6 g/1 U; 1.2M HNO₃. A major advantage of this technique is that relatively small volumes of highly active raffinate are produced, which although having limited evaporability, can be conveniently stored. Criticality in the dissolver is controlled by the mass-limitation technique. In the rest of the plant it is concentration controlled.

Solvent extraction is carried out in stirred mixer–settlers. TBP is used as the extractant in odourless kerosene as diluent.

Because of the small quantity of uranium present and the large quantity of aluminium nitrate, which is an excellent salting-out agent, 'dilute TBP' can be used for extraction. Thus, in the original flowsheet, 5 per cent TBP–OK is used compared to 20 per cent TBP–OK in use at Windscale. The high degree of salting-out gives decontamination factors from fission products in the first cycle of one or two orders of magnitude greater than in the second separation

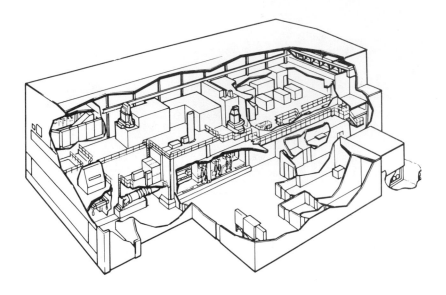

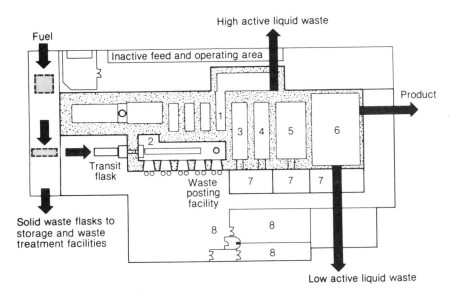

1. Dissolver centrifuge and accountancy.
2. PFR fuel breakdown cave.
3. Solvent extraction cycle 1.
4. Solvent extraction cycle 1.
5. Solvent extraction cycle 2.
6. Solvent extraction cycle 3.
7. Active cell maintenance access area.
8. Manipulator and plant component servicing area.

Fig. 14.22 The PFR reprocessing plant, Dounreay

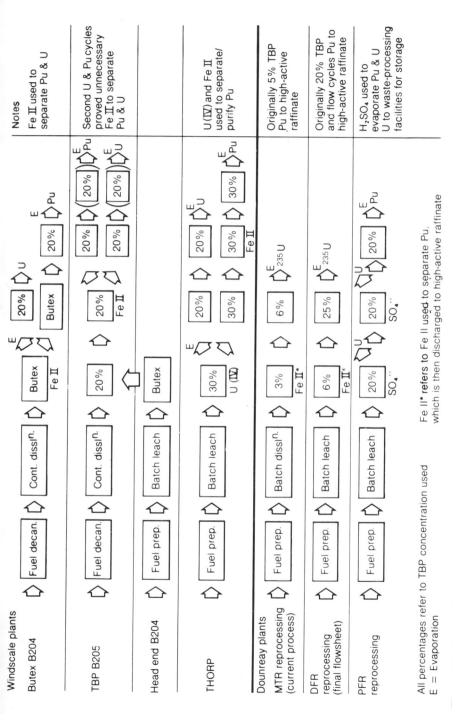

Fig. 14.23 Process flow sheets used in UK reprocessing plants

plant at Windscale. Hence only two cycles of extraction and strip are required to decontaminate the uranium. Owing to the high enrichment of the uranium, very little plutonium is formed during reactor irradiation of MTR fuels, and this is rejected to the highly active, raffinate stream in the first cycle by reducing the plutonium to the inextractable Pu(III) state. A further advantage of the high degree of salting-out in the first cycle is that only six stages are required for extraction compared to eight stages in the Windscale plant. The second cycle of the MTR process also originally operated on 5 per cent TBP–OK. Subsequent process modifications resulted in a change to the first cycle TBP concentration to 3 per cent TBP, which led to further improvements in decontamination from fission products.

The uranyl nitrate product from the process is evaporated to about 300 g/l U and after adjusting the enrichment by the addition of ^{235}U, the uranium is processed to make more MTR fuel which is then recycled back to a reactor.

The two Dounreay plants (see §6.5.2 for a description of the fast reactor fuel reprocessing plant) are operated on a *campaign* basis rather than continuously. In 1966 the head end of the MTR plant was coupled to the solvent extraction section of the fast reactor plant and sequential MTR/fast reactor fuel processing was operated with a consequent saving in overall costs. The fast reactor plant was closed down in 1972 for modification (as described in §6.5.2) and the MTR plant was restored to its original status. This plant continues to operate. Figures 14.20 and 14.23 show views of plant items and the flowsheet.

6.5.2 *The Dounreay fast reactor (DFR) fuel reprocessing plant (1960–1975)*[24]
The Authority's development of the fast reactor system, mentioned earlier, resulted in the construction of the 60 MW(t) Dounreay fast reactor which was fuelled initially by uranium metal fuel of 46 per cent ^{235}U enrichment. Since highly enriched uranium was in short supply it was decided that the DFR fuel inventory should be minimized by building reprocessing, product conversion, and fuel fabrication plants alongside the reactor at Dounreay. This meant that the entire fuel cycle was thus located at a single site. The fast reactor commenced operation in 1958 and the fuel reprocessing plant was commissioned in 1960 (Figs. 14.21 and 14.22).

The development of the original process flowsheet and equipment for DFR fuel reprocessing was carried out by the Research and Development Section at Windscale Works so that full advantage would be taken of the Magnox fuel reprocessing experience there. The solvent extraction flowsheet adopted 20 per cent TBP–OK, following its successful use in the second plutonium purification plant at Windscale. Because of the high fissile material content of fast reactor fuel the solvent extraction contactors were designed to be geometrically limited to avoid the possibility of an accidental criticality excursion.

A novel feature of these mixer–settlers was that the mixing of solvent and aqueous solutions in them was carried out by the use of oscillating air pulses rather than by stirrers or turbo-mixers (see Fig. 14.13).

At a later stage there were economic and logistic pressures to reduce the highly enriched uranium inventory required for DFR and its fuel cycle and a series of plant trails was carried out progressively to reduce the decay time before reprocessing to achieve a once-off reduction in uranium inventory. Ultimately, DFR fuel was processed routinely at 90 days' cooling with occasional reductions down to 80 days' cooling. This, together with the increased fuel burn-up achieved by the molybdenum alloying addition, meant that the specific activity fed to the reprocessing plant ultimately reached an order of magnitude greater than the original design parameters. Since the uranium conversion plant and the fuel fabrication plant, to which the recovered ^{235}U was to be fed, were unshielded glovebox facilities, the reprocessing plant product specification could not be relaxed and the required fission-product decontamination factor had to be increased by a factor of ten.

The original flowsheet and plant were designed to process fuel consisting of 46 per cent enriched uranium alloyed with 0.5 atomic per cent (^{a}/o) chromium irradiated for 8800 megawatt days per ton(ne) and cooled for 120 days before reprocessing.

As experience of fast reactor fuel performance and endurance was obtained, the fuel composition was changed to an enrichment of 75 per cent ^{235}U alloyed with 7 per cent by weight (^{w}/o) of molybdenum. this enabled the irradiation to be increased to 25 000 MW d/t and also required a significant process flowsheet change. To prevent the precipitation of molybdenum salts in the dissolver, ferric nitrate was used to complex molybdenum. Inclusion of molybdenum in the fuel meant that the concentration of uranium in the dissolver product had to be reduced from 200 g/l (U–Cr fuel) to 120 g/l (U–Mo fuel). Further, the addition of ferric nitrate to the dissolver to prevent precipitation of molybdenum necessitated a change from a stainless steel dissolver which would be rapidly corroded by the ferric nitrate at 100 °C to a titanium dissolver. With the original solvent extraction process using 20 per cent TBP–OK, the plant tended to become hydraulically unstable during long campaigns or runs. This was caused by the accumulation of niobium-bearing solids which led to a stabilization of emulsion at liquid–liquid interfaces in the mixer–settlers. It was sometimes necessary to suspend processing to flush these solids out of the first cycle extractor in order to restore hydraulic stability to the plant.

The use of the solvent extraction part of the DFR plant for MTR reprocessing has been noted. In an attempt to increase the throughput and the decontamination factor in the DFR plant when processing DFR fuel, it was decided to examine the consequences of increasing the TBP concentration. A 30 per cent TBP–OK flowsheet was, however, found to be difficult to control, particularly when high solvent loading was attempted. A 25 per cent TBP–OK flowsheet was successful in giving a higher throughput and better decontamination factors than the 20 per cent TBP–OK process and, although this flowsheet was used for some time, there was still some instability of the liquid–liquid interface. Experience with the reprocessing of MTR fuels led then to consideration of the use of solvent containing less than 20 per cent TBP for processing Dounreay fast reactor fuel. Pilot plant tests showed that with 6 per cent TBP as the first-cycle solvent, the enhanced differences in the specific gravity, viscosity, and surface tension between the aqueous and solvent phases eliminated all traces of interface instability and gave a significantly improved hydraulic performance resulting in a greater throughput.

However, if dilute nitric acid was used for stripping the uranium, the volume of first-cycle aqueous product was too large for the rest of the plant (which operated on 25 per cent TBP–OK). The use of dilute sulphuric acid for stripping gave an acceptably low volume for the first-cycle product. Sufficient nitric acid is stripped to prevent undue plant corrosion. After intercycle acidity adjustment with nitric acid, the process used a second cycle of 25 per cent TBP–OK followed by sulphuric acid stripping, and a third cycle of 25 per cent TBP–OK with nitric acid stripping.

When used in the plant on fully radioactive liquors, the fission-product decontamination factors in the first two cycles were sufficiently high for the third cycle of operation to be omitted. The second-cycle strip was then carried out using nitric acid in order to obtain a uranyl nitrate product. Table 14.6 lists typical decontamination factors for fast reactor fuel reprocessing using different TBP concentrations. As forecast from pilot-plant experiments, the interface stability was excellent in the plant and there has been no suggestion of blockages during plant operation. Use of low TBP concentrations has the added advantage that the residence time per mole of TBP per pass is reduced. As a consequence of this the radiation dose per pass to the TBP is reduced and hence the total radiation damage (or radiolysis) to the organic phase is reduced.

A feature of highly enriched uranium reactor fuels is that only very small quantities of plutonium are formed as compared to the quantity of plutonium that would be formed in an equivalent natural or near-natural uranium fuel having the same burn-up. Thus in both the MTR and DFR process, plutonium

Table 14.6 Comparison of decontamination factors as a function of TBP concentration

Overall decontamination factor	6% TBP cycle I, 25% TBP cycle II, cycle III not operated	20% TBP all three cycles	25% TBP cycles I & II, 20% TBP cycle III	30% TBP cycle I, 20% TBP cycles II & III
β	1.5×10^7	5.6×10^5	1.2×10^6	1.5×10^5
γ	7×10^7	ca 7×10^6	1.1×10^7 (approx.)	2.7×10^6
Ru	4.4×10^7	2.7×10^7	not measured	2.6×10^6
Zr/Nb	6.3×10^7	3.5×10^6	2.9×10^7	3.2×10^6

Note
Data are for Dounreay fast reactor fuel reprocessing. Feed to the plant was standard irradiated DFR fuel cooled for about 90–120 days prior to reprocessing. Standard flow sheets for a given TBP concentration used in all cases

is reduced to Pu(III) by ferrous sulphamate in the first cycle and is rejected to the highly active raffinate. Both MTR and DFR highly active raffinates contain high salt concentration (aluminium nitrate or ferric nitrate–molybdenum complex, respectively) and in consequence are not significantly evaporable. The high first-cycle decontamination factors of both processes means that the second-cycle raffinate can be discharged to sea without treatment after monitoring to check the radioactivity content.

In 1975 the Dounreay fast reactor fuel reprocessing plant was shut down for a major refit to enable it to reprocess fuel from the Dounreay Prototype Fast Reactor (PFR).

6.5.3 *Prototype fast reactor (PFR) fuel reprocessing*[25]

A prototype fast reactor (PFR) of 600 MW(t) output was sanctioned in 1966 as the next stage in the development of the fast reactor system in the UK. The policy of reprocessing and refabricating the fissile-material content of irradiated fast reactor fuel to demonstrate the feasibility and reliability of the total fuel cycle was continued, and, in 1972, it was decided that the irradiated PFR fuel should be reprocessed at Dounreay by modifying and extending the existing DFR fuel-reprocessing plant for the purpose[26] (Fig. 14.22). Flowsheet and equipment development to support the design of the plant modifications and additions for this fully active demonstration plant were carried out over the period 1976–80 (Fig. 14.23). The modification programme was completed in 1980 with the plant becoming active in the autumn of that year.

The decommissioning and decontamination of the old DFR plant is fully described elsewhere.[16] It is worth noting here that, in complete contrast to the original philosophy of 'no entry/no maintenance', the whole plant was entered, and was fully decontaminated and dismantled. All the old equipment was removed, only the mixer–settlers being reused; shielding was pulled down and rebuilt, and new equipment was installed, all under closely controlled personnel radiation limits.

The PFR plant comprises a head-end section where fuel sub-assemblies, minus their top and bottom end sections, are received, having been cleaned of sodium at the reactor. A section of wrapper is removed by cutting it with a laser beam, thus exposing the fuel pins which are then individually pulled from the sub-assembly. The wrapper carcase and the internal structure (*honeycomb grids* and *egg-box grids*) remain as waste. The individual fuel pins are fowarded to a cropping machine where they are cropped into the dissolver basket which is then placed in the dissolver. After fuel leaching has occurred, the basket containing the hulls is withdrawn for reuse and the dissolver liquor plus any solids pass via a centrifuge into a *conditioning vessel*, which is also used to enable a measurement of the plutonium in the dissolved (UPu)O$_2$ (uranium–plutonium dioxide) to be made.

The clarified, conditioned liquor is then fed to the solvent extraction plant. Because of the high plutonium content of the liquor (30 to 40 g/l) criticality restrictions must be applied not only in the solvent extraction process, but also over the whole plant, wherever fuel or liquor is likely to be present. Criticality control in the new sections of the plant (fuel-handling, dissolver, centrifuge, tanks, etc.) is achieved by geometrical means as in the solvent extraction equipment.

In order to obtain the required uranium–plutonium throughput to match the output of irradiated fuel from the reactor it was necessary to use two extraction sections in parallel with the first cycle, and from then on to combine the solvent products into one stream. In the first-cycle strip units all the plutonium, but only part of the uranium, is stripped from the organic phase for further decontamination. The balance of the uranium is stripped from the organic phase after the 'partial strip' and is not a product stream. Complete separation of plutonium from uranium occurs in the second cycle of decontamination; this uranium is also a non-product stream and the plutonium is finally purified in the third cycle evaporated, and returned to Windscale to be remade into fuel for the PFR.

Because of the high Pu/U ratio in PFR fuel, uranium–plutonium separation cannot be achieved in a reprocessing plant such as that at Dounreay by using ferrous sulphamate. The quantity of ferrous sulphamate required would make the plant inoperable; uranium tetranitrate $U(NO_3)_4$ cannot be used for similar reasons. A reductant such as hydroxylamine could be used, but the kinetics of the reaction are too slow to achieve a high throughput in a small plant.

In the PFR reprocessing plant sulphuric acid is used for both the partial strip and the complete separation. The plutonium is not reduced to Pu(III), but is converted to Pu(IV) sulphate, $Pu(SO_4)_2$, which is not soluble in the organic phase. The uranium present is partly converted to uranyl sulphate, which is similarly insoluble in the organic phase but, as the equations below show, the reactions are reversible. Hence, by controlling the nitric acid/sulphuric acid ratio in the process, the required separations can be achieved:

$$Pu(NO_3)_4 + 2\,H_2SO_4 \;\rightleftharpoons\; Pu(SO_4)_2 + HNO_3$$

$$UO_2(NO_3)_2 + H_2SO_4 \;\rightleftharpoons\; UO_2SO_4 + 2\,HNO_3.$$

In the PFR process the highly active raffinate is stored in underground tanks in the usual way, with or without evaporation. Owing to the high burn-up of the fuel (8–10 per cent), the fission-product content of this liquid is high and only limited evaporation is possible if precipitation of solids is to be avoided. The medium-and low-active aqueous wastes (including all the uranium) are collected and the uranium is precipitated by ammonia together with any fission products and plutonium present. This solid is retained on the site and the decontaminated liquor is discharged to sea after monitoring.

A centrifuge collects the fission-product insolubles, together with any small fragments of sub-assembly debris. Leached fuel hulls are then placed into stainless steel drums and stored in an *engineered store* (Fig. 14.16).

The PFR fuel-reprocessing plant is currently operating on irradiated fuel from the PFR. The development programme and design studies have identified areas in which direct scale-up of the PFR reprocessing plant would be inadequate or inappropriate for a large-scale commercial fast reactor (CFR) reprocessing-plant design. Consequently, a parallel development programme has been initiated to provide data for the design and active demonstration of the size and type of equipment which will be required to recycle the quantities of irradiated fuel produced from an expanding programme of CFR installation in the UK. The size and timing of the fuel cycle plants required to support a CFR programme is entirely dependent upon the initial reactor installation and the programme build-up in relation to plutonium availability. For planning purposes, however, the early demonstration of fully active, full-scale equipment in a plant sized to support a commercial demonstration fast reactor (CDFR) may be envisaged.

Acknowledgements

The authors would like to acknowledge the assistance of the staff of the Reprocessing Development Group, Dounreay, in collecting information for this chapter.

In particular, they are indebted to Mrs. J.M. Porter, Mr. B. Steele, and Mr. J.B. Spence (Dounreay), who played a considerable part in producing and checking the original drafts, and to Mr. J.A.G. Heller (London Office, UKAEA) for his help with the final draft of this chapter.

References

1. BRITISH NUCLEAR FUELS LTD. *Fuel services from BNFL*. BNFL, Capenhurst (1974).
2. WARNER, B.F., MARSHALL, W.W., NAYLOR, A., and SHORT, G.D. The development of the new separation plant, Windscale. In *Proc. 3rd UN Conf. on Peaceful Uses of Atomic Energy, 1964*, vol. 10. United Nations, New York (1965). Pp. 224–230.
3. INTERNATIONAL ATOMIC ENERGY AGENCY. *Non-proliferation and international safeguards*. IAEA, Vienna (1979).
5. SPENCE, R. Chemical process development for the Windscale plutonium plant. *J. R. Instn. Chem.*, **18**, 357–368 (1957).
6. GOWING, M. *Independence and Deterrence*, vol. 2 (Policy, execution). Macmillan, London (1974).
7. SMYTHE, M.D. *Atomic energy for military purposes*. Princeton University Press (1946).
9. INTERNATIONAL UNION OF PURE AND APPLIED CHEMISTRY. *Definitions of technical terms used in solvent extraction*. Information Bulletin 63. IUPAC, Oxford (1977).
10. INTERNATIONAL ATOMIC ENERGY AGENCY. *Techniques for the solidification of high level Waste*. Technical Report Series No. 176. IAEA, Vienna (1977).
11. *Nuclear power and the environment*. The Government's response to *The Sixth report of the Royal Commission on Environmental Pollution Cmnd 6820*. HMSO, London (1977).
12. INTERNATIONAL ATOMIC ENERGY AGENCY. The management of radioactive waste from the nuclear fuel cycle. Session VII Conditioning medium-level waste, vol. II, pp. 97–186. IAEA Vienna (1976).
14. ICRP. Recommendations of the International Commission on Radiological Protection. ICRP Pub. 26 (1977).
15. IAEA. *The management of radioactive waste from the nuclear fuel cycle*. IAEA, Vienna (1976).
16. BARRETT, T.R. and THOM, D. *The decommissioning and reconstruction of the fast reactor fuel reprocessing plant, Dounreay*. IAEA *SM* 234/9. IAEA, Vienna (1978).
17. IAEA. *Proc. Int. Symp. on Management of Gaseous Wastes from Nuclear Facilities, Vienna, February 1980*. IAEA, Vienna (1980).
18. WHITMELL, D.S., NELSON, R.S., WILLIAMSON, R. and SMITH, M.J.S. Immobilization of krypton by incorporation into a metallic matrix by combined ion implantation and sputtering *Nucl. Energy*, **18** (5), 349–352 (1979).
19. HINTON, C. *Chem. Ind.* 700–711 (14 July 1956).
    NICHOLLS, C.M. and SPENCE, R. *Trans. Inst. Chem. Engng.*, **35**, 380–393 (1957).
    HOWELLS, G.R., HUGHES, T.G., MACKEY, D.R., and SADDINGTON, K. The chemical processing of irradiated fuels from thermal reactors. In *Proc. 2nd UN Conf. on the Peaceful Uses of Atomic Energy*, Geneva 1958 vol. 17. United Nations, Geneva (1958) pp. 3–24.
20. CORNS, H., CLELLAND, D.W., HUGHES, T.G. and de LISLE, J.W. The new separation plant Windscale: design of plant and plant control methods. In *Proc. 3rd UN Conf. on the Peaceful Uses of Atomic Energy* vol. 10. Geneva 1964. United Nations, New York (1965) pp. 233–240
21. *Report by the Chief Inspector of Nuclear Installations on the incident in Building B204, at the Windscale Works of BNFL (26 September 1973)*. Cmnd. 5703. HMSO, London (1974).
22. PARKER, THE HON. MR. JUSTICE. *The Windscale Inquiry*. HMSO, London (1977).

23. BUCK, C., HOWELLS, G.R., PARRY, T.A., WARNER, B.F. and WILLIAMS, J.A. Chemical processes at the UK Atomic Energy Authority Works Dounreay. In *Proc. 2nd UN Conf. on the Peaceful Uses of Atomic Energy*, vol. 17 Geneva 1958. United Nations, Geneva (1958) pp. 25–45

24. MILLS, A.L. and LILLYMAN, E. In A review of metallic fast reactor fuel reprocessing at Dounreay. In *Proc. Int. Solvent Extraction Conf., 1974*, pp. 1499–1518. Society of Chemical Industry, London (1974).

25. MILLS, A.L. The Dounreay Prototype Fast Reactor Fuel Reprocessing Scheme–flowsheets and chemistry. In *Proc. Symp. on Fast Reactor Fuel Reprocessing, May 15–18, 1979*. Society of Chemistry and Industry, London (1980).

26. BARRETT, T.R. The reconstruction of the fast reactor fuel reprocessing plant at Dounreay. In *Proc. Symp. on Fast Reactor Fuel Reprocessing, May 15–18, 1979*. Society of Chemistry and Industry, London (1980).

15

The management of radioactive waste from civil nuclear power generation

N. J. KEEN

The total radioactivity of wastes produced by the generation of electricity by nuclear power is extremely high initially, but it begins to decay immediately the fuel is withdrawn from the reactor. For the first few hundred years, the radioactivity is dominated by the fission products, but after these have decayed the remaining radioactivity is mainly due to long-lived actinide elements. The nuclear wastes arises in a variety of chemical and physical forms in widely differing volumes and levels of activity. The total volume of the waste is very small, some thousand times less than that produced from coal, in generating an equal quantity of electricity.

When the irradiated nuclear fuel is reprocessed to separate uranium and plutonium for recycle, more than 99 per cent of all the activity in the dissolved fuel is concentrated in one liquid waste stream. This liquid is stored in tanks at present. For long-term storage or eventual disposal it is preferable to solidify it, since storage of a solid would require much less continuous supervision to maintain an even higher degree of safety because of its complete immobility. Conversion of the highly active liquid to a stable insoluble solid has now been demonstrated on an industrial scale, and the method is likely to be adopted in most countries using nuclear power. Methods are available for converting all the other wastes of lower activity to solid forms packaged suitably for disposal.

Gaseous wastes and low-level liquid wastes are discharged to the environment in a regulated manner, as authorized by the government departments responsible. Certain solid low-level wastes are disposed of by burial in shallow trenches at an authorized site, or are incorporated in concrete and dropped into an internationally agreed part of the Atlantic ocean. All wastes of higher levels of activity are currently held in stores pending the approval of appropriate disposal routes, which are currently under development.

Methods are being studied in many countries for the final disposal of all wastes in a manner that will assure that their radioactivity remains isolated from the biosphere until they have decayed to levels where they pose no threat to man. The methods include emplacement of waste containers in deep geological formations on land, on to the ocean bed, or into the ocean bed sediments. The highest-level waste would be stored for at least some decades to allow the initially high heat output to decay. All other wastes can be disposed of as soon as the route and the preferred solid package have been agreed with the relevant authorizing government department.

Studies of the long-term safety of wastes containing substantial quantities of long-lived radionuclides indicate that the radiation to man from any radioactivity that returns to the biosphere will most probably be very small.

Contents

1 Introduction

There have been nuclear power programmes, in the UK for more than 25 years. Many of the civil reactors constructed during the programme will be in operation up to the turn of the century, and the Government has recently announced plans for the construction of further stations to meet the country's electricity requirements. To the present time, various low-level wastes have been disposed of, but the vast majority of all the waste with higher levels of radioactivity that has been generated is held in stores on the reactor and fuel-reprocessing sites pending decisions on final disposal. There is a widespread apprehension that nuclear waste constitutes an insoluble problem and will be a threat to man for millions of years. This chapter seeks to correct this misconception. It describes briefly the way in which radioactive waste is currently being managed, and the methods now under development to ensure that it can be stored or finally disposed of with no threat to man in the short or long term.

From the very beginning of the development of nuclear energy, scientists and engineers appreciated the potential hazards and recognized the need to deal with the waste. In this respect the situation is quite different from that of many traditional industries, which had developed over long periods before the deleterious effects on workers and the environment were recognized. Knowledge of the biological effects of ionizing radiation, and the setting-up of the International Commission on Radiological Protection, preceded the advent of the nuclear power industry by many years so that from the outset there has been an established basis for setting the standards of safety.

This chapter is based on the premise that nuclear fuel will continue to be reprocessed in the UK in order to separate and recycle unburnt uranium and plutonium. The plutonium will probably be recycled in fast reactors, where it can be burnt efficiently, since this is the best course to follow if uranium stocks are to be efficiently utilized. Therefore, the wastes considered in the chapter are those arising from nuclear fuel cycles with reprocessing, although passing mention is made of certain implications of the use of the *once-through* nuclear fuel cycle, in which spent nuclear fuel would be regarded as a waste.

Wastes from uranium mining are also important, but as there is no direct UK experience in the management of these, the question is not dealt with in this chapter.[1] Also wastes arising from the complete dismantling and removal of reactors and other nuclear plant have not been included. These are dealt with in Chapter 18.

ın §2, an outline is given of the general principles underlying the control and management of radioactive wastes in the UK. Section 3 summarizes the type and quantities of the principal radionuclides involved and their characteristics of importance to waste management. Section 4 deals with the ways in which different categories of waste arise at various points in the fuel cycle, and the measures currently employed in their management. Research and development at present under way is directed to methods for converting all categories of waste to a state which will enable them to be disposed of, and this is also outlined in this section. The final sections outline the options being considered for the disposal of wastes, including disposal into deep subterranean formations and on to or into the deep-ocean floor. It indicates that there is confidence in finding a final method of isolating these wastes where they can safely be left to decay to safe levels before they can return to man's environment.

In the article an attempt has been made to indicate the possible arisings of various types of waste in the UK up to the year 2000. These figures are intended only to provide a rough guide. Many of the factors which could substantially change the data are uncertain. These include the size of the nuclear programme, the mix of reactors employed and the time at which they become operational. The figures quoted in the article are based on the reference case used in Energy Policy Green Paper (*Cmnd* 7101): namely installed nuclear capacity of 40 GW(e) by the year 2000. A similar figure was used in the 'high reference case' in the UK input to the International Nuclear Fuel cycle Evaluation Study.[5] It should be noted that the programme now seems unlikely to reach more than half this size.

2 The objectives and principles of radioactive waste management

There is wide international agreement that the objectives of radioactive waste management must be

    (a) to comply with radiological protection principles for present and future generations;

    (b) further, to minimize any impact on future generations to the maximum extent practicable;

    (c) to preserve the quality of the natural environment; and

    (d) to avoid pre-empting present or future exploitation of natural resources.

The UK policy for managing radioactive wastes is laid down in the Radioactive Substances Act (1960) which gave force of law to the recommendations of the white paper, *The control of radioactive wastes, Cmnd.* 884. This requires the industry

    (1) to ensure, irrespective of cost, that no member of the public shall receive more than one-tenth of the maximum radiation dose prescribed for occupational workers by the International Commission

on Radiological Protection (ICRP), which (in brief) stipulates a maximum dose of 0.5 rem (5 mSv) per year for the whole body;

(2) to ensure, irrespective of cost, that the whole population of the country shall not receive an average dose of more than 1 rem per person in 30 years; and

(3) to do what is reasonably practicable, having regard to cost, convenience, and the national importance of the subject, to reduce doses far below these levels.

In 1977, the ICRP revised their recommendations for radiation dose limits and *Cmnd.* 884 was reviewed.[2] The review committee restated the basic objectives for UK practice in the light of the revised ICRP recommendations as follows:

(a) that all practices giving rise to radioactive waste shall be justified, i.e. the need for the practice must be established in terms of its overall net benefit;

(b) that radiation exposure of individuals and the collective dose to the population arising from radioactive wastes shall be reduced to levels which are as low as reasonably achievable, economic and social factors being taken into account; and

(c) that in present circumstances the average effective dose equivalent (i.e. the measure of the risk from exposure to radiation which takes account of the different sensitivity of different organs of the body) from all sources, except natural background radiation and medical procedures, to representative members of a critical group shall not exceed 0.5 rem (5 mSv) in any one year.

2.1 Licensing and authorization

With the growth of nuclear power for industrial purposes the UK has evolved a system of licensing, control and authorization for all nuclear installations and operations, enforced by government departments or their agencies to ensure that the above objectives are met. For example, civil nuclear sites, such as power stations and reprocessing plants, are licensed by the Nuclear Installations Inspectorate, which specifies conditions under which plants (including waste treatment plants and waste stores) may be operated. Radioactive waste can be disposed of only under authorizations by appropriate government departments (the Department of the Environment, the Scottish Office, and the Welsh Office, together with the Ministry of Agriculture, Food and Fisheries in England and Wales and the Department of Agriculture and Fisheries in Scotland). Radioactive wastes may be transported only according to the terms of licences issued by the Department of the Environment, the Scottish or the Welsh Offices.

2.2 Waste management principles

The ways in which the objectives of radioactive-waste management can be converted into a practicable scheme can be set out in broad principles for design of plants and their operation as follows:

   (1) minimization of arisings of waste at source,
   (2) dilution and dispersion of low-level waste to the environment,
   (3) retention in store until activity has decayed to low levels, and
   (4) isolation of the long-lived radioactive components.

2.2.1 *Minimization of waste arisings*

The actual volume of the hazardous components of the waste—the radio-nuclides—is very small. Even in vitrified highly active liquid waste (the most concentrated category) the radionuclide component will be about 1 per cent by weight. Intermediate-level wastes contain only a few parts per million.

The bulk of the waste consists of common non-radioactive materials such as would be found in any laboratory and industrial operation though these will be associated with radionuclides. To simplify the management of the waste, it is very desirable to maintain a high standard of housekeeping and keep the bulk of the waste as low as possible. For example, operators will aim to avoid introducing any unnecessary inactive material into the active area of the reactor or reprocessing plant, since thereafter it would all be suspect and would have to be regarded as a low-level radioactive waste.

In the selection of processes for dealing with nuclear material, the consequences of all side-streams leading to wastes must be considered, and the use of reagents that create problems in dealing with the waste should be avoided. Also, in the design of radioactive plant and equipment, emphasis is placed on high reliability since maintenance and replacement inevitably generate waste, much of it bulky.

An extension to this principle is that unnecessary operations on radioactive waste should be avoided. Until such time as the form in which wastes can be finally disposed of can be clearly defined, it is better to provide safe storage in an interim condition than to have to rework or back-track at some future date, thereby adding to the bulk of waste and incurring the risk of additional radiation dose to operators.

2.2.2 *Dilution and dispersion to the environment*

The safety of this procedure relies on dilution being sufficiently large and rapid that the concentration of activity presents no appreciable risk to human populations by any route. In practice it is used only where the total quantity of activity is low. It is a very important route since it permits the disposal of large volumes of waste that contain little or no radioactivity.

In the UK, wastes are disposed by dispersal only within limits which are closely controlled by the licensing department. Gaseous effluents are discharged to the atmosphere, low-activity liquids to the sea, estuaries, or rivers, and low-activity solids are buried at shallow depths on selected sites or disposed of in the deep ocean after incorporation in concrete.

2.2.3 *Retention in storage until radioactivity has decayed*

This approach can be particularly useful in dealing with wastes in which all the radionuclides have relatively short half-lives. It is often advantageous to collect and hold such wastes in stores until the passage of time has reduced the activity to levels where the waste can be safely dispersed into the environment, thus avoiding the dose to operators if the waste were to be processed or to the public at large if the materials were released to the environment immediately.

2.2.4 *Isolation*

Waste containing large quantities of radioactivity must be isolated from man's environment until the activity has decayed to a level at which there is no significant risk. *Storage*, by which is meant retention in engineered facilities under supervision, is one means of isolation and is practicable as long as the stores are maintained. Eventually, it may be decided to *dispose* of such wastes in a permanent manner that does not depend for its safety on even the minimal degree of supervision and maintenance that a store would require. In radioactive waste management, disposal is defined as the release or emplacement of waste materials without the intention of being able to retrieve them.

As a first step towards isolation, the waste is immobilised, i.e. converted to a suitably stable, insoluble, monolithic, dust-free, and non-inflammable solid form. Isolation is then further secured by the appropriate location of the wastes, so that natural barriers will also impede the accidental return of activity to man.

3 The characteristics of the radionuclides in nuclear waste

An appreciation of the potential hazard presented by nuclear waste and of the measures that can be taken to meet them requires a general understanding of the nature of the various wastes and of their radioactivity. For an account of the biological effects of radioactive substances, the reader is referred to Chapter 22. It is important to appreciate that nuclear wastes are not poisonous chemicals but are variants (isotopes) of ordinary everyday atoms from which α-, β-, or γ-radiation is emitted as they return to an inactive form. It is these radiations which can be harmful.

In a nuclear reactor, the 'burning' of uranium and other similar heavy elements such as plutonium leads to the formation of three types of radioactive isotope:

    (1) the fission products,

    (2) the actinides, and

    (3) the neutron-activation products.

The characteristic feature of these radionuclides are briefly summarized in §§3.1–3.3.

3.1 The fission products

The fission process produces some thirty different chemical elements and over a hundred different isotopes. The proportions of each vary somewhat with the type of reactor and fuel, but not crucially, and there is little difference in this respect between thermal and fast reactors. Most of the fission products are initially radioactive and decay with the emission of beta particles and gamma rays until a stable isotope is produced. The principal fission products are shown in Table 15.1. The most important nuclides from the point of view of their potential hazard have half-lives lying predominantly in the range up to 30 years, although there are a few of small fission yield—i.e. quantity produced per unit of energy generated—that have very long half-lives. The fission products require heavy shielding or isolation by distance to protect man from the penetrating gamma radiation.

3.2 The actinides

Elements in the actinide group are produced in nuclear reactors as the result of neutron capture by uranium. The most important actinide is plutonium, which is itself a valuable fuel and in the UK is recovered and stored for future use in fast reactors. The other actinides, neptunium, americium, and curium have only limited uses at present and most of them are consigned to the waste. The actinides decay mainly by emission of alpha particles until a stable isotope of lead is formed. The quantities and proportions of the

Table 15.1 Output of some selected fission products from an AGR (for production of 1 gigawatt-year of electricity)

Nuclide	Half-life	Output at time of discharge of fuel (curies)
Group 1: half-lives less than 10 y		
Iodine-131	8 d	1.7×10^7
Cerium-141	32.5 d	3.0×10^7
Zirconium-95	65 d	3.1×10^7
Cerium-144	284 d	2.5×10^7
Ruthenium-106	1 y	1.0×10^7
Caesium-134	2.1 y	3.3×10^6
Promethium-147	2.6 y	4.9×10^6
Europium-154	8.6 y	1.1×10^5
Group 2: half-lives 10–30 y		
Krypton-85	10.8 y	2.8×10^5
Tritium	12.3 y	1.2×10^3
Strontium-90	29 y	2.3×10^6
Caesium-137	30 y	2.8×10^6
Group 3: half-lives greater than 100 000 y		
Technetium-99	2×10^5 y	3.9×10^2
Zirconium-93	1.5×10^6 y	8.3×10
Caesium-135	2.3×10^6 y	1.2×10
Iodine-129	1.7×10^7 y	7.4×10^{-1}

actinides in waste depends on the type of reactor fuel. Thus in Magnox reactors plutonium is produced, but americium and curium are hardly produced at all because the uranium is relatively lightly irradiated and there is little opportunity for the higher atomic number actinides to be formed. Greater quantities of the higher actinides are produced in the reactors such as the PWR and AGR, where the fuel is subjected to a much larger neutron dose.

In the fast reactor, there is scope, if desired, for producing plutonium in the blanket by the irradiation of the ^{238}U contained in the breeder elements. At the same time, the core of the reactor serves as a net incinerator of plutonium, and the balance between the processes of production in the blanket and incineration in the core is one which, to an extent, can be determined by the operator of the plant. Thus by adjusting the quantity of ^{238}U in the blanket, the reactor can be operated either as a net producer (*breeder*) or a net incinerator of plutonium. Even in the breeding mode, however, the fast reactor will produce less plutonium per unit of electricity generated than most thermal reactors.

Alpha particles are very easily stopped by matter, and therefore actinide-contaminated materials do not require thick shielding to protect man. However alpha particles are very energetic, and alpha emitters are very toxic substances if inhaled as dusts. Where significant quantities of actinides occur in waste, they must be immobilized or contained to minimize risk of inhalation or ingestion. Table 15.2 shows that the half-lives of the significant nuclides vary from a few years to billions of years. It frequently worries people that a toxic material can last for so long. However, it should be remembered that toxic materials such as lead and mercury have infinite half-lives. With regard to the specific problem of radioactivity, it should be

Table 15.2 Half-life and activities of selected actinides produced in generation of 1 gigawatt-year of electricity

Nuclide	Half-life	Radiation emitted	Activity AGR†	(curies) FBR‡
Uranium-235	7.1×10^8 y	α	—	—
Uranium-238	4.5×10^9 y	α	—	—
Neptunium-237	2.1×10^6 y	α	4.6	1.6
Plutonium-238	87 y	α	2.9×10^4	5.4×10^4
Plutonium-239	2.4×10^4 y	α	1.0×10^4	1.4×10^5
Plutonium-240	6.6×10^3 y	α	1.9×10^4	1.7×10^5
Plutonium-241	15 y	β	3.1×10^6	1.2×10^7
Plutonium-242	3.9×10^5 y	α	3.6×10	8.6×10
Americium-241	433 y	α	3.9×10^3	2.2×10^4
Americium-243	7.4×10^3 y	α	1.7×10^2	2.4×10^2
Curium-242	163 d	α	8.2×10^5	2.5×10^6
Curium-244	18 y	α	1.0×10^4	1.6×10^4

† irradiation time 1370 days, mean burn-up 17 250 MW d/t
‡ core fuel only (plutonium ex Magnox after 1 year storage), irradiation time 372 days, mean burn-up 80 000 MW d/t

noted that the longer the half-life of an element is, the less radiation is emitted. The isotope uranium-238, which occurs naturally in many common rocks, has a half-life of 4.5 billion years.

3.3 The neutron activation products

These are produced when the neutrons are absorbed (captured) by certain structural materials in the reactor, the coolant, or the nuclear fuel cladding. They decay with the emission of beta and gamma radiation. The total activity due to them is very much less than that of the fission products and they are mainly quite short-lived, but there are a few with longer half-lives. Nickel-63 (half-life = 92 y) is formed when steels are irradiated, and carbon-14 (half-life = 5730 y) is formed by irradiation of certain isotopes of carbon and nitrogen in moderators, steels and air.

3.4 Quantities of fission products and actinide activity in nuclear waste

Practically all the fission products and actinides are retained within the fuel elements and, as well be discussed in the next section, the vast majority of this radioactivity will be retained in one waste stream—usually termed the highly active liquid waste—when the fuel is reprocessed to separate uranium and plutonium. From the moment the fuel elements are withdrawn from the reactor, the initial very high radioactivity decays rapidly since much of it is due to species with short half-lives. The fuel is therefore stored for some time to allow this to happen before any further operations are undertaken; thus fuel is not likely to be reprocessed until half a year to several years has elapsed.

Figures 15.1 (a) and (b) show the fission-product and actinide radioactivities in the waste produced by the generation of 1 GW of electricity for one year in

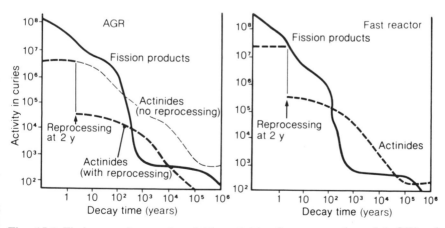

Fig. 15.1 Fission product and actinide activities from generation of 1 GWy of electricity

a thermal (advanced gas-cooled) and in a fast reactor; the curves for a water-cooled thermal reactor such as a pressurized-water reactor would be very similar to that for the AGR. In Fig. 15.1 (a) the actinide radioactivity is shown for two cases—the fuel without reprocessing and the waste after fuel has been reprocessed. The following points should be noted.

1. The total radioactivity decays steeply; it does not last at high levels for millions of years. In the first month after removal from the reactor it drops by a factor of 20; it is reduced by a further 100 000-fold in 500 years and by a million-fold in one million years.

2. Figure 15.1 (a) shows that the actinide radioactivity of the waste stream is reduced more than 100 times by reprocessing during which all but about 0.1 per cent of the plutonium and uranium is removed; the remaining actinide radioactivity at this stage is due mainly to isotopes of americium and curium. In the longer term the difference between total actinide radioactivities in unreprocesed fuel and reprocessing waste is not as large—about a factor of 10; this is due to the differences in rates of radioactive decay (and growth of certain decay products) in the different mixtures of nuclides.

3. The radioactivity of the waste is dominated for the first few hundred years by the fission products, mainly ^{137}Cs and ^{90}Sr. ^{241}Am is the dominant radioactivity between 300 and 3000 years. Thereafter, in succession the main contributors to the declining total radioactivity are the fission products ^{99}Tc, ^{93}Zr, ^{135}Cs, and, at 100 000 years, ^{129}I. The principal actinide component at these times is ^{237}Np, which is formed by decay of ^{241}Am, in addition to some which is present from the beginning.

4. There is little difference between the total quantities of radioactivity from the generation of the same amount of electricity in a thermal reactor such as an AGR and a fast reactor; there are minor differences between the relative proportions of the nuclides that contribute to the total radioactivity and these are mainly in the actinides, but none of the differences affect the management of the wastes.

The decay of fission products and actinides in high-level waste results in the generation of a considerable amount of heat as the energetic alpha, beta, and gamma radiation is emitted. The heat output for the wastes from the same two reactors as before, is shown in Figs 15.2(a) and (b). The following three points should be noted.

1. The fission products generate the greater part of the heat for the first few hundred years.

2. The actinides generate heat over a very long period but at a lower rate; the heat output by various radionuclides depends not only on the disintegration rate of the atoms (curies), but also on the energy of radiation characteristic of each of them. In general, the energy of alpha-particle radiation is much greater than that of beta and gamma radiation, and hence there are no exact parallels between the curves in Figs 15.1 and 15.2.

3. If plutonium and uranium were not removed and spent fuel were disposed

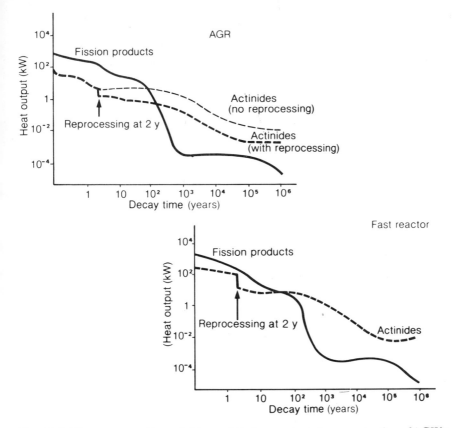

Fig. 15.2 Heat generated by actinides and fission products from production of 1 GWy of electricity

as waste, the long-term heat and radioactivity would be more than ten times greater (this is particularly important to disposal in rock formations).

4 The origins, physical form, and treatment of radioactive waste

The wastes arising from the various points in the nuclear fuel cycle are very diverse in volume, physical form, chemical composition, and radionuclide composition and concentration, and no single means of dealing with them would be sensible. Several schemes of categorization are used:

 (a) The origin of the waste—this gives a useful indication of the radionuclides present and their approximate concentration;

 (b) the physical and chemical form—gases, liquids, and solids, which require very different procedures; and

(c) the total radioactivity—the terms 'high', 'medium', and 'low' are frequently applied to denote levels of radiation emission per unit mass of waste, but there are no generally agreed definitions.

The main points at which waste originates in the nuclear fuel cycles, are summarized in Fig. 15.3.

The uranium ore concentrate imported into the UK is essentially free from radium-226, and the only significant radioactive constituents of the

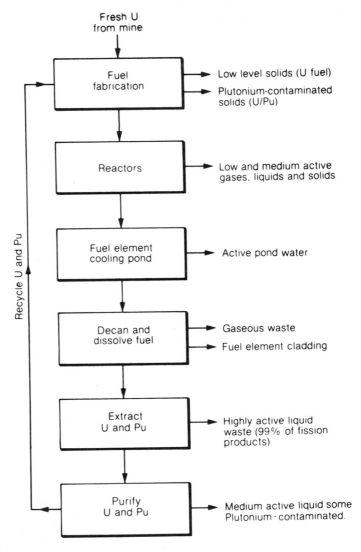

Fig. 15.3 Schematic outline of nuclear fuel cycle

wastes arising at the factory where concentrate is processed, the uranium-235 content enriched, and fuel fabricated for thermal reactors, are the naturally occurring uranium isotopes (mainly ^{238}U) and their immediate decay products. The wastes are of low specific activity and are readily dealt with under current authorizations by shallow burial of solids, and dilution and dispersion of liquid effluents. The fabrication of fast reactor fuels gives rise to plutonium-contaminated wastes (see §4.5).

During operation of the reactor, a very large proportion of the radioactivity is produced in and retained by the irradiated nuclear fuel elements which contain the plutonium, higher actinides, and all the fission products that have been formed, together with much unburnt uranium. Such minor quantities of these as escape from any defects in the cladding of fuel elements eventually appear as radioactive contaminants in the miscellaneous gaseous, liquid, and solid wastes of low- and intermediate-level radioactivity. The internal structure of the reactor also becomes radioactive, largely with neutron-activation products: this will constitute a waste when the reactor is eventually decommissioned.

When the fuel is no longer serviceable it is withdrawn from the reactor as *spent fuel*, and placed in a large tank of water (*pond*), where heat is removed and the shorter-lived radionuclides decay. The pond water becomes somewhat contaminated due to minor releases from any fuel with defective cladding, and is continually treated to remove such activity.

The spent fuel is then transferred to the reprocessing plant, where the metal cladding that contains the fuel is removed and the fuel is dissolved in nitric acid. In both stages, gaseous radionuclides that have been contained within the fuel are released. The uranium and plutonium are then extracted from the nitric acid solution of the fuel, leaving more than 99 per cent of the non-volatile fission products behind in a solution termed the *highly active liquid waste*. The separated uranium and plutonium are further purified and either stored for future use or fabricated into fresh fuel. These operations give rise to a number of medium-level active wastes, some of which are predominantly beta–gamma active, others mainly plutonium-contaminated (i.e. alpha active).

If fuel were not to be reprocessed but were disposed of intact as waste, some of the waste streams indicated in the outline fuel cycle would be avoided, but as has been pointed out in §3.4, a considerably increased quantity of long-term activity and heat has then to be disposed of.

The following sub-sections enlarge on the characteristics of the principal waste categories and describe the techniques in use or under development to convert them to form suitable for their storage or final disposal. In general, it may be said that there are methods for the conversion of all types of waste. These are essentially based upon the principles discussed earlier: the conversion of the most highly active waste into the smallest practicable volume, and the dilution of the remainder to a sufficiently low concentration to allow for its safe release to the environment. The methods themselves are simple. The main task remaining is the engineering of these techniques into safe and reliable operations for routine use in factories, and a decision as to precise forms in which the wastes are required to meet the criteria for safe storage or disposal.

4.1 Gaseous wastes

Gaseous effluents from reactors and reprocessing plants can become contaminated with certain fission products. The principal species that may be involved are krypton, argon, tritium, iodine, and carbon-14. Radioactive dust particles entrained in plant off-gas (mainly ventilation air) have also to be considered.

The principal radioactivity in airborne discharges from nuclear reactors in the UK at present is argon-41, which is short-lived and comes mainly from the earlier Magnox reactors, where it is formed by neutron activation of air used to cool the steel pressure vessel. This problem is associated exclusively with air-cooling, and thus does not occur on recent Magnox stations or on AGRs, PWRs or fast reactors.

The main radioactivity discharged to atmosphere comes from the reprocessing plant when the fuel is dissolved and all the krypton-85 is released together with some tritium and iodine-129; the remainder of the last two enters the liquid waste streams. The much more abundant and toxic iodine-131 is absent because the period of storage of the fuel before reprocessing allows it to decay completely. A small quantity of carbon-14, probably as carbon dioxide is also released.

The effect of the discharges has to be considered at two levels. Firstly the main radiation dose would be to people living in an area near the plant. Effluent is therefore discharged from high stacks to ensure that the activity is diluted to safe levels by dispersion. The other level relates to the very low global dose as these species become distributed throughout the atmosphere. They are removed by radioactive decay (^{85}Kr) or, in the case of tritium (as water), iodine-129, and carbon-14, may be removed to rivers and sea or the earth by rainfall. Assessments have shown that the global radiation dose, even from a much larger world nuclear programme than exists at present, does not warrant action being taken to restrict these discharges before the year 2000.[3] Consideration of the small dose to individuals living near to reprocessing plants is likely to be of greater importance in assessing the need and timing of restrictions in releases of ^{85}Kr, ^{129}I and, possibly further in the future, ^{14}C. The gaseous-waste discharges from nuclear power generation in the UK in 1976 and a forecast of possible discharge in the year 2000, assuming no measures are taken to remove active species from the off-gases, are given in Table 15.3.[2]

Technology for the separation of ^{85}Kr from process plant off-gas and for immobilizing the separated krypton in a stable solid for safe disposal is under development in Europe and the USA. Tritium, iodine-129, and carbon-14 are of lower priority but are under study.

Table 15.3 Estimated annual release of gaseous effluent and low-activity liquid waste; activity in the UK in curies[2]

Year	1976	2000[1] [2] [3]
Reactors		
Gaseous: mainly ⁴¹Ar, ⁸⁵Kr, ¹³³Xe	400 000	240 000
Liquid: mainly tritium	800	100 000
others	600	2 000
Reprocessing plants		
Gaseous: mainly ⁸⁵Kr	1 000 000	14 000 000[4]
tritium	12 000	50 000
others (mainly ¹⁴C	250	600
Liquid: tritium	54 000	1 000 000
other emitters	183 000	20 000
alpha emitters	1 600	1 800

Notes
1. Assumes release of all arisings
2. For Oxide fuel throughput of 1200 t/y, consistent with a maximum 'reference case' of 40 GW(e) capacity by year 2000
3. Whether materials would actually be discharged at the levels indicated is conditional on the terms of the licence from the authorizing department at that time.
4. Plant for krypton removal may be installed by year 2000, which would reduce krypton-85 emissions

Before discharge, gaseous effluents are filtered to remove entrained radioactive dusts and aerosols. The filters are replaced periodically and require disposal, usually as part of the medium-level solid waste.

4.2 Low-level solid wastes

A large quantity of miscellaneous trash from laboratories, reprocessing plant and reactor operation areas, consisting of clothing, paper, equipment, scrap metal, tools, etc. is often only suspect or slightly contaminated with radioactivity. Wastes are monitored and those containing less than 20 mCi/m³ of alpha-activity and 60 mCi/m³ of beta-activity, with a surface radiation dose-rate less than 0.75 rad/h, are disposed of by burial in shallow trenches under a minimum cover of 1 m of soil. BNFL has a disposal site for such low-level waste at Drigg, a licensed fenced site near Windscale. This has the capacity to receive low-level waste for a number of years though a second site may be required next century. The safety of the site has been endorsed by the Department of the Environment following a detailed survey. A similar site is operated on the UKAEA Establishment at Dounreay. Clearly, the main requirements for long-term safety are that the quantity of long-lived radioactivity must be strictly limited, that a careful survey is made to ensure that radioactivity leached by groundwater cannot enter drinking water and that the site remains under suitable control until regulatory authorities deem it safe to release it for unrestricted use.

Other solid wastes of slightly higher radioactivity are disposed to the deep-ocean floor, some 800 km off Lands End in an internationally agreed area and within the consultation/surveillance arrangement of the Nuclear Energy Agency of the OECD. The UK operation is conducted according to conditions laid down by the Ministry of Agriculture, Fisheries and Food jointly with the Department of the Environment. The wastes are cast into cement in steel drums, using internationally recommended package designs and with radioactivity limits in conformity with the definition and recommendations for disposal to sea of radioactive waste made by the International Atomic Energy Agency under the London Convention.[4] The annual disposal by the UK has recently averaged 1000–3000 tonnes gross incorporating up to 300 tonnes of solid waste. Such a consignment would contain some 500–1000 Ci of alpha acitivity, about 50 000 Ci of tritium, and 20 000 Ci of other beta emitters.[2]

Although it is likely that these packages will retain their integrity on the sea floor for many years, this is not regarded as vital to the safety of disposal. The prime function of the package is to ensure that the waste reaches the sea floor intact. The safety of the method lies in the vast dilution of the activity as it disperses at the bottom of the ocean. The quantities of activity are very small indeed compared with the natural radioactivity of the ocean and there is no evidence of exposure to man following these disposals. Radiologically, the disposal of waste in this category to sea seems to be the best option to avoid risks due to unnecessary handling and prolonged storage. The present level is well below that allowed in the London Convention and looking forward it seems sensible and safe to consider increased disposals to sea, to include (with international agreement) bulky articles, such as pieces of decommissioned plant, following suitable treatment.

The annual limits of radioactivity to be disposed to the sea-bed agreed internationally are based on very conservative safety factors. The Fisheries Laboratory, Lowestoft, of the Ministry of Agriculture, Fisheries and Food, have some 30 years' experience of these, and the Institute of Oceanographic Sciences has joined in investigating the possible fate of radioactivity dispersed on the sea-floor; this will provide an even firmer basis for defining safe limits for a disposal area without recourse to arbitrary imposition of large additional factors.

4.3 Intermediate-level solid wastes

A number of waste streams arise at various points in the fuel cycle whose activity is lower than the highly active wastes but higher than those that are presently consigned to sea. In general these streams are treated in such a way that the bulk of the material can be discharged safely to the environment as low-level waste while the residual material, containing most of the activity, is retained in a concentrated form. These concentrates, and other intermediate (medium) active arisings include:

(1) air filters from treatment of gaseous wastes,
(2) ion-exchange resins from water-reactor coolant clean-up,
(3) ion-exchange resins and sludges from fuel element pond water clean-up and other liquid effluent treatment, and
(4) large items of contaminated equipment, such as plant components, reactor components, fuel transport flasks, etc.

Wastes of this category arising from reactor operations are usually contaminated mainly with neutron activation product isotopes, together with some fission products. Long-lived actinides are only rarely involved. Generally the main problem they present is their bulk rather than the level of activity. Some wastes in this category, particularly those originating from the fuel element pond water treatment plant and from reprocessing plants, can however contain considerable fission product and actinide activity.

Processes have been developed on a laboratory scale, and are being further developed, for reducing the volume as far as is practicable or necessary, immobilizing the activity, and packaging the waste. The main techniques that have been employed so far are incorporation into cements or bitumen, or combinations of the two. Thermosetting resins and polymer impregnated cement are also likely to be increasingly used. To protect the immobilized waste during handling and transportation it is set in steel drums in each case.

Cement has the advantage of being cheap, non-combustible, and providing shielding against external exposure. It is adequately strong and stable to radiation. On the other hand, wet sludges are not readily incorporated to give a high loading, in which case the final volume is greater than with bitumen or resins. The ability of cements to resist leaching by groundwater varies with the radionuclides. Some are strongly retained, e.g. the actinides, but some such as caesium are leachable. Improved cement mixes and impregnation with plastic polymers are measures that can be used to improve the resistance of cements to leaching.

Bitumen has been extensively used for treatment of waste in Belgian, French, and German establishments. The waste material is mixed with hot, fluid bitumen which is then allowed to cool and set. Its advantages over cement are its larger capacity (meaning a smaller volume) and its lower weight, which is advantageous for transport. Additionally, it shows greater resistance to leaching. However, it provides only limited radiation shielding and its combustibility and propensity to flow when hot requires careful attention to fire prevention in production, transport, and storage/disposal. Experience on incorporation into thermosetting resins is limited at present to pilot-scale work. Bitumen could be very well suited to certain materials such as ion-exchange resins that are difficult to incorporate satisfactorily in cement.

4.4 Fuel element cladding waste

Nuclear fuels used for electricity generation consist of uranium metal or pellets of uranium dioxide, or in the case of fast reactors pellets of uranium and plutonium oxide, clad in thin-walled tubes. The cladding waste is the second most highly active waste. It consists of scrap magnesium alloy from Magnox fuel cladding, zircaloy (a zirconium alloy used in water reactor fuels), and stainless steel (from AGRs and FBRs), which remains when the fuel is dismantled. It is highly alpha and beta–gamma radioactive due to

(1) neutron activation of constituents of the alloys (e.g. iron-59, nickel-63, cobalt-60) and

(2) adhering fragments of fuel, containing uranium, plutonium and fission products.

Typically after the spent fuel from a thermal reactor has been stored for 1 year, the activity of the neutron-activation products in the cladding waste amounts to about 10 000 curies/tonne of fuel processed, and fission-product activity amounts to about 2000 curies/tonne, assuming that 0.2 per cent of the fuel remains in the cladding.

Processes are under development for the removal of these fragments of fuel from the cladding waste, enabling the recycle of the uranium and plutonium and the removal of the fission products to the highly active liquid waste. Treated cladding will still be highly active from the activation products, some with fairly long half-lives, e.g. ^{63}Ni. If the removal of the fuel fragments has been thorough, there will be no significant amount of long-lived actinides in the waste; however, it is probable that some actinide contamination will remain. Hence at present, this waste should be included with the *intermediate-level wastes with some long-lived activity*. It will be compacted mechanically, and in common with other intermediate-level wastes, will be immobilized in a suitable matrix and packaged for disposal. Cement is being considered, as well as lead alloys (being researched in Belgium) and glasses (France).

4.5 Plutonium-contaminated solid wastes with low beta–gamma activity

A considerable volume of miscellaneous solid wastes, both combustible and non-combustible, arise at the tail end of the reprocessing plant, where the separated plutonium is concentrated and converted to its solid oxide. Similar sorts of waste arise in the plant for making uranium–plutonium oxide fuel. These wastes include rubber gloves, swabs, paper, plastics, glassware, ceramics, metal tools, pieces of equipment, and glove boxes. They also include chemical residues from the treatment of liquid effluents and ventilation filters from plutonium plants.

These wastes differ from those discussed in §4.2 in being contaminated with plutonium, thus presenting an alpha-contamination hazard. Against this, the content of beta–gamma emitters is low. Hence the hazards of this material arise from inhalation or ingestion, not direct radiation. The toxic hazard is very long-lived owing to the long half-lives of the actinides.

Processes are under development to recover the plutonium as far as economic or environmental considerations require it, and to immobilize and package the waste for disposal, with such volume reduction as is possible. Combustible wastes may be incinerated to reduce the volume and produce an ash from which the plutonium may be recovered and returned to the fuel cycle.

The ash, after extraction of plutonium, can be immobilized by incorporation in cement or ceramic compositions. Non-combustible items would be treated to remove or fix any loose contamination and packaged for transport to a disposal site.

4.6 Low-level liquid wastes

Typically these consist of tritium in liquid wastes from nuclear power stations, treated water from fuel storage ponds, the low-activity streams from reprocessing plants, laboratory waste, and active laundry waste.

Radioactive liquid wastes are treated by various means. Methods include (1) ion exchange, (2) filtration, (3) reverse osmosis, and (4) flocculation. Reverse osmosis is a process essentially akin to filtration, involving the use of extremly high-quality membranes for the removal of dissolved salts from solution. Flocculation (the process conventionally employed in water treatment plants) involves the removal of solutes by precipitation, a *floc* agent being added to the precipitate to improve its cohesion, and to facilitate the separation of liquid and solid by filtration.

Following treatment, low-active effluent is discharged within the limits permitted under the authorizations. The standards required differ according to the assessment of safe limits for each particular site.

The treated effluents from power stations are discharged into the sea or estuaries, except in the case of Trawsfynydd, where the effluent enters a fresh-water lake. Effluents from the reprocessing plant at Windscale are discharged 2 km out from the Cumbrian coast into the Irish Sea via a pipeline. Effluents from the UKAEA establishments at Dounreay and Winfrith are similarly discharged. Effluents from the uranium fuel factory at Springfields are discharged into the tidal reaches of the Ribble Estuary.

The safety of the disposal relies on the principle of dispersion and dilution, and the limit authorized by the departments responsible for the environment is defined on the basis of a careful assessment of all potential mechanisms for reconcentration of critical radionuclides, the pathways back to man, and the consequent possible radiation dose received by the *critical group* (i.e. the most highly exposed members of the population).

As the scale of operations increases, the overall efficiency of the effluent treatment plants may need to be increased to maintain the discharge within the current authorization, and possibly to meet any reduction in authorized levels that might be decided. Table 15.3 shows the amounts discharged in 1976 and a forecast is given of the liquid effluent that might be discharged in the year 2000 from a plant reprocessing 1200 t/y of oxide fuel at an irradiation level of 37 000 MW d/t and cooled for one year prior to reprocessing. This is roughly consistent with the needs of a nuclear power programme of 40 GW installed electrical capacity in the year 2000 (the *high* projection for the UK, as submitted to INFCE Working Group 1.[5]

4.7 **Highly active liquid waste**

This is the liquor remaining after the fuel has been dissolved in nitric acid, and the uranium and plutonium have been extracted. The process is highly efficient and the liquid contains 99 per cent of all the remaining non-volatile fission products from the dissolver liquor and almost all of the higher actinides from the irradiated fuel. It also contains small amounts of unextracted uranium and plutonium; the quantities of these depend on the efficiency of the process but typically they are about 0.1–0.5 per cent of the original contents in the fuel. The liquor also contains some inactive constituents derived from dissolution of small amounts of fuel cladding and corrosion products from plant vessels. The UK practice has always been to use separation processes that avoid addition of large quantities of inactive reagents since that would increase the mass of solids in the waste and would interfere with the evaporation of the liquid. The aim is to reduce the volume of wastes in this category arising from thermal reactor fuels to the order of 10–20 m^3 per gigawatt-year of electricity generated.

One of the features of this waste which requires consideration at every stage of the process is its very high heat output. The basic sequence of operations in the planning for the long-term management of this waste adopted in the UK and most other countries with nuclear power programmes is:

(1) interim storage in liquid form,
(2) solidification and packaging after a period of cooling,
(3) engineered storage of the solidified waste for further cooling, and
(4) disposal of the solidified waste.

4.7.1 *Tank storage of highly active liquid waste*

In the UK the evaporated liquid waste is stored on the site where it is produced, in high-integrity stainless steel tanks contained in concrete vaults. The tanks are cooled by water circulation to remove decay heat. The concrete vaults are also lined with stainless steel to provide a further line of containment in the improbable event that the primary containment should fail. Provision is made for transfer of the contents of a tank to one of several spare tanks that are always kept available should the need arise. At present, some 990 m^3 of highly active waste are contained in 12 tanks at Windscale; this represents the accumulation from virtually our entire nuclear programme, both defence and civil over the past 25 years. Liquid wastes from reprocessing fuels of materials-testing reactors and the Dounreay fast reactor are stored in a similar way at Dounreay. Similar tanks are used in other countries for waste from civil nuclear fuel, and experience has demonstrated that the practice is very safe.

Storage of the waste as liquid has the advantage of operational flexibility and leaves open the widest range of options for future treatment and disposal, but the practice can be regarded as no more than an interim measure. Although experience has shown that the practice is safe, the tanks and cooling systems require supervision and eventual replacement, and there is a wide international consensus that these liquids should be converted to a solid. This would:

    (1) reduce their mobility in the event of an accident,

    (2) require less supervision,

    (3) enable the wastes to be transported for disposal,

    (4) enable some further reduction in volume to be achieved, and

    (5) be more economic.

4.7.2 *Solidification of highly active liquid waste and properties of the solid*

Solidification of the liquid waste immediately after production would at first appear to be desirable, since it offers potentially more safety in storage and a possible economy by elimination of the need for liquid-storage tanks. However, there are two reasons why it will be desirable to retain some liquid storage. Firstly, it provides a buffer stock to enable the solidification and reprocessing plants to be operated flexibly at a higher overall efficiency. Secondly, a period of cooling permits the use of larger glass blocks without exceeding the desirable centre temperature limits. A careful balance must be struck between the competing factors.

The waste needs to be converted to a solid whose properties, in conjunction with the characteristics of the selected disposal system, will ensure that the radioactive constituents remain isolated from man's environment. Ideally, the requirements of the solid form are that it should:

    (1) have a good capacity to accommodate all the elements in waste and cope with variations in waste composition;

    (2) be capable of being made safely and reliably in a remotely operated and maintained plant;

    (3) have a good resistance to leaching by water since this controls the rate at which radioactivity could disperse;

    (4) have a high thermal conductivity to dissipate the heat produced by radioactive decay;

    (5) have good mechanical integrity; and

    (6) have good resistance to radiation damage.

Vitrification—i.e. the incorporation of radioactive waste oxides into glass[6,7]—has been satisfactorily demonstrated as meeting all the requirements indicated above, and work on the development of industrial-scale processes for the manufacture of suitable borosilicate glasses is at an advanced stage in the UK, France, the Federal Republic of Germany, the USA, Japan, India, Canada, and the USSR. Work is being carried out in parallel on the study of environmentally significant properties of such glasses.

Borosilicate glasses can accommodate a wide range of elements because a glass has no very definite structure. Compositions have been found that will incorporate all the non-gaseous radionuclides and *tramp* elements in waste, and which can contain up to 25 per cent by weight of the waste oxides. The melting-points of glasses are in the range 900–1000 °C, which is within the experience of existing glass-making industry and acceptable in the design of radioactive plants.

The selected glasses have proved very resistant to effects of heat and radiation. All glasses are subject to a phenomenon known as devitrification (i.e. reversion to small crystals) when held at elevated temperature in the dry state for long periods. The borosilicate glasses that have been developed do not recrystallize readily and, when induced to do so by retention for a long time above 500 °C, the crystalline phases produced do not significantly reduce the strength in the glass block, and in some cases may indeed increase it. The block remains coherent and shows no change in the leach-resistance.

No effects on the glasses have been found in experiments to give the full lifetime dose of beta–gamma radiation, which would, in any case, not be expected to be particularly damaging. The most damaging radiation would be that due to the alpha decay of actinides. Specimens of glass, which have included the very radioactive plutonium-238, have now accumulated an accelerated radiation dose equivalent to that which would accrue over 100 000 years from Magnox waste, with no change in physical properties and no more than a halving of the leach resistance.[9]

However, two recent French and Australian papers[11,12] both claim to have found evidence to suggest that there is a critical radiation dose, beyond which the resistance of the glass to attack by water deteriorates sharply.

Obviously, it is essential that any evidence concerning a matter as important as the long-term properties of waste be properly examined. The papers report on experiments carried out by bombarding inactive glass samples externally with beams of energetic charged lead and argon ions to simulate the effects of alpha decay of atoms within the glass. However, the particles used have low penetrating power and result in damage which is concentrated in a surface layer about 5 and 0.25 millionths of a centimetre thick for the respective lead and argon ions.

In contrast, the method of simulation favoured by research workers at Harwell[9] and other laboratories in Germany[13] and the USA,[10] is to incorporate alpha-particle-emitting radionuclides into the glass as mentioned above. This irradiates the whole sample internally and gives an exact simulation of what will occur in practice, but on a greatly reduced timescale. Calculations suggest that the dose found necessary in the Australian paper to affect borosilicate glass adversely will never be reached in the lifetime of the radioactivity incorporated in it. Moreover, the critical dose suggested in the French paper has been surpassed in experiments at Harwell with no significant effects. The work described in these two papers[11,12] does not therefore cast doubt on present ideas for incorporating radioactive waste into glass.

Many of the results on leaching of glass relate to rapid tests in a large excess of boiling distilled water to provide data to assist the developments of suitable glass compositions. These tests are not typical of conditions that might be expected in a final waste repository where the flow of water might be restricted, and many other tests including the effects of temperature, water composition, and rate of water movement are being conducted. The leach resistance of the UK borosilicate glasses is very good, and by any normal standards the material would be regarded as insoluble. Over the practically significant range of water temperatures, from 10 °C to 100 °C, the quantity that dissolves per square centimetre of exposed surface is between one millionth and one ten thousandth of a gram per day.[9] These rates are similar to the dissolution rate of some granites and basalts. As has been pointed out by a number of workers, at higher temperatures, say 300 °C, water under pressure attacks the glass quite rapidly, and the block could be disintegrated in a matter of a few months. However, there is no question of allowing waste glasses to be subject to such conditions as will be discussed further in the section on disposal.

A number of other waste forms have been or are being studied in various laboratories.[7,8,10] These include ceramics, glass-ceramics, composite materials such as glass pellets embedded in a metal matrix, and synthetic crystalline minerals. The purpose behind these studies is to seek even greater thermal stability and resistance to leaching. One interesting concept being studied in Germany, Belgium, and the USA is the incorporation of beads of glass or ceramic oxide containing the waste in a metallic lead matrix. The metal would provide an additional leach-resistant barrier, and have good thermal conductivity and thereby facilitate dispersal of heat. Another attractive idea being studied at the Catholic University in the USA is the preparation of high-silica porous glass beads which are soaked in liquid waste. The beads are then fired to immbolize the activity. The active beads are then coated with an inactive layer of silica and refired. The leach resistance of the product is extremely high, as would be expected of high-silica glasses which have high melting-points. Yet another concept, pioneered at Pennsylvania State University by Roy and co-workers is the conversion of wastes to ceramic forms composed of highly insoluble crystalline substances. Ringwood's SYNROC[11] is a further development of this idea, in which the waste elements would be incorporated into a mixture of three crystalline structures which are known to be stable entities in many ancient rocks.

Some of these materials would be very attractive matrices for containing the elements of high-level waste, and several of them have very good resistance to leaching or hydro-thermal alteration. However, none of these concepts has been taken beyond the laboratory bench scale, and often not with real waste. Furthermore, it is not known whether any of them can meet satisfactorily all the other requirements mentioned at the beginning of this section. In particular, all of the concepts involve considerable process complexity and, some of them, the use of temperatures and pressures considerably higher than desirable in a production plant handling radioactive materials. It is possible that one of them may supersede glasses eventually and the possibilities should continue to be explored, but it would be mistake to delay the solidification of wastes by a good technique in the hope of having a better one.

4.7.3   *Industrial-scale conversion of highly active liquid waste to glass*

The feasibility of vitrifying highly active liquid waste is now well established on an industrial scale. The processes under development in various laboratories take a number of different forms. Essentially, these are different engineering approaches to the same type of end-product. These developments range in status from laboratory bench-scale active work to large-scale inactive and, in one case, industrial-scale fully active operation. Two types of processes will be outlined here.

The first glass-making process was developed on the semi-pilot scale in Canada.[7] Radioactive waste was fused with a nepheline syenite (a naturally occurring silicate mineral) to produce a glass, samples of which have been buried and their leaching behaviour studied since 1960. Work was stopped when no fuel processing was contemplated in the Canadian programme.

In the UK FINGAL (fixation in glass of active liquid) was developed and operated from 1962 to 1966 at Harwell on a pilot-scale plant using up to 15 000 curies of Magnox waste per batch. Naturally, the shielded cell in which the plant was situated became known as FINGAL's Cave. The main feature of this process is that the cylindrical vessel in which the glass is manufactured is also used as the final storage container. This type of process has become known as *in-can melting*. The liquid waste together with glass-forming materials are fed continuously into a heated stainless steel vessel in which the mixture fuses to a homogenous glass. When the vessel is full, it is removed from the furnace, cooled, sealed, and becomes the storage vessel. A unique feature of FINGAL was the method used for trapping active dust that is carried out of the process vessel with the off-gases. These passed through a filter installed in a second stainless steel cylinder. This cylinder, with filter was used in the next glass-making cycle so that the filter was incorporated in glass.

The FINGAL plant produced blocks of glass about 50 kg in weight. In the early 1970s, British Nuclear Fuels in collaboration with the UKAEA began the development of HARVEST (highly active residues vitrification engineering study) as a process suitable for routine industrial operation. The principle of HARVEST[7,15] is shown in Fig. 15.4. It is based on the FINGAL process, but the scale is increased and the internal dust filter omitted, the dust being collected in a separate condenser system, from which all radioactivity would be concentrated and recycled.

Both FINGAL and HARVEST are characterized by the emphasis on engineering simplicity. They are batch processes, with no moving parts and the glass-making vessel has to endure only one cycle at temperature—the most demanding stage in terms of potential for failure during operation.

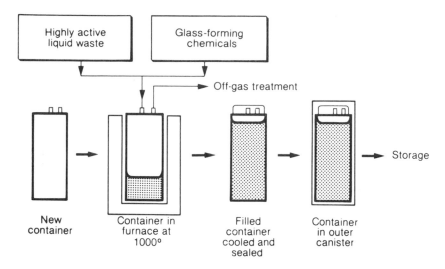

Fig. 15.4 Principle of the HARVEST process

In France, a very ambitious development has been brought to fruition with the successful operation of the AVM (Atelier de Vitrification Marcoule) plant[14] at Marcoule. This has been operating on a commercial scale since June 1978, converting long-stored Magnox wastes to glass. By April 1979, 143 containers had been produced, holding a total of 48 tonnes of glass prepared from liquid wastes arising from reprocessing of 2500 tonnes of uranium metal fuel. The AVM process (Fig. 15.5) makes glass in two stages continuously. In the first stage, liquid waste is dried to a free-flowing powder in a rotating kiln, and in the second, the dried waste is fed, together with glass powder, into the glass-making furnace where the mixture is melted. The molten glass is cast periodically into the storage container, which is cooled and closed with a welded lid. Thus, the engineering is more complex, and the glass-making furnace has to last for much longer periods than in the *in-can-melting* processes. However, the advantage is that the type of plant has greater potential for scale-up to a larger production rate per unit.

During the next decade or so, several countries plan to bring fully active plants into operation. The French AVM process is broadly accepted within the reprocessing industry internationally as an established prototype pro-duction operation, and it is expected to be scaled up in plant size for operations at the French Cap la Hague reprocessing site. HARVEST has reached the stage where it could be an alternative industrial process, although in terms of radioactive plant development it is some years behind. Further development would not alter the basic characteristics of HARVEST, namely that it is a batch process in which the size of the glass storage vessel, once chosen, cannot easily be varied. AVM is more flexible in this respect. British Nuclear Fuels, who have a technical liaison agreement with the French CEA, have announced that their first glassification plant at Windscale will

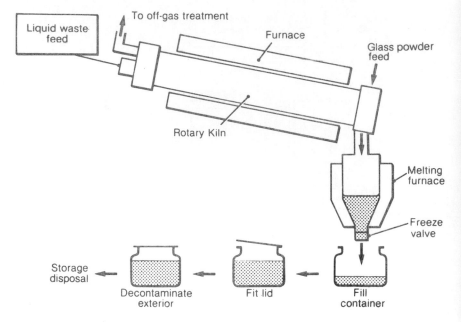

Fig. 15.5 Principle of the AVM process

probably be based on the AVM process. The availability of information on French technology, together with earlier experience in the UK on the HARVEST process development, will enable provision of the first plant to be brought forward by a few years.

5 Storage of treated waste

5.1 Volume of waste for disposal

From the foregoing, it can be seen that methods are available and plants can be in operation, within the next decade or so, to convert all types of waste into solid forms, in containers appropriate to the level of activity in each category of waste and suitable for the disposal method eventually chosen. The cumulative quantities of solid waste of all categories that would arise in the UK by the year 2000 from a nuclear programme with an installed capacity of 40 GW(e) in the year 2000 is summarized in Table 15.4. This gives the approximate volumes of wastes treated and packaged using the techniques outlined in §4 of this chapter.

The volumes of waste listed in Table 15.4 are very small when one considers that they are the output from the cumulative generation of 400 gigawatt-years of electricity, enough to meet the electricity needs of 30

Table 15.4 Indicative cumulative nuclear wastes arising in the UK[1] by year 2000

Waste category	Volume after treatment (m³)	Activity (Ci) alpha	Activity (Ci) beta–gamma	Characteristics	Possible disposal options
1. Highly active waste (as glass blocks)	900	2×10^7	4×10^9	High initial heat output Long-lived actinide activity	1. Several decades retrievable air-cooled storage followed by 2. deep geological or deep ocean emplacement
2. Intermediate-level waste with significant actinide activity:					
(i) fuel cladding and associated associated residues	21 000	6.8×10^5	9×10^7		Deep geological or deep ocean emplacement
(ii) sludges and resins from liquid effluent treatment	3 000	5.4×10^4	1.4×10^6	Low heat output Long-lived actinide activity	No prior need for cooled storage
(iii) plutonium–contaminated wastes[2]	12 000	1.5×10^5	4.5×10^6		
Total[3]	36 000	8.8×10^5	9.6×10^7		

Table 15.4 (*cont.*)

3. Intermediate-level wastes with no significant actinide activity:						
(i) misc. fuel element components ex reactors	5 000	—	—	3.6×10^8	Low heat output Very little long-lived activity (1000 y)	Emplacement in excavation at intermediate depths
(ii) discarded plant and reactor components	10 000	—	—	6.5×10^7		
(iii) sludges, resins, etc. from liquid effluent treatment	11 000	—	—	1.7×10^6		
Total	26 000	—	—	4.3×10^8		
4. Low level solid waste[4]	600 000	low	low		Suitable for shallow land burial	Burial in shallow trenches

Notes

1. This assumes a nuclear capacity of 40 GW(e) in the year 2000 (the high reference case, Ref. 5). All numbers should be taken as indicative, since the final figures will depend on actual technological methods used.
2. The volume of category 2(iii) could be reduced about three times by incineration of combustible waste and compaction of non-combustible waste.
3. Figures given assume no special measure used to remove the plutonium and americium from category 2 wastes. If removal to the maximum extent practicable, alpha-activity would be reduced ten times and beta–gamma activity three times. Recovered plutonium would be recycled and americium transferred to category 1.
4. Untreated volume could be reduced about eight times if incineration and compaction were used.

million people for 30 years. To give a sense of perspective, the production of 1 gigawatt-year of electricity leads to the production of about 4 m³ of high-level waste glass in about 10 containers, containing more than 95 per cent of all the activity. The total volume of all the other categories of waste, including the low-level solids, is about 1800 m³. (A modern coal-fired station delivering the same amount of electricity produces 0.5 million tonnes of ash with a volume of 400 000 m³.)

The low-level solid waste, by far the largest volume category in Table 15.4, will not accumulate since it can be baled, with some volume reduction, and disposed of as it arises by burial in shallow trenches, or incorporated in concrete and disposed to sea as described in §4.

The levels of activity in the other four categories require a much greater degree of isolation from man's environment. At present no disposal routes providing this degree of isolation are available, and until they have been developed and approved by the licensing authorities these wastes must be held in stores under supervision. The design and operation of suitable stores poses no novel technical or safety problems, and the length of time for which the wastes are held in store before disposal is a question of the chosen strategy for final disposal and of economics. The requirements differ between the categories in this respect as will be discussed in §5.2 and §5.3

5.2 Storage of containers of highly active waste

An essential step in the long-term management of the containers of glassified highly active waste is likely to be storage, at least until the heat emission associated with radioactive decay has fallen to a level that will not impose major problems if the waste is finally disposed into a suitable repository, which can be sealed and which would require no further surveillance.[16]

As an example, the reference design for the HARVEST process envisages the production of cylindrical blocks of glass in stainless steel vessels about 0.5 m diameter and up to 3 m long, containing 1 tonne of glass of which 25 per cent would be waste elements. This would accommodate the waste from the production of 0.11 GW y of electricity. A typical vessel in the inactive experimental plant at Harwell is shown in Fig. 15.6. The initial radioactive heat emission from such a block, manufactured four years after the spent fuel had been withdrawn from the reactor, would be about 10 kW, but this would decrease by an order of magnitude over the first 75 years of storage. The dissipation of this heat is one of the major considerations in the design of stores and final repositories to avoid damage to the glass, its container or the repository.

Initially, the cylinders of glassified waste will be stacked in a pool, rather like a swimming pool, or in concrete vaults with forced air or natural convective air cooling, as used now by the French for waste containers from the AVM plant. The purpose of the water in the pool, apart from providing cooling, is to shield man from the intense beta–gamma radiation. In the case

Fig. 15.6 HARVEST waste container in the Harwell inactive pilot plant

of concrete vaults, the thickness of the concrete provides the necessary shielding. In either case the technology is very simple, and the designs can draw on much experience of storage of spent fuel elements.

The approximate centre temperature of a HARVEST block stored in a cooling pond as described above would decrease with time as shown below. The surface of the block would at all times be a little above the water temperature of about 30 °C. The maximum temperature of the glass would always be well below that at which recrystallization (devitrification) would occur. The relationship between time of storage and temperature is as follows:

Time of storage (years)	At manufacture	10	20	50
Centre temperature (°C)	270	110	75	55

It can be argued that, radiologically, a period of storage in a simple, cooled store is the safest course to follow. The store can be constructed above or below ground. The activity is extremely well contained in a heavy inert solid in a corrosion resistant container, most unattractive to terrorists or saboteurs and very difficult to disperse. The probability of activity getting back to man is extremely small. While in the store, the containers can be retrieved for examination and over-clad if necessary with another corrosion resistant container designed to meet disposal conditions. Simple storage of this type requires only minimal supervision and can certainly be maintained as long as society is capable of industrial operations.

5.3 Storage of intermediate-level wastes

The wastes, such as fuel element cladding, the miscellaneous intermediate-level active solids, and plutonium-contaminated solids, that emit only a little or negligible heat but whose total activity or long half-life require long-term isolation from man, can be conditioned and safely stored indefinitely if that is what society wants. Alternatively, in principle, they can be sent for final disposal as soon as practicable after they have been suitably immobilized and packaged. There are few technical or environmental advantages in a deliberate policy of storage prior to disposal, as in the case of the heat-generating, high-level waste.

Indeed, because of the relatively much larger volume of this category of waste it might be much better housekeeping to dispose of it as it arises in future to avoid the construction and maintenance of unncessarily large storage facilities. At present, all of this waste is held in various stores, such as silos, tanks, drums, in non-immobilized condition. No route for final disposal has been defined at present, except for the small fraction that is currently included in the annual sea-disposal operation. As soon as the route is defined, decisions can be taken as to the appropriate method of immobilization and packaging. Some expert opinion is strongly in favour of immobilization being carried out even before the disposal route is known, to increase the safety of wastes in storage.

The main requirement of an intermediate waste store is a relatively simple building in which the containers can be kept dry, free from fire risk and with suitable shielding from gamma radiation such as concrete walls.

6 Final disposal of waste—general

In the previous section, it has been argued that retrievable storage of glassified highly active waste in suitable containers is a very safe method of management and could be conducted indefinitely. There remains the possibility that people sometime in the future may want to consign this waste to somewhere remote from the surface of the earth, rather than continue with even the minimal monitoring of wastes in stores. Although this is not a requirement that has been placed on any other industry, it is incumbent on the generation that has developed nuclear power to show that there are options other than indefinite storage of the waste which could be employed if and when required. Indeed, one of the recommendations of the Royal Commission on Environmental Pollution in its sixth report was that at least one technique for final disposal, not requiring any further human surveillance, should be demonstrated.

Thus, the objective in disposal would be to place the waste in a situation such that, with no further action by man, the radionuclides cannot return to his environment in concentrations regarded as hazardous. The criteria for safety have not yet been formally ratified. One possibility is to assume that it will be required to limit any dose to man resulting from released radionuclides at any time in the future to the current ICRP-recommended limits, but there are obvious difficulties in taking any criteria to very long times in the future.

We will deal first with conceptual schemes for disposal of highly active waste, since this has received most attention to date and is the most challenging problem. Then comment will be made on the extent to which intermediate active waste disposal is different.

Various schemes have been suggested for placing waste outside man's environment. The most comprehensive review covering these schemes, ranging from esoteric to down-to-earth, has been undertaken by the US Department of Energy.[17] Ideas that have been discarded for reasons outlined in the remainder of this section include despatching waste into outer space, destruction of radionuclides by transmutation, deposition of waste packages in the Antarctic icecap, and emplacement of liquid or solid waste into very deep boreholes, permitting the radioactive decay heat to fuse them into the surrounding rocks.

6.1 Space disposal

It has been suggested by a number of people that one could rid the earth of wastes by shooting them into the sun, or even out of the solar system. This ideas was attractive enough to merit serious appraisal by NASA for the US Department of Energy. They concluded that selected wastes could be so disposed of, using technoloy that is already in development, and identified the more likely concepts. Typically, this would be a space shuttle to lift waste packages into near-earth orbit, from which the package would be propelled into an appropriate solar orbit that would be stable for millions of years. However, costs would be high and the payload, dictated by the number of launchings possible, would limit disposal to separated components of wastes, so that a terrestrial repository system would be needed for the leftover materials. Most important of all, a fail-safe ejection system and a waste container able to survive accidental re-entry and impact would be required. The US Department of Energy do not seem to be doing more than continuing the appraisal, and there is no evidence that the scheme would receive any support at all in other countries with waste management research and development programmes in hand.

6.2 Icesheet disposal

The disposal of packages of waste in continental icesheets, in particular, the Antarctic, was suggested by a group of scientists in 1973. The advantages offered would be remoteness from human activities, the simplicity of the technology for emplacement, and the properties of ice (self-sealing through plastic flow, impermeable), which appeared likely to contain the radioactivity for thousands of years. Several concepts have been proposed, including the following.

1. The containers could be emplaced in boreholes in the icecap; the heat from the containers would cause them to melt their way down to the land surface, some 4000 metres beneath the ice sheet, the ice re-sealing above the containers. They would reach bed-rock in about 10 years.

2. Containers might be anchored below 'rafts' on the surface of the ice or placed in repositories supported above the icecap surface. Retrieval of the containers would be possible with these options for about 200 years, and eventually, new snow and ice would accumulate over them. They would reach bed-rock in about 30 000 years.

However, there are considerable uncertainties regarding the long-term stability of icesheets and their interaction with the waste. There is reason to believe that liquid water might exist at the base of an icesheet, which could transport dissolved radionuclides directly to the oceans. At present, there is very little information on which to judge the safety of the concept Furthermore, there has been international agreement (the Antarctic Treaty) not to introduce radioactivity into the area. Hence, no one is pursuing this concept of disposal.

6.3 Partitioning and nuclear transmutation

Various suggestions have been made that the long-lived radionuclides might be destroyed by nuclear transmutation processes leaving no need to consider any possible environment effects in the long term. The transmutation of long-lived fission products, except possibly technetium-99, does not appear to be remotely practicable on an industrial scale with present technology, but in principle, all the actinides could be destroyed by irradiation in neutron fission reactors in a manner parallel to that of the use of plutonium as a fuel. The actinides would be replaced by fission products in substantially the same proportions as those resulting from the burning of uranium and plutonium in our present reactors. It would be necessary, in the first instance, to separate the actinides from all wastes and fabricate them into suitable fuel elements. The chemical separation processes, the fabrication of special fuel elements, and the reactor physics of their transmutation are regarded as feasible to the extent that potential solutions to the technical problems involved have been identified. However, a very large research and development effort would be required over several decades before industrial-scale operation could be considered as practicable.

The separation of the 'waste' actinides (neptunium, americium, and curium) from the other waste to the required degree calls for the use of additional, very complex separation processes, different from those already well developed for uranium and plutonium. This would produce its own set of wastes.

At a recent international meeting[18] the results of studies in several laboratories of radiological benefits of the practice were presented. The conclusions were that the reduction in risk in the long term was marginal, compared with the small risk associated with disposal into suitably selected geological formations, and the cost would be very high. This small reduction in long-term risk would almost certainly be offset by increased risk to the present population by increased exposure of operators and by a proliferation of low-level wastes. Hence it was generally agreed that there is no safety incentive to pursue the concept.

Three practicable options are now generally recognized and in the UK are under active consideration following the Government white paper (*Cmnd.* 6820) on *Nuclear power and the environment.*[19] These options are the emplacement of solid waste in suitable containers:

(1) in deep geological formations under the land,
(2) *in* sediments on the deep ocean floor, and
(3) *on* the deep ocean bed.

It is the present UK government policy to research the feasibility of all three options for disposal to the point where a choice can be made and one of them developed fully for use by the end of the century, if the government of the day so wishes. Research on the land-disposal option has progressed much further than the ocean-disposal programmes. However, since about 1978, studies relating to the safety of deep-ocean disposal of high-level waste has accelerated in several countries, including the UK.

7 Disposal in geological formations under the land

The concept of *geological disposal* has been accepted internationally in principle as a practicable, safe, long-term method of nuclear waste management. The philosophy behind this acceptance is that many regions of the earth's surface, including many parts of the UK, have been stable for extremely long periods, running to hundreds of millions of years. It is expected therefore, that suitable encapsulated waste, emplaced deep into a carefully chosen geological formation, can be effectively isolated from the biosphere.

Given that a repository is constructed in such a stable rock formation and sealed off, there is no way in which radionuclides can reach the surface through natural processes, except by dissolution of the waste and transport in ground water. Therefore the water content of the rock, its rate of movement, and its isolation from surface water are some of the more important considerations.

Guidelines which might be used within the UK in the selection of areas containing geological formations potentially suitable for the location of sites for disposal of high-level, heat-emitting radioactive wastes have been published by the Institute of Geological Sciences.[20] The aim is to select a rock formation and area that meets the following criteria:

(1) sufficient thickness and area to house the repository, and at sufficient depth to isolate it from the surface,

(2) simple and determinable hydrogeological conditions, low hydraulic conductivity,

(3) long geological history of stability, away from active faults and abnormally high seismic activity,

(4) adequate thermal conductivity,

(5) suitable chemical characteristics, high ion adsorption or exchange capacity, adequate thermal stability,

(6) absence of potential resources that might encourage man's intrusion,

(7) absence of previous mining activities, and

(8) suitability for mining, construction, and operation of a repository for the required period and, subsequently, for being back-filled and sealed.

7.1 Preferred rock formations

The types of geological formation that have been considered [21] [22] for the containment of high-level waste are (1) salt, (2) argillaceous deposits such as clays, shales, and slates, and (3) crystalline hard rocks such as granites and gneisses. Each has its advantages and disadvantages as a location for a repository, and these are described briefly in the following numbered paragraphs.

1. *Salt.* Salt formations are many millions of years old and their very existence indicates the absence of water-flow. Salt is a relatively good conductor of heat, which is advantageous in heat dissipation. Under the pressure of the overlying rock, salt *creeps* so that any fissures produced by earth movements are self-sealing—the feature that has prevented the entry of surface water. This property also ensures that salt will rapidly close round an emplaced hot-waste container, thereby isolating it effectively from moving ground water. Salt is a mechanically strong rock and easily sustains un-supported cavities, avoiding the need for roof supports in excavation. The water content is not absolutely zero, and small quantities are associated with small inclusions of brine in salt crystals and with certain hydrated minerals such as carnallite ($KCl.MgCl_2.6H_2O$). It has been observed that brine in-clusions can migrate under the influence of a temperature gradient, which might cause the accumulation of concentrated brine round hot-waste con-tainers. One part of the present EEC research programme in the Federal Republic of Germany is aimed at finding the extent to which this would occur and the consequences. It seems likely that problems from the formation of free water by the dehydration of carnallite can be avoided by careful surveying of the salt formation to ensure that waste is emplaced in a region where this mineral is absent. Salt formations of potential interest occur in two very different ways: deposits more or less horizontally bedded, and salt 'domes' which have been formed when earth movements have squeezed or folded bedded salt deposits upwards into structures often kilometres in

extent and more than 1000 m in depth. In Germany and Holland salt domes are plentiful, and in the USA both salt domes and vast thick bedded formations exist, and hence are receiving the greatest attention in these countries. In the UK there are no salt domes under the mainland, and the bedded salts are in the main not thick enough to house a repository. Thicker deposits exist, but these are frequently associated with potash deposits that would constitute a resource that might be exploited. Some deposits of alternating salt and shale layers of potential interest occur in Cheshire and Somerset.

2. *Argillaceous formations (clay and clay-based rocks).* These are sedimentary rocks varying from unconsolidated, plastic clays to the harder rocks such as shales and slates produced when clays have been under great pressure for long periods. The main attraction of these is their very large ion exchange capacity which enables them to adsorb radioactive species leached from the waste by flowing groundwater. The plastic clays are self-sealing, but the shales and slates are prone to crack and fissure if subject to earth movements. The rate of water movement through these formations is relatively low and virtually zero in plastic clays. However, the quantity of water inherent in their constitution can be considerable. It is greatest in clays and progressively less in the shales and slates. It is the behaviour of this water under the influence of heat that constitutes the most important question in relation to the emplacement of hot wastes; but these formations appear to have very good potential for plutonium-contaminated and other intermediate-level wastes with low heat emission. In the UK, areas in which there are argillaceous formations that merit more detailed examination include Southern Scotland, Cumbria, Gwynedd–Powys, Leicestershire–Nottinghamshire, and Somerset. The presence of the salt bands in the salt/shale formations mentioned above is clear evidence of freedom of moving water for a long geological time.

3. *Crystalline rock formations (granites, gneisses).* In the unfissured state these rocks are impermeable to water flow. In addition they are quite good absorbers/adsorbers of materials that might be leached out of waste. Near the surface they can be extensively fissured and considerable water movement can occur through the fissure system. These fissures are known to decrease with depth and a major part of exploratory work is to establish their extent and the rate of water movement at depths to 1000 m. Granite is not as good a thermal conductor as salt and hence dissipation of heat will be slower. It is less susceptible to alteration by heating than the clay-based rocks. It is a strong rock, suitable for engineering operations, but being brittle, may be subject to cracking if stress is induced by earth movements or the thermal expansion of a large volume of rock. In the UK most of the crystalline rocks occur in Scotland, and a few in Northern England. Those in Devon and Cornwall do not appear very promising.

Study of the geological maps of the United Kingdom by the Institute of Geological Sciences[23] using the guidelines mentioned earlier in this section has resulted in the identification of over 100 areas containing potentially suitable rock formations of all three types. These range in area from less than 5 km^2 to 6000 km^2 and the total area amounts to about 16 per cent of the land of the UK. It may be noted that in preliminary design studies of a repository, about 0.1 km^2 would be needed to accommodate all the high-level waste generated in the UK this century. There seems, therefore, every likelihood that a suitable location might be found.

However, the above selection was made largely on the basis of knowledge of surface geology, and much more detailed information of the geological conditions at depth is required before proposing any site for a repository. The aim of current research and development programmes is to verify the whole concept by conducting detailed studies of the various factors indicated in the 'guidelines' that must be taken into account in an assessment of the safety of geological disposal. Present plans are for such studies to be carried out at a number of sites covering a range of geological environments in crystalline and argillaceous formations. This would require the drilling of a pattern of boreholes at each research site to elucidate the geological features at depth and to measure rate of water movement. Only when this phase of the research has been completed would be geologists be in a position to recommend the most likely type of rock mass for containing a repository, and if the government of the day so decided, one or more sites could be selected for construction of a repository. The site selection process would not be limited to the research sites that had been studied in the initial programme, but would include a review of all areas containing rocks of the preferred types. In addition to the geological factors, a number of important social and logistical factors would have to be taken into account including *inter alia* the question of land use, transport, and economics.

Any proposal for building even an experimental demonstration repository would be the subject of planning approval under the Town and Country Planning Act (1971).

Any site selected would require a very detailed geological and hydrological investigation before construction of a repository began.

7.2 Repository design considerations

The design of a repository for high-level waste depends on the type of rock in which it is to be built and the constraints imposed by the temperature limits to protect the waste and the integrity of the repository. It is necessary to take account of:

(1) the maximum rock temperature adjacent to freshly emplaced waste containers;

(2) the bulk rock temperature, i.e. regional effects; and

(3) the temperature of the waste containers and of the glass within them.

Criterion (1) will control the extent to which the walls of the hole or chamber

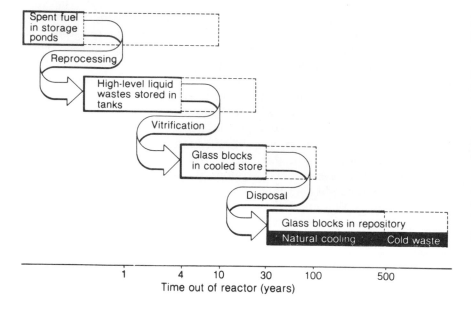

Fig. 15.7 Sequence of operations in storage and disposal of highly active waste

in which waste is placed will break down. Criterion (2) will determine the effect on the stability of the rock formation as a whole, i.e. whether the dispersion of heat could so alter the movement of water within the formation as to affect the isolation of the waste. Criterion (3) will determine the rate of corrosion of the container and the rate of leaching of glass. For granite rocks it is intended for the present to aim for a target maximum temperature of about 100 °C. This provides one of the important design criteria for a repository. To meet the temperature limit indicated, containers would need to be spaced at minimum distances from each other of 15–20 m, and the heat output per container at time of emplacement would need to be limited to 1 kilowatt unless the repository were to be cooled, e.g. by air circulation, in the earlier years. A large HARVEST container with AGR waste would require 60 years' cooling before the heat output was down to 1 kilowatt. In general there is considerable flexibility in the options, including the amount of waste put into a container (and therefore size and numbers of containers produced) and the length of cooling in storage before placing in the repository. The sequence of choices in operations in the storage and disposal of high-level waste is shown in Fig. 15.7.

Figure 15.8[16] shows one of several conceptual designs of a repository constructed in a hard rock (e.g. granite). It utilizes a three-dimensional configuration for the emplacement of containers, to minimize the amount of excavation necessary. A similar design, but with a planar array, would be equally possible and might be more desirable on grounds of temperature distribution, but would occupy a greater area. The repository would be

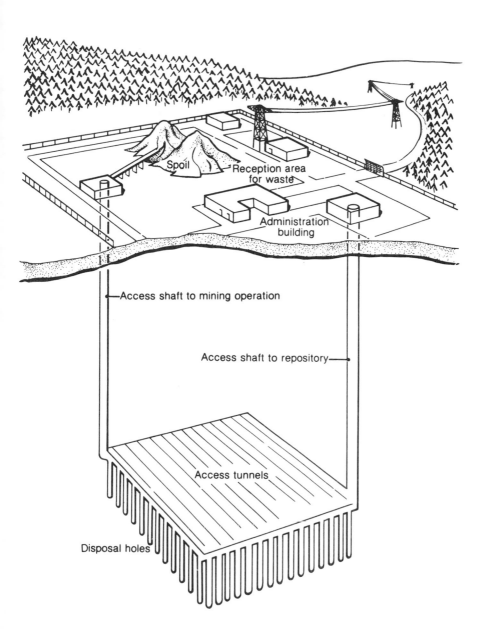

Fig. 15.8 Conceptual design of a repository for high-level waste

constructed at least 300 m below the surface to avoid rocks that have been subject to cracking by permafrost during previous ice ages. At any specific site, the depth chosen will depend on the hydrogeological conditions found.

The underground part of the repository would consist of a system of parallel tunnels, forming a grid with about 20 m separation between each. The tunnel floors would contain boreholes into which the waste containers would be placed, stacked on top of each other. The containers would have packed round them a clay-like material chosen for its ability to restrict the ingress of water to the bore-hole and to adsorb strongly any dissolved radionuclides that emanate from waste containers that have been penetrated. The access tunnels would be progressively back-filled as the repository is filled. Rock spoil would be used, possibly together with cement or material with adsorption properties. Finally, the openings connecting the repository to the surface would be effectively plugged and sealed at the end of the repository's life.

The only features at the surface would consist of a reception area for the waste, a shaft head gear for removal of rock spoil from mining tunnels, and for lowering waste containers to the repository level, and an administration building. The total land area required for a major waste repository is very small. For illustrative purposes, the repository to accommodate the high-level waste arising to year 2000—given in Table 15.4 (about 3500 HARVEST containers)—requires a volume of rock with an area of about 0.1 km^2 and a thickness of 150 m. The technology for constructing it is available now.

The design described here is only one of several concepts at present being studied.[24] Others, in which waste containers are placed in galleries off the access tunnel, packed and back-filled as before, might be more economic. Another variant is the storage of shielded containers in tunnels, air-cooled, with the ability to re-enter and inspect as necessary until a decision is taken to back-fill and declare the waste disposed of. However, the present studies are enough to provide an estimate of the size of the operations, the environmental impact, and the approximate costs of dealing with the quantity of waste that might be generated.

The visual intrusion of the operation would be very small in relation to the quantity of power the waste represents. Excavation of the repository would result in the net production of about 500 000 m^3 of rock spoil, since half could be used in back-filling. This quantity can readily be landscaped, if indeed, a market is not available to take away the material for ballast. To generate the same quantity of electricity in coal-fired stations would require 1400 million tonnes of coal, the mining of which would lead to the production of about 200 million m^3 of spoil—400 times as much. Thus, the magnitude of nuclear waste disposal in terms of its physical size is such that considerable trouble can be taken to see that it has only a very small visual impact, and indeed, once the repository was closed there would be much less to see than the landscaped spoil from a very small conventional quarry.

7.3 Geological disposal of intermediate-level waste

The intermediate-level waste may be segregated into the following two streams which have different disposal requirements:

(a) stream 1—those wastes containing activated materials, fission products, and actinides, i.e. long-lived activity;

(b) stream 2—those containing activated materials, or fission products, or both and only minor quantities of actinides, i.e. no long-lived activity.

The repository design requirements for stream 1 (α-, β-, γ-activity) is very similar to that for highly active waste insofar as radionuclide paths back to man is concerned. The important difference—heat generation in the first 100 years is very much less—modifies the design constraints. The requirements of rock stability, low water movement, and ability to accept climatic and topographic changes over long periods of time are still of prime importance.

One solution for this stream would be to dispose it in containers alongside the highly active waste, making use of the spare excavated volume of the tunnels. This is some ten times that occupied by the containers of highly active glass. The additional cost above that for the highly active waste repository would be negligible. However, unless the repository were air-cooled, it would be necessary to await 50 years or more, so that emplacement could be in phase with the highly active waste-disposal programme.

Another solution would be the construction of a repository specifically dedicated to intermediate-level waste containing long-lived alpha activity. It might be acceptable to dispose of the waste containers in large caverns (since there is no requirement to spread the heat load) located at adequate depths in argillaceous rock formations.

For stream 2 (active lifetime less than 1000 years) it might be possible to use underground caverns excavated at a shallower depth than for stream 1, or possibly caverns that already exist as a result of quarrying slate or other materials. 'Dry' regions might be selected and back-fill arrangements devised which have capacity for absorbing radionuclides and providing a barrier against human access.

7.4 The safety of a geological disposal repository

Before a geological repository can be constructed and operated, the regulatory authorities will require to be satisfied (on behalf of the public) that it would be safe during the period of loading the waste, and in the long term after it had been closed. The safety of a geological repository long into the future cannot be 'demonstrated' by direct experiment, and can only be assessed by considering all the factors that control the return of radioactivity to man's environment, and the radiation dose he would receive from this.

At present safety criteria for judgement of acceptability of consequences of today's actions into the very far future have not been decided or agreed upon either nationally or internationally. ICRP recommendations that doses to individuals should not exceed current ICRP radiation limits are directly applicable to planned releases of radionuclides directly into the environment (e.g. routine discharges from reactors and plants to atmosphere or the sea). The ICRP does not recommend an approach that is directly applicable to situations where there is probability, but not certainty, that a release into the environment will occur. Risks associated with a disposal option in which the waste is isolated from the biosphere for a long time require two components to be assessed:

(1) the probability that a release will occur, and
(2) the probability that subsequent radiation doses will give rise to deleterious effects.

The debate on this topic is proceeding at a scientific level in the international community, but in the end the agreement of criteria will be as much social and political as scientific and will therefore have to be endorsed by government.

There are two types of process whereby radioactive material buried deep in the earth might return to the surface. The first of these is catastrophic disruption, by which the waste could be exposed or brought up to surface. Examples of such catastrophic events that have been assessed are giant meteorite impact, extremely violent earthquakes, volcanoes, glaciation, nuclear weapons, or accidental intrusion by man. Assessments show that, provided the repository is located at an adequate depth in a suitably selected area, the probabilities of any of these events is so extremely low that the risk entailed can be regarded as negligible. The other type of process, and the only one by which radionuclides might find their way back to the biosphere, would be via solution or suspension in moving ground water, following breaching of the container and leaching of the solid waste.

The safety of a repository of the type outlined in §7.2 depends on the existence of a series of barriers designed to arrest the movement of radionuclides, so that they will have decayed before they reach the biosphere. The barriers will have the following features.

1. *Choice of site*. The site chosen should be one in which access of water to the waste is very limited and the rate of water movement and any mixing with surface waters is very slow.
2. *The waste canister*. Material and thickness would be chosen to minimize corrosion rates. With present technology, materials are available that could prevent access of water to the waste for more than 500 years, and if necessary, for thousands.
3. *The stability and leach resistance of the solidified waste*. Borosilicate glasses could take thousands of years to dissolve completely, even at high pressure, provided temperatures are not allowed to exceed 100 °C.
4. *Packing materials round the waste canisters (to exclude water and adsorb any radionuclides leached*. Materials based on bentonite clays offer very low permeability and high cation exchange capacity. They would have

sufficient plasticity to accommodate minor movements of the rock in which the canister is placed.

5. *Sorption of radionuclides in the rocks deep underground.* The term *sorption* covers a variety of chemical and physical processes whereby dissolved and entrained radionuclides can be removed from the water.

6. *Dilution in surface waters of any ground water reaching the surface.*

Studies have been made in several countries of the rate and time at which radionuclides from buried waste canisters would reach the biosphere and the subsequent radiation dose to man, either directly or by ingestion from food and water. This has involved introducing the values of the various parameters of the barriers and processes into complex computer models.[25] In such exercises, it is necessary to allow for possible changes in conditions in the future that could alter the hydrogeological system—e.g. changes in climatic conditions. Most of these studies have concerned hypothetical and generalized cases of repositories in various rock types.

One of the most comprehensive studies of a waste disposal system is that carried out by the Swedish organization Kärn-Bränsle-Säkerhet (KBS).[26] This was based on measurements of important parameters in an actual site in crystalline rock in Sweden. This work showed that the radiological impact on man from the waste would be small. In the most unfavourable case, that of an individual drinking water from a deep-drilled well near the repository, the increase in radiation dose to an individual using the water would be less than 2 per cent of the ICRP recommended limit and less than local variations which occur in the natural background radiation in various parts of Sweden. The assessment concluded that the *most probable* radiation dose to an individual from the repository would be less than 1 per cent of this *worst-case* value.

Preliminary results of similar studies conducted elsewhere also conclude that the radiation received by man from waste disposal in various formations would be very small—a small fraction of natural background radiation—and most of the exposure occurring over a long time, starting far into the future.

Working Group 7 of the International Nuclear Fuel Cycle Evaluation (INFCE) reviewed the management and disposal of wastes that arise from several nuclear fuel cycles. The Group concluded that wastes from any of the fuel cycles examined were capable of being managed and disposed of with a high degree of safety by application of existing technology. They commented that safety analyses and calculations of future doses are limited by the accuracy of available models to describe natural phenomena, but that the uncertainty is not such as to affect the conclusion that disposal can be carried out without undue risk to man or the environment.

In the UK, the NRPB[27,28] have conducted a preliminary assessment of the radiological impact of the hypothetical repository in granite. The results showed that the maximum radiation doses would be very small from nearly all the radionuclides, and that the main contributors are the very long-lived fission products (e.g. ^{129}I and ^{99}Tc) and ^{237}Np, which have been assessed, on basis of available data, to be poorly *sorbed* on geologic media. Plutonium isotopes do not appear because this element is strongly sorbed; neither do the very toxic ^{90}Sr and ^{137}Cs, since they would decay before they were released from the glass.

The numerical values of the radiation dose to man obtained in the NRPB study cannot be taken as definitive. They depend on the values of the parameters used, and in the study, values of many of the parameters had to be assumed since data are not available for actual rocks at depth, particularly the hydrogeological conditions and the sorption properties of some radionuclides under the conditions that would be obtained. As the authors point out, the purpose of their analysis is to draw attention to the factors which are of the greatest importance in controlling releases, and to identify the information that must be obtained so that a definitive assessment of the radiologial consequences of geological disposal can be made. A significant conclusion is the identification of the importance of sorption in delaying or preventing the movement of long-lived radionuclides; this proved to be of greater importance than improvements in durability of the waste container and glass, the value of which does not appear to be critical provided they survive intact for 300 years.

The actual figure for estimated dose to individuals from certain radionuclides in some of the earlier assessments[26, 27] may need revision in the light of more recently revised data in *ICRP*-26 for the radiotoxicity of all radionuclides. The changes are unlikely to alter the general conclusions. To take the studies forward from the present idealized generic models it is essential that better data, as identified by NRPB, are determined. For this it is necessary to conduct work in the field in actual typical rock formations.

It is worth noting that after about 1000 years the total potential toxic hazard is of the same order as that of many naturally occurring toxic ones. This simplistic comparison does not replace the need for rigorous analyses involving assessments of pathways back to man for every critical radionuclide, but it helps to put the potential risk into perspective.[29]

An interesting Canadian experiment that provides additional grounds for confidence was the burial of small radioactive glass blocks 4 m deep in wet ground for 20 years. These have been examined, and at the present rate of dissolution, it would take over 100 million years to dissolve them completely. These tests also showed that the radioactivity that was leached out had moved more slowly than the ground water, owing to sorption processes; its activity is decaying faster than it migrates.

Finally, nature itself has provided a demonstration that the products of nuclear fission can remain isolated underground for very long times. At Oklo in Gabon, West Africa, the remains of six natural nuclear reactors in a uranium ore body have been found.[30] Nuclear fission continuing for several hundred thousand years resulted in the formation of several tonnes of fission products and significant amounts of plutonium. These have decayed but can be identified by their daughter products. The plutonium that was formed was still there but has long since decayed in place to uranium-235. Remarkably, at least half of the thirty-odd fission-product elements have remained immobilized in the ore. These include the rare-earth elements (lanthanides), most or all of the zirconium, ruthenium, rhodium, palladium, niobium, and silver, and some of the molybdenum and iodine. Many of the elements with valencies of one or two (rubidium, caesium, strontium, and barium) have disappeared. On the other hand, there is no deficit of zirconium-90; this isotope is produced by decay of strontium-90. It can be concluded that little of the strontium was transported from the vicinity of the reactor. Although no data exist on elements heavier than plutonium, the similarity in chemistry of the actinides and the lanthanides suggests that they would behave similarly, and the Oklo observations support the conclusion that all heavy elements remain relatively firmly fixed.

This natural demonstration embodied the principles which have been adopted in geologic disposal of nuclear wastes, and provides assurance that with careful implementation the risk to future generations could be negligible.

8 Disposal to the deep ocean

The current practice of disposing of some low-level solid waste, after incorporation into concrete, into the deep ocean within the terms prescribed by the London Convention has been described in §4.3. This treaty, of which the UK is a signatory, expressly prohibits the disposal of highly active waste to the oceans, and any proposal to do so would require renegotiation of its terms. There are, however, reasons to suggest that the deep ocean could accept much higher levels of waste suitably emplaced, without risk to man or the environment, and therefore the possibility is worth careful investigation.

There are two concepts for deep-ocean disposal of high-level waste: the emplacement of waste canisters into the ocean bed sediments, and placing them on the ocean bed.

8.1 **Disposal into the ocean bed sediments**

This is in many ways similar to disposal into geological formations on land, but with the certainty that the formations are permanently water saturated. The principle is, again, one of multiple barriers between the waste and man's environment. The first barrier is the canister. Current knowledge suggests that it would be possible to design it to remain intact for 1000 years. This of course depends on the material of the canister, its thickness, the effect of temperature, radiation, and sediment pore water composition. High temperatures would be undesirable, and the radiogenic heat from the canister would almost certainly require to be limited, either by restricting the loading per canister or by storage for a period before emplacement in the sediments. The second barrier is the solid form of the waste itself—glass or other suitable solids. It has been estimated that provided the temperature was limited, it could take between 1000 and 5000 years or longer before all of the radionuclides had been dissolved out into the surrounding sediment. The third barrier is the sediment, which would be selected to possess high power for adsorption of radionuclides. The intention of sub-sea-bed disposal is to prevent radionuclides reaching the sea bed; but if they do, the ocean itself provides a fourth barrier in dispersing and diluting the activity.

8.2 Disposal on the ocean bed

The first two barriers, the canister and the waste form, are similar to those discussed in §8.1. The difference is that there would be very good heat removal, thus limiting temperatures and corrosion rates, and hence less need for limiting waste concentrations or for long land storage. However, the sediment barrier is no longer there and leached radionuclides will escape directly into the sea water. The levels of radioactivity will depend on the processes of dispersion by bottom currents and dilution by mixing. Some radioactivity may be scavenged from the water by adsorption on to suspended particles, which would be either deposited locally or transported elsewhere, depending on conditions.

8.3 Emplacement technology and choice of disposal sites

Techniques for placing canisters on the ocean bed are clearly possible with present-day knowledge. It could be by free fall of individual canisters, or by the controlled lowering of a large number of them in complex engineered structures. For emplacing canisters to depths of, say, 30 m into uncon-solidated sediments, it has been suggested that a free fall technique could be used. The canister would be streamlined to form a projectile which would penetrate the sediment, and the plasticity of the sediment would be relied upon to seal the entry hole. For harder sediments, wide-diameter, deep-sea drilling techniques are not far beyond the techniques already used success-fully by the Deep-Sea Drilling Project.

The site or sites chosen would be in the most stable and quiescent areas of the deep ocean floor in 'mid-plate' regions of low topography, where there is a good thickness of sediment of the preferred type. The areas to be avoided would be (1) the mid-ocean ridges, which are steep mountains with very little sediment; (2) continental margins where the sediments are too steeply sloped and subject to strong turbidity currents (underwater avalanches); (3) earthquake and volcanic zones (deep ocean troughs) because sediments would be unstable; and (4) high northern and southern latitudes to avoid glacier-borne boulders on the sea floor and the rapid vertical mixing of ocean waters.

8.4 The safety of deep-ocean disposal

Research aimed at determining the feasibility of ocean disposal, the selection of suitable areas, and the long-term safety is being conducted in close collaboration in an international programme involving, at present, the USA, UK, France, Japan, and Canada under the auspices of the Nuclear Energy Agency of the OECD.[31]

A crucial part of these studies is a total system study to assess the interactions between the parts of the system, in order to determine which are most critical and which are unimportant in isolating the radioactivity from man. A preliminary assessment of the radiological consequences arising from the disposal of high-level waste from a hypothetical very large world nuclear programme (72 000 HARVEST containers) on to the deep-ocean floor has been published by the National Radiological Protection Board,[32] and another is in preparation on sub-sea-bed disposal. These studies assess the various pathways back to man that radionuclides might take given a number of assumptions. The results of the first study showed that the maximum radiation dose to man by any route would be far below the current ICRP recommended limits. However, such studies are only indicative and their validity depends on the identification of all relevant factors and on experimental confirmation of their numerical value. Much of the research programme that is currently being undertaken is directed towards establishing these values, in particular, the durability of the container and its contents, the properties of the sediments, the movement and mixing of waters in the deep ocean, and the complex biological chains between the deep ocean and the fisheries.

It seems highly probable that disposal of the higher categories of waste into the deep ocean would be safe and practicable, but any decision to use the route would clearly require international agreement. The research of the next decade will be directed towards provision of sound scientific and technical evidence on which a case could be made. Meanwhile, it is clear that some of our intermediate-level wastes could quite safely be disposed of on to the ocean floor.

9 The cost of radioactive waste management

The question is often raised as to whether the very costly measures that are being developed for the treatment, storage, transport, and eventual disposal of all nuclear wastes would make nuclear power uneconomic. This is a crucial question if one is to follow the principle that the polluter must pay. A very interesting cost study has been carried out as part of the International Nuclear Fuel Cycle Evaluation.[33]

The purpose of the study was primarily to compare the waste management costs of seven different fuel cycles to determine whether any of them presented any particular advantages over the others. The study covered various options, including *once-through* with spent fuel disposal, and uranium–plutonium reprocessing and recycle. The reactor types did not include Magnox and advanced gas-cooled reactors, which are presently in use in the UK. However, the range of cycles and reactor types included in the study was sufficiently wide to cover a range of costs that would encompass those of the present UK reactors. The exercise did include the light water reactor (LWR) and the fast reactor, both of which are under consideration in the UK at the present time.

Costs were included for the conditioning (treatment and packaging), transport, and disposal of wastes from every step in the cycle: uranium mining and milling, refining and enrichment of uranium, fuel fabrication, power plant, and spent fuel disposal and reprocessing. The assessment took account of the following factors:

(1) capital costs for all facilities, excluding land costs;
(2) operation and maintenance costs, including salaries;
(3) cost of financing;
(4) research and development costs;
(5) cost of transportation of conditioned waste;
(6) no taxation was included, and
(7) no resource value for spent fuel, nor for recycled material.

The detailed assumptions concerning depreciation, interest rates, wages and salaries, repair and maintenance costs, insurance, etc. are spelt out in the INFCE Working Group 7 report.

The total costs of waste management, including disposal into a repository in *hard rock* ranged from 10.7 to 17.3 million US dollars per gigawatt-year of electricity sold (1978 prices), depending on the fuel cycle. In terms of energy costs, this range corresponds to 1–2 mills/(kW h). In the basic cost assumptions, the price of electricity was assumed to be 100 mills/(kW h); on this basis the cost of waste management would amount to only about 2 per cent of the total cost of electricity. Even assuming a financial error in the costing ground assumptions, it would seem likely that the cost would be within 5 per cent of the cost of electricity.

The following points which arose from the analysis might be noted.

1. No allowances were made for optimizing the technology or the waste management strategy in terms of searching for the least expensive solutions. For example, it was assumed that all categories of waste would be consigned to a deep repository, whereas simplified schemes for low-level wastes such as shallow burial or ocean disposal would reduce the costs for all fuel cycles.
2. The total costs included wastes from mining and milling of uranium. These costs, in fact, are already covered in charges made to electricity utilities when they purchase uranium.
3. The costs of treatment and disposal of reprocessing wastes and of spent

fuel are about equal and are the largest single component of the costs. The cost for power plant wastes are of the same order or slightly less for some fuel cycles.
4. The fast reactor wastes come into the lower echelons of costs for the fuel cycles studied.
5. Increasing predisposal storage times from 10 years (the base case taken) to 40 years would reduce costs by about 6.5 per cent.
6. Costs are sensitive to the size of the nuclear power programme. The increase in unit cost for a power programme that produces 10 GW instead of the 50 GW that was the basis of the study would be some 8–24 per cent, depending on the fuel cycle. The UK situation would fall within this range.

In conclusion, the study shows that the costs of waste management amount to only a few per cent of the value of electricity generated.

10 Conclusions

1. The nuclear fuel cycle generates a range of radioactive wastes, some of which are undoubtedly extremely hazardous.
2. The technology for converting the most active category into a glass (vitrification) possessing adequate properties to safeguard the environment has now reached the industrial scale.
3. The vitrified product should be placed in stores with simple cooling arrangements for many decades at least. The engineering and inspection requirements of such stores are straightforward. Such storage would be a very safe method for managing these wastes for as long as any industrial activity can be maintained.
4. Other waste categories are of much lower activity but much greater volume. Methods are known for their conversion to forms for storage with as much security as those developed for high-level waste; it requires only the determination and decision to develop and use them.
5. Low-level solid wastes will be disposed of by shallow land burial or sea dumping as at present, but all the wastes of higher activity can safely remain in stores until it is decided that a repository requiring no further human surveillance should be established. The safety of a final repository would be assured by the existence of a multiplicity of barriers: the properties of the solid waste, the container, the design of the repository, and geologic barriers. Several analyses have indicated that the long-term radiation dose to the population arising from properly sited repositories should be well below acceptable limits. The visual impact of a repository will be small compared with many other installations used by modern communities.
6. While the nuclear industry is right to approach the final solution to its waste with caution, evidence already exists to show that the margin of safety

built in to the strategy outlined in this article is likely to prove very large indeed. The risks would be very small, and future generations are likely to be critical of the present if it delays decisions in the hope that something even better may be developed.

References

1. O'RIORDAN, M.C. *Radiological protection in uranium mining and milling*. Miscellaneous Paper No. 7. Geology and Society, London (1978).
2. DEPARTMENT OF THE ENVIRONMENT. *The control of radioactive waste* (a review of *Cmnd*.884). DOE, London (1979).
3. KELLY, G.N., JONES, J.A., BRYANT, P.M. AND MORLEY, F. *The predicted radiation exposure of the population of the European Community resulting from discharges of krypton-85, tritium, carbon-14 and iodine-129 from the nuclear power industry to the year 2000*. Report, Commission of the European Communities, Luxembourg (1975).
4. ‡ *Cmnd*. 5169. HMSO, London (1972).
5. INTERNATIONAL NUCLEAR FUEL CYCLE EVALUATION. *Report of Working Group 1: fuel and heavy water availability*. *INFCE/PC/2/1*. IAEA, Vienna (1980).
6. GROVER, J.R. *High-level waste solidification–why we chose glass*. UKAEA Report No. *AERE* 9432. UKAEA, Harwell (1979).
7. INTERNATIONAL ATOMIC ENERGY AGENCY. *Techniques for the solidification of high-level wastes*. Technical Reports Series No. 176. IAEA, Vienna (1977).
8. NATIONAL TECHNICAL INFORMATION SERVICE. *Proc. Int. Conf. on Ceramics in Nuclear Waste Management, held in Cincinnati, Ohio. CONF*-790420. US Department of Commerce, Springfield, Virginia (1979).
9. BOULT, K.A., DALTON, J.T., HALL, A.R. HOUGH,'A., AND MARPLES, J.A.C. *The leaching of radioactive waste storage glasses*. UKAEA Report *R*9188. HMSO, London (1978).
10. INTERNATIONAL ATOMIC ENERGY AGENCY *Characteristics of solidified high-level waste products*. Technical Reports Series No. 187. IAEA, Vienna (1979).
11. RINGWOOD, A.E., KESSON, S.E., WARE, N.G., HIBBERSON, W., AND MAJOR, A. Immobilisation of high level nuclear reactor wastes in SYNROC. *Nature*, **278**, 219–223 (1979).
12. DRAN, J.C., MAURETTE, M., AND PETIT, J.C. Radioactive waste storage materials: their alpha-recoil ageing. *Science*, **209**, 1518–1519 (1980).
13. HIRSCH, E.H. A new irradiation effect and its implications for the disposal of high-level radioactive waste. *Science*, **209**, 1520–1522 (1980).
14. COSTI, J.A. JOUAN, A.F., PAPAULT, C., AND PORTEAU, C.H. Vitrification of high-level waste solutions at Marcoule, France. In *Proc. European Nuclear Conf. Hamburg, (May 1979)*. Vulkan Verlag, Essen (1979).
15. CLELLAND, D.W. A review of European high-level waste solidification technology. In *Proc. Int. Symp. on the Management of Nuclear Waste from the LWR Fuel Cycle. Denver Colorado, USA (11–16 July 1976)*. Report no. CONF 760701 National Technical Information Service, Springfield Virginia (1976) pp. 137–165.
16. ROBERTS, L.E.J. Radioactive waste–policy and perspectives. Lecture at British Nuclear Energy Society, 9 November 1978. *Atom*, no. 267, pp. 8–20 (1979).
17. US DEPARTMENT OF ENERGY. *Draft environmental impact statement, management of commercially generated radioactive waste*, vols. 1 and 2. *Doe/EIS*-0046. US Department of Energy, Washington DC (1979).
18. COMMISSION OF THE EUROPEAN COMMUNITIES. *Proc. 2nd Technical Meeting on Nuclear Transmutation of the Actinides. Ispra, Italy, 21–24 April 1980. EUR 6929 EN*. CEC, Brussels (1980).
19. *Nuclear power and the environment.*† *Cmnd*. 6820. HMSO, London (1977).
20. GRAY, D.A. *Disposal of Highly active solid radioactive waste into geological formations–relevant geological criteria for the United Kingdom*. Institute of Geological Sciences Report No. 76/12. HMSO, London (1976).
21. IAEA *Site selection factors for repositories of solid high-level and alpha-bearing wastes in geological formations*. Technical Report Series No. 177. IAEA, Vienna (1977).

22. CHAPMAN, N.A. *Geochemical Considerations in the choice of a host rock for the disposal of high-level radioactive waste.* Institute of Geological Sciences, Report 79/14. HMSO, London (1979).
23. MATHER, J.D., GRAY, D.A. and GREENWOOD, P.B. *Burying Britain's radioactive waste–the geological areas under investigation. Nature,* **281**, 332–334 (1979).
24. HAYTINK, B. *Conceptual design of radioactive waste repositories in geological formations.* European appl. Res. Reports. *Nucl. Sci. Technol.,* **2** (1, 179–215 (1980).
25. NUCLEAR ENERGY AGENCY. *Proc_Conf. on Risk Analysis and Computer Modelling in Relation to Disposal of Radioactive Wastes into Geological Formations, held at Ispra, Italy, March 23–27, 1977.* Nuclear Energy Agency, Paris (1977).
26. *Handling of spent nuclear fuel and final storage of vitrified high level reprocessing waste* (5 parts). Kärn-Bränsle-Säkerhet Reports. AB Teleplan, Solna, Sweden (1978).
27. HILL, M.D. and GRIMWOOD, P.D. *Preliminary assessment of the radiological protection aspects of disposal of high-level radioactive waste in geologic formations. NRPB-R*69. National Radiological Protection Board, Harwell (1978).
28. HILL, M.D. *Analysis of the effect of various in parameter values on the predicted radiological consequences of geologic disposal of high-level waste. NRPB-R*86. National Radiological Protection Board, Harwell, (1979).
29. TONNESSEN, K.A. and COHEN, J.J. Survey of naturally occurring hazardous materials in deep geological formations. In *A perspective on the relative hazard of deep burial of nuclear wastes. UCRL*-52199. Lawrence Livermore Laboratory, Livermore, Cal., USA (1977).
30. COWAN, G.A. A natural fission reactor (Oklo) *Scient. Am.* **235**, no. 1 (1976).
31. § ANDERSON, D.R. In *Proc. of the 4th Annual Seabed Working Group meeting, Albuquerque, New Mexico, March 5–7, 1979. SAND* 79-1156. US Department of Commerce, Springfield Virginia (1979).
32. GRIMWOOD, P.D. and WEBB, G.A.M. *Assessment of the radiological protection aspects of disposal of high-level radioactive wastes on the ocean floor. NRPB-R*48. National Radiological Protection Board, Harwell (1976).
33. INTERNATIONAL NUCLEAR FUEL CYCLE EVALUATION. *Report of Working Group 7. Waste mamagement and disposal: INFCE/PC/2/7.* IAEA, Vienna (1980).

‡ Final Act of the Inter-Governmental Conf. on the Convention of Dumping Waste at Sea.
† The Government's response to *The Sixth report of the Royal Commission on Environmental Pollution: Cmnd.*6820 (1977).

§ Available from National Technical Information Service, US Department of Commerce, 5285 Port Royal Road, Springfield, Virginia, USA.

Suggestions for further reading

1. NUCLEAR ENERGY AGENCY. *Objectives, concepts and strategies for the management of radioactive waste arising from nuclear power programmes.* HMSO, London (1977). (This is a very useful, compact book covering all aspects of radioactive waste management.)
2. SIMON, R. and ORLOWSKI, S. Editors *Proc. 1st European Community Conf. on Radioactive Waste Management and Disposal, Luxembourg, May 20–23, 1980.* Harwood Academic Press, New York (1980). (This gives the results of research conducted under the EEC programme)

16

Decommissioning of nuclear facilities

K. SADDINGTON

The disposal of obsolete nuclear facilities, generally termed 'decommissioning', is a requirement which will grow with the world-wide expansion of nuclear energy programmes. It is therefore attracting increasing international attention.

After all fissile material has been removed from the plant there is a number of options open for disposal of the remaining plant and buildings. As a minimum the plant may be closed down, made safe, and left intact on a care and maintenance basis. At the other extreme, all plant and buildings may be demolished and removed from the site which, after any necessary clean-up, is then available for unconditional reuse.

The factors which govern the choice of option are identified together with the consequences which stem from the decision. Technical aspects are discussed under the headings of *Decontamination, Demolition,* and *Disposal.* Other aspects include costs and safety requirements.

The current status of decommissioning is outlined in terms of practical experience, paper studies, and present legislation. The need for further information in all of these areas is recognized. Nevertheless, the conclusion is reached that the decommissioning of redundant nuclear plants can be safely accomplished, even to the cleared-site condition.

Contents

1 Introduction

As in the case of conventional industrial plants, much can be done extend to the normal lifetime of nuclear plant. Regular maintenance has an important role to play here, as does the use of improved operational techniques. In some cases it may be judged desirable for major engineering modifications to be carried out during the life of the plant, to take account of improvements in component design. There is nevertheless a limit to the cost-effectiveness of maintenance and modification procedures of these kinds. In particular, significant advances in nuclear engineering are likely to be made during the operational life of a commercial nuclear reactor, and it is thus virtually certain that a point will come at which it will be economic for newly designed, more efficient plants to be installed as replacements for their by now obsolescent predecessors.

When this point is reached, and nuclear plants are made redundant, the same problem will arise as with their non-nuclear counterparts. What is to be done with them? The practice that would once have been followed with industrial installations—to recover any valuable materials, and to leave the plant and buildings derelict—is one which public opinion now universally rejects. Environmental standards now demand that derelict sites be rendered not only safe but also visually acceptable. Some effort of landscaping is a minimum requirement. Ideally, it is felt, the site should be carefully cleared for reuse.

It is these environmental standards which will need to be met in dealing with redundant nuclear plants. Given the problems of residual radioactivity, it is indeed especially important in these cases that adequate action be taken to avert possible environmental damage. The treatment of redundant nuclear plant has therefore been studied over the past fifteen years, the aim of these studies being to provide assurance that an expanding nuclear programme will not result in an ever-growing number of derelict plants being left scattered throughout the country. For a responsible nuclear industry can clearly not aim to press forward with the world-wide expansion of nuclear power without demonstrating its ability to deal with the end-product in a publicly acceptable manner. This is the basis of the work being carried out under the general heading of decommissioning.

2 Decommissioning

The term 'decommissioning' refers specifically to the steps to be taken when a nuclear facility has ceased operation completely, without intent to restart at any stage. To date, practical experience of decommissioning as so defined is very limited since few commercial nuclear plants have completed their normal operational lives. Many paper studies have been carried out, and these are valuable in setting out the issues involved, the available options, and the technical and financial consequences of the policy decisions taken. Much of the published information of a practical nature stems from the maintenance and remedial work carried out on plants which have been closed down temporarily by accident or design and which are scheduled to restart thereafter. The value of such experience is limited since the different objectives of life-time remedial work and end-of-life decommissioning often lead to differences in the techniques to be employed.

3 Magnitude of the problem

During the last decade it has been possible for the nuclear industry to examine the problem of dealing with nuclear plants which have completed their operational lives. The problem is by no means a remote and futuristic one. Current conservative estimates of the lives of commercial nuclear power reactors suggest that in the United Kingdom alone about thirty will be retired from active service by the end of the century. The world-wide figure will be several times greater, and will begin to accelerate rapidly thereafter. To these must be added the redundant chemical plants, and pilot and research facilities, some of which have already been shut down and left in a safe temporary condition awaiting decisions as to their future final state. Reliable policies and plans need to be formulated well in advance of any widespread need for work of this kind, and the time is therefore ripe to carry out the necessary studies, backed where desirable by practical development work. The value of real experience gained during the complete decommissioning of a few typical plants will be invaluable in this context.

From a technical standpoint, the decommissioning of a nuclear plant poses no radiological problems different from those incurred in maintenance and operational work. Indeed, once a nuclear reactor has been shut down and its fuel has been discharged, the residual induced activity in the component structure will be lower by several orders of magnitude than the equilibrium activity contained in the fuel rods. For a 1000 MW(e) pressurized-water reactor (PWR) the latter will be of the order of 10^9 Ci, and this will, after reprocessing, appear in the form of highly active chemical waste. In contrast the induced activity of the reactor structure and components after final shutdown will be approximately 10^6 Ci.

The new problems which arise in decommissioning are therefore not so much those of radioactivity as such, but rather those of the large-scale demolition and the transport and disposal of large and unwieldy sections of radioactive steel and concrete.

4 Decommissioning options

Once a nuclear facility has been closed down and the nuclear fuel has been removed from it, decisions will need to be taken on the extent to which it should be decommissioned, and the time-scale within which this should be achieved. The ultimate target would be to remove everything from the site, which would be left in a condition suitable for unqualified reuse. Short of this target there are a variety of options which spread across three well-defined stages. Any of the options can be regarded as a permanent end point, or as a temporary situation from which to progress further at a later date.

While some commonality of approach can be expected with plants of the same generic type (e.g. PWRs), the decision in any particular case is likely to be influenced by features specific to the plant in question. As well as general and economic factors, consideration will be given then to such specific factors as the operational life-history of the plant, its location, the presence of other reactors on the site, and the potential alternative use of the facilities and site released.

The following three main decommissioning stages may be identified.

Stage 1. This stage is reached once the nuclear facility has been shut down, nuclear fuel has been discharged, operational materials removed, and the control system disconnected so as to prevent resumption of operation. The facility is thereby rendered safe and is kept under continuing surveillance in order to ensure that it remains so.

Stage 2. In decommissioning to Stage 2, the installation is reduced to the minimum size possible without penetration into the most highly radioactive or contaminated areas (e.g. the core of a reactor). The structural integrity of the primary containment and the biological shield continue to be assured, in order to avoid hazards to personnel or to the environment. This residual structure, occupying only 5–10 per cent of the original area, is maintained under surveillance.

Stage 3. In decommissioning the facility to this third and final stage, all remaining plant, components, and materials are removed from the site. Following decontamination, if necessary, the site is released for redevelopment or for general use by the public. There is no further requirement for surveillance.

Figures 16.1-16.3 show artist's impressions of the Windscale advanced gas-cooled reactor (WAGR) after decommissioning to Stages, 1, 2, and 3, respectively.

5 Radioactive inventory

The specific difficulties inherent in the decommissioning of a nuclear facility are due above all to the problems associated with the appreciable radioactivity of areas of the plant. It should, of course, be remembered that these are problems affecting only a small proportion of the facility (no more than 20 per cent in the case of a nuclear power station) and that the decommissioning

Fig. 16.1 Windscale Advanced Gas-Cooled Reactor before decommissioning

Fig. 16.2 Windscale Advanced Gas-Cooled Reactor as it would appear after stage 2 decommissioning

Fig. 16.3 Windscale Advanced Gas-Cooled Reactor as it would appear after Stage 3 decommissioning (cleared site)

and disposal of non-radioactive materials can proceed without the need for special techniques, using conventional equipment. The attention of those concerned with nuclear plant decommissioning has been focused therefore on the 'difficult 20 per cent', and it is with these activated and contaminated areas of nuclear plant that the remainder of this chapter will be primarily concerned.

The formulation of development programmes and of practical decommissioning plans demands knowledge of the total radioactive inventory of the plant, its distribution through the system, and its decay characteristics. For a chemical plant the information may be provided from an analysis of wash-liquor samples, following primary decontamination with suitable stripping solutions. This must be supplemented by direct instrumental monitoring, the aim of this being to detect the presence of solid residues, which may have accumulated in inaccessible parts of the plant and may resist attempts at decontamination.

For a nuclear reactor, the compilation of the radioactive inventory is more complex. It includes:

(1) the neutron-induced activity in the fixed structure of the plant;
(2) the neutron-induced activity in removable components, e.g. control rods remaining in the plant after discharge of the fuel; and
(3) contamination around the primary coolant circuit arising from activated corrosion products (from cans and components).

The value of (3) is determined by the operational history of the reactor and can be estimated only on the basis of sampling. Estimated values of (1) and (2) are obtainable, by calculation, however, this being based upon the following two sets of data.

1. *Data on the abundance within structural materials of those isotopes which are activated by neutron irradiation.* The most important structural materials in this respect, for UK reactors at present under study, are stainless steel, mild steel, concrete, and graphite. The composition of the last of these is well defined due to the 'nuclear' specification required for its use and to the analytical control procedures employed to maintain this. The specifications for the steels (which include the reinforcement steels within the concrete) are not normally so well defined and must be supplemented by analyses for trace elements on representative samples. The abundance of trace elements in concrete is determined principally by the composition of the aggregate, this being dictated by the geographical source. Here again analysis of samples is necessary.

2. *Data on the mean reactor flux, the nuclear load factor, and the time over which the reactor has opearated at this load factor.* This allows neutron irradiation conditions to be calculated and the level of isotope irradiation to be estimated. Account will need to be taken of the precise location of host materials within the reactor core and of possible shielding and neutron attenuation by internal components, which will affect the degree of activation.

By means such as these a radioactive inventory can be calculated which can be regarded as adequate as a basis for technical judgements. Wherever possible however the results must be validated and corrected by physical measurements on samples taken from within the reactor.

The integrated inventory data thus obtained can be used to estimate contact radiation dose levels at the surface of each component or structure; simplified geometries will generally be used in the estimation of these doses. These estimates can be used only as a guide, and need to be verified by means of direct measurement of exposure rates at sample points on or near the surfaces in question, portable radiation survey equipment being employed for this purpose. Individual contact dose-rates for reactor core components vary according to the nature of the material and the position in the core. The variations are considerable, ranging in a typical case from millirems per hour (mrem/h) to several thousand rems per hour (rem/h). For example, the core and shroud baffle plates of the Elk River boiling water reactor (ERR),[1] which was decommissioned during the period 1971–74, had contact dose levels in excess of 6000 rem/h.

However, since the proximity of other components provides a measure of self-shielding, the environmental dose level will be significantly reduced. Indeed by careful planning of the demolition sequence the self-shielding effect can be maximized so as to bring area dose levels below 100 rem/h. This is further reduced for the protection of the workers by carrying out the work from a distance with the aid of remotely operated tools and equipment, by the use of television cameras, and by the provision of local shielding. By such means, the radiation exposure of the personnel can be brought below statutory limits. Thus in the ERR project (Appendix A) the average exposure was 0.8 rem per man. In the decommissioning and rebuilding of the Dounreay chemical plant used to reprocess the irradiated fuel (see Appendix B), the average dose (1976) was 0.86 rem. The Working Group 4 report from the International Nuclear Fuel Cycle Evaluation (INFCE)[2] noted that the decommissioning of an active reference reprocessing plant could be undertaken with essentially no increase in exposure of the general public and a minimal occupational exposure. The INFCE report noted that, on present evidence the occupational exposure resulting from decommissioning would be 'a small fraction of the overall occupational exposure during the whole operating life of the plant' and could be accomplished 'with only a small increase in the average occupational exposure'.

6 Discussion of decommissioning options

Stage 1
The measures taken at Stage 1 render the installation safe and substantially intact on a care and maintenance basis. Decommissioning if taken only to this stage involves a continuing requirement for monitoring over a period of many decades, entailing a considerable financial outlay.

Stage 2

In terms of capital outlay Stage 2 is a relatively inexpensive exercise, since during this phase we are dealing with plant and materials which are largely non-radioactive and uncontaminated and which can be treated by normal demolition practices. During this phase, valuable materials and components may be removed for use elsewhere or sold as marketable scrap. The requirement for continuous monitoring and surveillance of the residual structure entails financial penalties as in Stage 1. Additional cost would be involved by a decision to mound over the residual structure. In addition to involving a financial penalty, such a decision would have the added disadvantage of significantly reducing the area of the site that could subsequently be released, and the aesthetic benefit to be gained may well be regarded as being outweighed by these disadvantages.

Stage 3

Decommissioning to Stage 3 ('green field' or cleared site) involves the demolition and disposal of those structures remaining at the completion of Stage 2. These structures include the most radioactive and contaminated parts of the nuclear plant and the work involved is in consequence more difficult and costly than that required for Stages 1 and 2. The successful decommissioning of plant to Stage 3 demands that attention be directed to a number of issues, including those set out in the following list.

1. Conventional engineering problems inherent in the demolition of a complex plant structure.
2. Problems specifically arising from the radioactive and contaminated character of many of the components.
3. Any conflict between requirements imposed by (1) and (2). The sequence and timing of operations dictated by the radioactivity may for instance interfere in part with the engineering dismantling logic.
4. The need for the definition and development of the special demolition methods and equipment demanded by such a complicated and potentially hazardous task.
5. The need for definition of disposal methods for the different types of waste material (including radioactive and contaminated waste). Where acceptable procedures are not at present available, there will be a need to initiate development work. The choice of disposal procedures may itself in part dictate demolition methods (dictating for example the limiting size, weight, and radioactivity content of the pieces after demolition).
6. The safe transport of the waste materials from the reactor site to the authorized disposal point. This includes definition of the routes to be taken and the types of container to be used. It may be necessary to develop new designs of container for special handling requirements, and this may affect handling procedures *en route* to disposal.
7. Legal requirements governing the removal and disposal of hazardous waste materials from the site. These include mandatory advance consultation with the authorizing bodies to obtain formal permission in respect of the procedures proposed.

8. Resource requirements, in particular, manpower. Evaluation of these will need to take into account limitations imposed for health reasons on manpower utilization. Account will have also to be taken of the need for back-up services to help ensure operational safety. The decommissioning project will lie outside normal site activities and there may be a consequential requirement temporarily to reinforce local manpower resources by the use of contract labour or specialist demolition services.

7 Time-scale

In an evaluation of options it is necessary to consider not only the extent of decommissioning but also the time-scale within which objectives are to be achieved. How soon after final plant closure should decommissioning start? Should work be carried on uninterruptedly to the defined end-point? Or should there be a policy of deliberate time delays, work being temporarily halted at various stages so as to allow radioactivity to decay, thereby easing radiation problems at later stages of the operation?

Figures 16.4-16.6 show the radioactivity decay curves for the Windscale advanced gas-cooled reactor, for a typical Magnox station[3] and for a pressurized-water reactor. The curves are drawn for the total activity in the system, expressed in curies, and for the associated beta and gamma energy levels, expressed as beta or gamma activity x energy. The latter curves are of greatest relevance in terms of radiation exposure hazards and clearly demonstrate the value of delay prior to the commencement of Stage 3. The curves show that the first 40–50 years following shutdown will be characterized by the rapid decay associated with the short-lived isotopes that dominate this period, ^{55}Fe (half-life 2.6 years) and ^{60}Co (half-line 5.27 years), which are present in the stainless steel structures. The activity spectrum over the ensuing period is dominated by the less-troublesome isotope ^{63}Ni (half-life 93 years), which decays much more slowly. The radiological problems of decommissioning, which arise almost entirely from the gamma activities of these radionuclides, are thus considerably alleviated by the decline in activity over the first 40–50 years, and considerable practical advantage may ensue from the deferral of work on Stage 3 for a period of up to 100 years after reactor shutdown. Activity levels decline only very gradually after this time however, and there is likely to be little benefit in further deferring decommissioning beyond this period.

Some technical advantage may be seen then in a policy of deliberate delay during decommissioning. The necessity for such a delay period is by no means established however, and, based on present evidence, it is entirely feasible that an uninterrupted decommissioning procedure could be achievable. It should be noted that the decay which automatically occurs during the period of fuel discharge and during Stages 1 and 2 will of itself greatly ease the radiation problems to be encountered during Stage 3. In

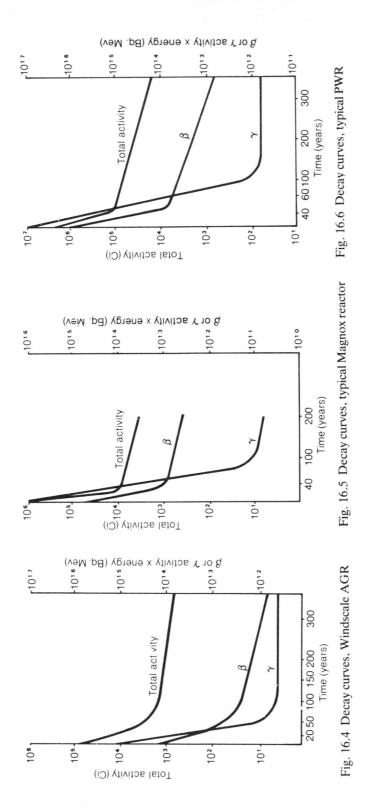

Fig. 16.4 Decay curves, Windscale AGR

Fig. 16.5 Decay curves, typical Magnox reactor

Fig. 16.6 Decay curves, typical PWR

favour of continuous decommissioning it can be argued that this will coincide with most complete understanding of plant characteristics and will ensure the continuing availability of experienced personnel. With this procedure moreover, the continuing integrity of structural components, of the biological shield, and of containment structures over a post-operative period of many decades will not be a matter of concern. A procedure of this kind also avoids the substantial costs of monitoring and surveillance over a protracted care and maintenance period.

The following estimated figures for WAGR serve to give a broad indication of the expected time involved in decommissioning continuously to the end of Stage 3, using a small team of skilled personnel:

discharge of fuel after closure	2–3 years
completion of Stages 1 and 2	1–3 years
completion of Stage 3	5 years
Total time approximately	10 years.

It may be noted that a procedure of continuous decommissioning commends itself particularly in the case of chemical plants. The relatively very low level of induced radioactivity in the structural and operational materials here means that the time between closure and final decommissioning need not be unduly prolonged, and that little advantage will accrue from a deferral of the final stage of decommissioning.

8 Decommissioning practice

The work of decommissioning involves a number of interrelated activities including decontamination, demolition, and disposal. Each of these may be partial or total, and it is clear that there may be some overlap between some or all of these activities. The choice of method in respect of any of these aspects of the decommissioning process may well impact upon decisions regarding other related activities. Decisions for example on the method of disposal of radioactive scrap and of the method of transport of the scrap to the disposal site may determine the size of the pieces into which the plant must be cut up during demolition. Again the demolition plan and the disposal method may influence the extent of prior decontamination.

8.1 Decontamination

The process of decontamination is undertaken in order to reduce the risk of occupational exposure during subsequent work. While in certain cases decontamination may be completed *in situ* before the commencement of demolition work, it is unlikely that this will always be judged appropriate. There is a trade-off between radiological protection and operational efficiency, and it will be necessary to optimize between the degree of protection accorded to the workforce on the one hand, and on the other hand considerations of economy and practical simplicity. Following demolition, further decontamination may be carried out on certain components in order to permit them to be handled and disposed of safely in the planned manner. On grounds both of cost and of waste minimization, it will be wise to restrict decontamination to the minimum demanded for the particular requirement.

The practicable extent of decontamination will tend to be greater in the case of chemical plants than in the case of reactors. Chemical plants are designed to operate with flowing liquids, so that *in situ* decontamination using the standard decontaminating fluids will be a relatively simple matter. Since the plant is not intended to return to service and the long-term integrity of pipework and vessels is no longer a relevant consideration, it will indeed be possible to make use of more aggressive decontaminating fluids, e.g. nitric acid–hydrofluoric acid mixtures.

The scope for *in situ* decontamination of reactor plant is more limited, the use of liquid decontaminants being restricted to a flushing out of the coolant circuits of light water reactors. A certain amount of practical experience exists with this, the procedure being one which is increasingly being used during the operational life of these reactors in order to reduce the accumulation of active deposits. For a gas-cooled reactor, *in situ* decontamination is mainly restricted to dry techniques, e.g. brushing and vacuum extraction of dusts. If necessary, further decontamination of some components can be carried out after removal from the reactor by one or more of the large number of available methods, which include chemical, physical, and electrochemical techniques.

8.2 Demolition

As has already been pointed out, a large fraction of any nuclear installation is free from radioactivity or contamination, and no specialized demolition techniques will be needed for this part of the plant. Components such as the inactive storage and feed tanks of a chemical plant, or the inactive pipework, steam-generating plant, and turbo-alternators of a nuclear reactor, can be dealt with by conventional demolition procedures. The methods to be employed with activated or contaminated components will depend on the level of residual activity. Experience to date has indicated that the methods employed are likely to be developments from existing techniques rather than completely new procedures. Established mechanical or flame cutting methods are likely to be employed for the demolition of steel sections, though some adaptation of the technology is likely in order to enable the work to be carried out remotely (for example from behind shielding), and to contain any airborne contaminated particles that may be released. Appendix A,[1] which describes the decommissioning of the Elk River Reactor, gives examples of the techniques that have been used. It is of paramount importance that all demolition operations should be planned in meticulous detail, and if necessary rehearsed under inactive conditions before work commences. This preparatory work, rather than the development of fundamentally new techniques, may be regarded as the *sine qua non* of successful nuclear plant demolition.

8.3 Disposal

Ideally the method and route for waste disposal should be determined and all necessary authorizations sought before the commencement of the particular demolition operations. The concept of 'long-term' temporary storage of waste following demolition is one which should, if possible, be avoided. The effect of a concept of this kind is to lead to the redundant double handling of active materials, a procedure which increases the chance of hazard to personnel. In the case of a redundant nuclear reactor it would certainly be better to delay demolition, components being retained in an entirely safe environment until such time as acceptable disposal plans can be formulated. This procedure would moreover have the advantage of permitting radioactive decay to occur, thereby lessening the hazard of subsequent handling.

The formulation of adequate disposal plans involves some categorization of future waste arisings in terms of their composition, size, weight, activity, and contamination. Where the current disposal options are not felt to be appropriate to decommissioning wastes, alternative methods will require to be evolved, and it will be necessary for these to be cleared with the appropriate regulatory bodies in advance. This categorization activity will be a continuing requirement as the detail of the demolition plan and of its integration with the disposal plan are worked out. The general objective will be to maximize the amount of material that can be recycled. Decontamination procedures will be used where economically justified in order to minimize activity levels for each class of waste, thereby alleviating the difficulties of handling and disposal.

Table 16.1 shows the aproximate tonnages of construction materials in the nuclear islands of two reactors under study. The pressure vessels and the reactor internals will all be highly activated and possibly contaminated also. The biological shields will also be activated and contaminated, though experience suggests that this will be limited to the inner 0.9–1.2 m the outer layers, which are more remote from the reactor core, not being subject to any significant degree of neutron bombardment. A large proportion of the biological shield—typically in the range of 42–68 per cent for a Magnox reactor—will then be inactive.

The wastes arising during decommissioning differ from operational wastes primarily in respect of their awkward shape and size, and while there will be scope during demolition for breaking down some of the more troublesome components, it is inevitable that there will be some items of waste which will prove particularly stubborn. Here again the solution to this problem will lie

Table 16.1 Quantities of structural materials (tonnes)

Item	WAGR (33 MW(e))				Typical Magnox reactor			
	Mild steel	Stainless steel	Graphite	Concrete	Mild steel	Stainless steel	Graphite	Concrete
Pressure vessels and internals	600	50	300		2500	200	2250	
Biological shield	200			4000	450			14 000

not so much in the devising of new disposal methods, but rather in adapting existing methods to these arisings. It will clearly be necessary for national and international acceptance to be obtained for the disposal procedures that are evolved.

An example of this problem is the case of mildly contaminated heat exchangers. The design of these is such that complete internal decontamination is difficult: the heat exchangers cannot be easily compacted, and their disposal in land-based burial sites would pre-empt a large volume of burial space, an uneconomic use of scarce disposal facilities. An attractive solution here may lie in the disposal of large sections or even complete heat exchangers to the ocean depths. Sections of activated steel plate might be disposed of in a similar way following decontamination, or packaged in a matrix such as concrete prior to disposal. The shielding could either be disposed of with the package or returned for further use, the choice depending upon economic considerations. The neutron-induced activity of these components is purely beta–gamma and will decay to insignificant levels in hundreds rather than thousands of years. Further details of disposal methods that may be appropriate to such components are given in Chapter 15.

9 Costs

The cost of decommissioning will clearly depend upon the option chosen. Cost projections have, at the present state of experience, to be essentially theoretical in nature, actual practical decommissioning experience to date being very limited. In the reactor field, apart from some early low-power experimental and prototype reactors, the only reactor fully to have been decommissioned so far is the Elk River boiling-water reactor in the USA. This reactor, which was of 22.5 MW(e) capacity was demolished during the period of 1971–4, the total project cost including technical support services being $6.15M.[1]

Numerous paper studies have been carried out since this time on a variety of reactor types. Given their essentially theoretical nature and the differences in ground rules, it is not surprising to find that there are wide variations in the cost estimates. For example, estimates for the cost of decommissioning of a PWR to Stage 3 immediately after shutdown vary from US $42M[4] to $92M[5] at 1978 prices. The figures relate to reference reactor types in the 1100–1200 MW(e) range. A more recent Swedish study[6] based upon the 600 MW(e) boiling-water reactors Oskarshamn II and Barsebeck I estimates the cost of immediate decommissioning to Stage 3 as the equivalent of US $120M (1979 prices) and equates this to 10–15 per cent of the installation cost of an equivalent (600 MW(e) BWR) new nuclear power plant. By extrapolation the order of cost for an 1100 MW(e) PWR is put at $130M (1979). Despite the variations in absolute costs there is increasing acceptance that the cost of decommissioning a nuclear power plant to Stage 3 will be 10–15 per cent of the cost of new construction at constant prices. This figure excludes the cost of defuelling (an operational charge) and makes no allowance for the value of recovered plant and scrap. These decommissioning costs, if spread over the operating life of the nuclear plant, are estimated to add much less than 5 per cent to the unit power cost.

Decisions on the extent to which decommissioning should be carried out will clearly involve a judgement as to the economic worth of this activity and some type of cost–benefit analysis is likely to be appropriate as a guide to decision taking. In considering the desirability in moving from Stage 1 to Stage 2 for instance, it will be relevant to take account of the value of any recovered materials, following decontamination if necessary. These will include valuable construction materials such as stainless steel, and could even include whole sections of plant which may have a second-hand value. The value of the site itself, up to 95 per cent of which may be vacated in the course of Stage 2 decommissioning, will also clearly need to be taken into account in an economic evaluation, as will the presence of other nuclear power plant. The overall worth of a site released in this way may well be deemed greater than its simple book value, particularly in a country where there is a shortage of land for construction purposes, and in particular where environmental and political difficulties exist with the opening up of new sites for nuclear installations.

These 'unquantifiable' aspects of the benefit of decommissioning operations will be of particular importance in any cost-benefit analysis carried out on a Stage 3 decommissioning. The justification here is likely to rest less on simple economic considerations, and more on the question of public acceptablity. The time scale of decommissioning may here be of particular importance to any analysis. Section 7 of this chapter has considered the advantages which a deferred demolition policy may have in respect of reduced radiation levels to the work-force. In practical terms, a long period for decay will permit worker access for demolition without the costly shielding and remote operation equipment required if dismantling were to be carried through on a shorter time-scale. A policy of deferral will on the other hand have the economic demerit of involving substantial costs over a period of many years, owing to the monitoring and surveillance which will be required.

10 Design for decommission

In the past nuclear plants have been designed for safety and performance, but without prime regard being paid to their subsequent decommissioning at the end of their working lives. Increasing attention is now being directed at the original design stage to maximizing the use of materials and design features which will be conducive to subsequent decommissioning. Considerable scope exists here for innovatory features. Some features falling within this category may be primarily directed at improving routine maintenance and prolonging plant life, and will to this extent help reduce the overall decommissioning waste arisings. Other design stage modifications may have a still more direct impact upon end-of-life operations, being directed at providing easier means of decommissioning: examples may include the provision of lifting lugs in strategic locations, or the incorporation of additional instrument access facilities to permit monitoring of residual activity levels after plant closure.

One aspect of reactor design that is likely to merit reconsideration from this point of view is the method of construction of the biological shield. As has already been mentioned, the induced activity in reinforced concrete falls off with distance from the reactor core. This points in the direction of the concept of a two-component biological-shield structure.[7] One version of this would consist of

    (1) an outer physically independent layer to provide containment and some shielding, and

    (2) an inner layer, which would be constructed to a thickness calculated to provide the major shielding requirements, and to contain all the significant induced activation expected during the operating life of the reactor.

An alternative approach to the construction of the biological shield would consist in the use of interlocking blocks, which might have advantages for some applications. Either approach would have the merit of facilitating the separation of activated and contaminated concrete from inactive material, thereby reducing the demand for active disposal facilities that would exist if this segregation were not carried out. There are similar opportunities for optimizing the size and shape of thermal shield and pressure vessel plates, so as to permit easier handling and disposal of activated components during and after demolition, without compromise to normal construction and operation.

11 Decommissioning plans

Stringent standards are imposed upon the owners and operators of nuclear facilities by regulatory bodies, serving as preconditions for the granting of an operating licence. These include stipulations regarding safety, personnel and environmental protection, and the safe transport and management of fertile and fissile materials. The advent of the large-scale decommissioning of commercial nuclear plants will lead to an extension of these requirements to this new field of activity. As in the case of normal nuclear plant operations, responsibility for the technical and financial aspects of the work can be expected to fall to the owners and operators of the plant. Logically therefore, as part of the documentation supporting an application for an operating licence, they should furnish an outline decommissiong plan or feasibility study to demonstrate their intention and ability to decommission the plant after the end of its operational life. A practice similar to this is already followed in a number of countries and is recommended for general application in a recent report on decommissioning compiled by international experts under the auspices of the International Atomic Energy Agency.[8] As well as providing licensing authorities with the data required for regulatory purposes, this outline plan would also have the function of providing valuable guidance to the designers of the facility. Such a preliminary plan would of its nature be sketchy and in time become obsolete in the light of continuing design studies. It would thus without doubt be necessary to refine and update it during the life of the plant, so as to take into account operational experience, development of new techniques, and other relevant information.

Prior to the start of the decommissioning, a detailed decommissioning plan would be compiled. The objective of this plan should be to provide sufficient information to all organizations involved in the decommissioning project, including the regulatory authorities from whom approval is required. Features to be incorporated in this final decommissioning plan should include *inter alia*:

(1) a clear description of the state of the plant before decommissioning commences;
(2) a description of the major activities of the decommissioning operation;
(3) a description of the techniques to be used in the decommissioning operation, mock-ups, and special training required, if any;
(4) an estimate of the type and quantity of radioactive wastes to be generated during the decommissioning operation and the plans for their disposal or storage;
(5) a safety analysis, including an environmental impact assessment and an analysis of the structural integrity of reactor buildings, during the decommissioning operation;
(6) a safety analysis, including an environmental impact assessment, for the state of the plant to be reached on completion of the decommissioning operation;
(7) a description of organizations involved and the role of each in the decommissioning activity;
(8) a description of the radiological protection measures; and
(9) a description of the surveillance inspection and test programme which will be implemented after decommissioning.

12 Current status

12.1 **Practical experience**

12.1.1 *Reactors*
The exploitation of nuclear energy on a commercial scale has not yet reached the point at which power plants have completed their operational lives and are ready to be decommissioned on a significant scale. Practical decommissioning experience is therefore limited in most countries to the treatment of early power, demonstration, and test reactors, which include some that have suffered premature closure owing to unsatisfactory operating experience. In most cases the extent of decommissioning has been limited to Stage 1 (known as 'mothballing' in the US), which merely ensures that the plant will pose no threat to public health and safety. Experience of a Stage 3 decommissioning is limited to the Elk River reactor (ERR) project described in Appendix A and that derived from the current decommissioning project on a 20 MW(t) sodium-cooled graphite-moderated thermal reactor, SRE, near Los Angeles.[9]

Much work currently being published under the title of 'decommissioning' is confined to the areas of decontamination and waste disposal, or again to experience relating to the maintenance, rehabilitation or modified reuse of nuclear plant. Decommissioning embraces much more than this, and the inclusion of these activities under this heading may be regarded as something of a misnomer. Such information is valuable nevertheless in helping build up a general awareness of possible techniques and equipment which may be of application in a decommissioning project.

Important though work to date has been, much more relevant experience will be needed before the industry can confidently proceed to any large-scale decommissioning operations. It is expected that the next five to ten years will bring much new important evidence to light on many aspects of decommissioning. Tentative plans are already in hand for decommissioning of a number of reactors, including the advanced gas-cooled reactor at Windscale (33 MW(e)) and the Shippingport PWR in Pennsylvania, USA, which has operated at power levels up to 150 MW(e).

12.1.2 *Fuel plants*
Present-day experience is again limited in this area. By their very nature plants of this kind occupy a small number of long-term sites, and such decommissioning as has occurred to date has usually been of a partial nature involving decontamination, partial demolition, and reconstruction for further use. The significant inventory in such plants is in the form of plant residues which lend themselves to a high degree of decontamination, radiological hazards during subsequent demolition operations therefore being much reduced.

A good example of what can be achieved in this area is given by the decommissioning and reconstruction of the fast reactor irradiated fuel re-processing plant at Dounreay in UK.[10] This plant was built originally in the 1950s to service the experimental Dounreay fast reactor fuel cycle. It was equipped with fuel-handling and cave breakdown facilities, dissolution equipment, and solvent-extraction systems. Here highly enriched uranium alloy fuel was reprocessed to high purity uranyl nitrate solution for sub-sequent refabrication into fuel. After 15 years of operation it was decided that the DFR reprocessing plant should be modified in order to permit the reprocessing of the mixed plutonium oxide-uranium oxide fuel from the prototype fast reactor (PFR), this calling for a more sophisticated and more extensive plant. The work involved in the decommissioning and reconstruc-tion of this plant is described in Appendix B. Similar techniques to these have been used in the decontamination and partial dismantling of the Eurochemic fuel reprocessing plant at Mol in Belgium.[11]

12.2 Paper studies

While the decommissioning of nuclear facilities to Stage 1 leaves them in a safe condition, in most cases this has been regarded as a temporary measure, doing no more than deferring the need for a determined assault on the problem of the handling and disposal of waste materials. While little practical experience exists of the later stages of decommissioning, the need for attention to be given to this problem has been recognized in the numerous paper studies that have appeared over recent years. Most nuclear nations have produced, or are in the process of producing, feasibility studies and detailed assessment of the possible decommissioning alternatives of large-scale plants. The volume of information derived from such studies is rapidly increasing, and while conflicting data have certainly come to light in some areas, real progress is nevertheless being made toward gaining a clearer consensus approach to the problem. There is a strong measure of agreement world-wide that the decommissioning of nuclear reactors to the Stage 3 cleared-site condition is technically feasible. Progress is being made with the compilation of reliable radioactive inventories, with the development of suitable demolition techniques and equipment, and with the study of possible new design features that may later help materially to ease radiation problems in decommissioning. The results of this work are being communicated to a wide audience, including the IAEA, which serves as a focal point for the exchange of information and views.

12.3 Legislation

A host of regulations and licensing requirements has already been imposed in most countries by national regulatory authorities to cover health and safety requirements for the construction and operation of nuclear facilities. These regulations serve to cover some aspects of decommissioning, for example, the disposal of at least some of the wastes arising during such operations. Most countries are still, however, some way short of having formulated a comprehensive set of regulatory controls for the decommissioning operation as a whole, and progress still remains to be made in the evolution of a regulatory code. The need for such a code is however widely recognized, and in parallel with technical decommissioning studies, efforts are being made within international forums to define the necessary administrative requirements. Studies of these kinds have of necessity had to await the completion of at least the initial stages of the technical studies of decommissioning, so that a sufficiently clear insight could emerge into the regulatory problems which these activities would raise. This point may however be regarded as having been reached and progress should now be made on this aspect, so that utilties and other bodies involved with decommissioning plans may be able to know in advance the kind of framework within which they will expect to work.

Of all the European countries the Federal Republic of Germany has the most specific legislation on decommissioning. The licensing procedures for new nuclear plants under the Atomic Energy Act require that plant operators furnish proof that the plant can be decommissioned and removed at the end of its operational life. It is indeed furthermore required that the design of the plant be shown to be one that is conducive to decommissioning. The precise interpretation of these legislative requirements has yet to be fully worked out. In particular, it is still uncertain whether a specific decommissioning study will need to be carried out for every nuclear plant that is licensed. Plant operators argue that the requirement should be for operators simply to demonstrate the general decommissioning feasibility of the type of reactor plant involved. It is also felt at present that some requirements imposed in respect of decommissioning might potentially conflict with operational safety requirements, and time may be needed for the resolution of these issues.

The regulations and guidelines enforced in the USA, while more comprehensive than those in other countries, cover the requirements and criteria for decommissioning only in a limited way. The current Nuclear Regulatory Commission (NRC) regulations are applicable to decommissioning only in relation

(1) to the financial competence of the licensee safely to operate and shut down the facility, and

(2) to an obligation upon the licensee to obtain approval for the decommissioning plan once the time for decommissioning has arrived.

The *NRC Regulatory Guide 1.86* provides non-binding advice on the acceptable methods for decommissioning and gives some criteria for residual activity levels following decontamination. It is however recognized in the USA that these requirements are inadequate and that there is at present too little guidance available on the subject of reactor decommissioning. Even less guidance exists on the decommissioning of other facilities. This has been recognized in the decision by NRC to place extensive contracts with the Battelle Pacific Northwest Laboratories for decommissioning studies on a large BWR, a large PWR, and a fuel-reprocessing plant.[4,12] The aim is to develop a library of information and analyses of decommissioning possibilities for different kinds of facilities. This will then serve as a basis for the further development of a general decommissioning policy and for the updating of the regulations applicable to decommissioning.

In the UK, as in many other member states of the Nuclear Energy Agency, there are no specific requirements regarding decommissioning. Under present legislation such work would be controlled by application of the provisions relating to the amendment or surrender of nuclear operating licenses. The treatment of waste would be controlled under existing regulations for radioactive waste treatment and disposal.

13 Future criteria

As in other fields of nuclear plant operations, the methods employed to carry out decommissioning will largely be determined by the industry on the basis of a growing body of experience in such work. But certain other criteria, such as those listed below, will need to be taken into account by the regulatory authorities when considering the method and degree of control which they will deem necessary to ensure that the work has been carried out satisfactorily.

(1) *The extent of decommissioning appropriate for the facility at the end of its operational life*
This could, in the limit, be the subject of a detailed study for each separate plant, and this would have the merit of taking into account features unique to that plant. These would include any design peculiarities, non-standard operational events, and the future plans for the vacated areas and the nuclear site itself. However, it has been argued that for nuclear facilities of the same type, e.g. all Magnox reactors or all fuel reprocessing plants, a great deal of duplication of effort could be avoided by the authorization of a standard practice for that type of plant. Local variants would then be cleared as modifications to the standard practice.

Apart from the extent of the decommissioning to be carried out it is for consideration whether the time-scale in which the work is done should also be the concern, amongst others, of the regulatory authorities.

(2) *Standards of safety, security, and structural integrity to be achieved as a result of the decommissioning activities*

(3) *Public health and environmental protection criteria to be adopted during and after decommissioning*
Although it would be possible for criteria to be based in part on those which

have been adopted for operational activities, new facets will be raised by decommissioning which will require attention from regulatory bodies concerned with these matters. Where, for example, a facility is left temporarily or permanently in an intermediate stage of decommissioning, criteria will need to be defined for the degree of surveillance and monitoring of residual structures required, and the length of time during which this will be necessary. Again, where a significant proportion of the products of decommissioning are judged suitable in principle for recycle, either for unrestricted or for qualified reuse, criteria will need to be set down governing the limits of residual radioactivity or contamination in such material.

The existing dose limits, which are based on ICRP recommendations, do not cover the full range of requirements. Where, for example, a site is to be cleared and to be returned for unrestricted reuse, limits will need to be set down governing the maximum permissible residual radioactivity of the soil. Again, there are at present no limits for induced radioactivity in materials subjected to neutron irradiation which are destined for intended recycle.

(4) *Financial arrangements to cover decommissioning and surveillance costs*
The funding of the capital and operational costs of a nuclear facility is generally regarded as a straightforward matter, and as one which necessitates no departure from normal commercial practice. Decommissioning costs constitute only a fraction of the lifetime operational costs of the plant, and it might therefore well be argued that no special arrangement is needed in their respect. Decommissioning costs do, however, raise a number of special problems which argue a possible case for involvement by national authorities in arrangements regarding the funding of these operations. These stem above all from the long time-scales involved. Modern nuclear reactors are expected to have an operating life in excess of 30 years. Immediate decommissioning to Stage 2 after shutdown may well take a further five years, and following this, decommissioning to Stage 3 may take anything between an additional few years and up to a century depending on the option to be adopted. The financial security of a commercial organization cannot be sufficiently well guaranteed over anything approaching the maximum period of time that may be involved, and the question therefore arises, particularly in countries where private industry builds and operates reactors, as to the means by which the continued availability of funding may be ensured. It is likely that public authorities will wish to formulate mandatory financing arrangements, for example by means of a sinking fund or some suitable alternative, so as to ensure that monetary resources are available when required.

(5) *Administrative arrangements for delayed decommissioning*
As with the question of finance, it is not possible to guarantee that the technical capability and management expertise of a utility will survive over the kind of extended time-scale that has been outlined. The suggestion has

therefore been mooted that long-term technical responsibility for surveillance and decommissioning operations should become the remit of a public organization dedicated to this purpose. The proposals discussed here, as in the preceding subsection, relate of course exclusively to those countries where nuclear generation is carried out by private industry. The need for such special national arrangements is clearly less pressing in a country such as the UK where national bodies already exist to carry out functions of this kind. These remarks relate furthermore primarily to reactors, and less to chemical plants, where, as has been shown in §7, the period of time between closure and final decommissiong is unlikely to be particularly prolonged. Chemical processing plants are in any case more often than not owned and operated by national bodies, thus obviating the need for any special arrangements.

(6) *Arrangements necessary to ensure the safe disposal of the decommissioning residues*

14 International co-operation

The growing importance of decommissioning has been recognized in the setting up under IAEA auspices of a series of meetings which have been held since 1973, at which representatives of member states have been able to exchange views and information. The IAEA has acted as a primary forum for such exchanges and has additionally been involved in the publication of technical literature in this area. It is intended that the IAEA should become a reference source for expertise on the subject of decommissioning.

Attention has also been given to decommissioning by the OECD Nuclear Energy Agency (NEA) Expert Group on the Management of Radioactive Wastes Arising from Nuclear Power Programmes. The NEA continues to take an active interest in this area and in March 1980 sponsored a specialist meeting on Decommissioning Requirements in the Design of Nuclear Facilities. The IAEA and NEA collaborated in arrangements for an International Symposium on Decommissioning in November 1978 at which wide-ranging discussions took place between representatives of the 26 nations represented.[13]

The commission of the European Communities (CEC) has initiated discussions on a suitable indirect action programme of research and development on the decommissioning of nuclear power plants.

15 Summary and conclusions

Numerous paper studies, backed by limited practical experience conclude that the decommissioning of redundant nuclear facilities to the cleared-site stage (Stage 3) is a practical possibility.

Decommissioning allows of a number of different options in respect of extent and time-scale, any one of which may be judged appropriate for a particular case.

Decontamination procedures normally employed during the operational life of nuclear plant are suitable for use in decommissioning operations; in addition, more aggressive decontaminants may be employed during decommissioning if desired.

With some adaptation, currently available demolition procedures are suitable for use in this field.

Considerable advantage will derive if adequate attention is given to future decommissioning needs during the initial design phase of a plant. A conceptual decommissioning plan, as well as providing necessary information to national regulatory authorities, can, if drawn up at this stage, serve as valuable guidance to designers.

Before decommissioning commences an updated and detailed decommissioning plan should be drawn up and submitted for approval by the regulatory authorities.

Decommissioning can be carried out without any significant additional environmental problems from waste arisings.

Current regulations governing the construction and operation of nuclear facilities and the treatment of operational wastes may need to be extended to cover any new requirements of decommissioning in terms of health and safety, environmental protection, etc., where these are not already covered by existing procedures.

Time is available to complete all necessary work in advance of any general need to decommission commercial plants presently in operation.

Acknowledgement

The author wishes to acknowledge the assistance given by staff of the Central Technical Service Group, Risley in providing the decay curves shown in Figs 16.4 , 16.5 and 16.6.

Appendix A The Elk River reactor (ERR)

This reactor, built as part of the US Atomic Energy Commission power reactor demonstration programme, was shut down after four years' operation in January 1968. In the period to September 1969, the fuel and control rods were discharged and removed from the site. Between 1971 and 1974, the reactor was dismantled and removed, leaving the site in the Stage 3 ('green-field') condition in which it had been prior to the original construction.

The ERR was a boiling water reactor (BWR), 58 MW(t) and 22.5 MW(e). The reactor vessel was 7.62 m high, 2.13 m internal diameter, and made of 75 mm thick carbon steel with a 2.54 mm cladding of stainless steel. The concrete bioshield/fuel element storage well was 13.7 m high with a maximum thickness of 2.74 m and steel reinforcement located at a depth of about 100 mm from all exposed surfaces on a 305 mm by 305 mm grid.

The decommissioning planning studies, carried out over many months, included, *inter alia*, definition of objectives, calculation of the radioactive inventory, estimation of the resultant maximum contact radiation levels on specified components, and formulation of the preferred general dismantling sequence.

The radioactive inventory calculations were cross-checked where possible against samples taken from the facility and showed good agreement. Over a total radioactive inventory of 10 000 Ci, 1200 Ci were in the pressure vessel and 8700 Ci were in the structural components within the vessel. Estimates of the resultant contact radiation levels were again cross-checked against actual measurements made at different locations in the reactor.

Typical values for components within the bio-shield were within the range 1–6000 R/h. The dismantling was divided into a number of major tasks which included:

(1) facility and site preparation;
(2) removal of all equipment and systems outside the bio-shield not required to support other dismantling activities;
(3) removal of the highly radioactive components contained within the reactor pressure vessel (the reactor internals);
(4) removal of the pressure vessel and segmentation of the outer thermal shield;
(5) removal of the outer thermal shield and of all radioactive/contaminated structures including the biological shield;
(6) removal of the reactor building and all remaining non-contaminated equipment contained within it.

Uncontaminated ancillary equipment outside the bio-shield was removed by standard demolition techniques as was the reactor building in its turn. Other equipment and systems external to the bio-shield which contained low levels of contamination were removed using normal reactor maintenance techniques, emphasis being placed on prior radiation surveys, control of contamination, and the minimization of personnel exposure. (This involved for example the use of wet cutting and bagging methods.)

The dismantling sequence adopted within the biological shield involved working outwards. Work commenced then with the removal of the reactor internals, followed by the inner thermal shield, the pressure vessel, the outer thermal shield, and the biological shield. To cope with high radiation levels the dismantling operations were carried out under water and special long-handled tools were developed. These included a sheet metal nibbler, and a series of hydraulically operated chisels and wedge tools.

The inner thermal shield was a 3.65 m high stainless steel cylinder, 2 m diameter and 25 mm thick. It was cut into 26 segments with a plasma cutting torch. (The larger segments which weighed 308.4 kg were approximately 1.5 m tall and had radiation levels as high as 1300 R/h on contact.) The use of a plasma torch has the advantage over oxyacetylene cutting of enabling stainless steel to be cut. Oxyacetylene cutting was used for carbon steels. Although plasma torches are currently used in industry, much development was needed to adapt the torch for remote operation. This involved the use of a manipulator actuated from a distant control panel. The panel was located outside the containment envelope, which had been erected to minimize the spread of contamination from the cutting operation.

A major operation in the decommissioning sequence involved the demolition of 1185 m³ of activated and contaminated concrete. To this end, a survey was made of all the available methods, and trials were carried out. The usual mechanical methods (drilling and rock-jacking, use of a demolition ball, wall sawing) were ruled out for this purpose, owing to the time involved and the consequent exposure of personnel. Flame cutting and thermal lance techniques were discarded owing to the generation of large amounts of toxic gases and smoke.

Finally, the use of explosives was considered. After considerable development using different procedures and a variety of explosives, a technique was established which was successfully used to demolish the bio-shield/fuel element storage well. Control of the radioactive contamination and debris was successfully achieved by the use of

(1) local contamination envelopes erected over the structures to be demolished,
(2) heavy rubber blasting mats to contain the debris, and
(3) a localized fog-spray system to damp down the dust.

Throughout the dismantling operations a cardinal requirement was to minimize the radiation exposure to personnel. Extensive use therefore was made of remote or underwater operations and the use of temporary shielding and other safeguards. It is however a commentary on the care taken to organize safely the work on the highly radioactive components, that while significant radiation exposure was incurred by the operational crew, the greatest portion of the radiation exposure was received during almost constant work in low-level radiation fields varying between 5 and 15 mR/h.

A total of 75 rem of whole-body exposure was received by upwards of 100 people connected with the dismantling project. The average exposure was about 0.8 rem while the maximum total exposure received by any workman was about 4.5 rem. Of the 75 rem total, approximately 12 rem were incurred during removal of the reactor internals, 45 rem during removal of the pressure vessel, 12 rem during removal of the biological shield, and 6 rem during removal of components external to it.

By this combination of detailed planning, development of optimum demolition techniques, extensive training of operational workers, and comprehensive health and safety control measures, the ERR was successfully decommissioned to the cleared-site condition at an overall cost of about US $6.15 million. Cars are now parked where the reactor stood.

Appendix B The fast reactor fuel reprocessing plant, Dounreay

The fast reactor irradiated fuel reprocessing plant Dounreay (FR/IFR) was built to reprocess the fuel from the experimental Dounreay fast reactor (DFR). It consisted of fuel-handling and cave breakdown facilities, dissolution equipment and solvent-extraction systems. The highly enriched (75 per cent ^{235}U) uranium alloy fuel was reprocessed to high-purity uranyl nitrate solution for subsequent reduction to metal and fuel refabrication. The active liquor effluents were segregated into 'high-active' and 'low-active' streams for storage or for discharge to sea after suitable monitoring control.

After 15 years operation it was decided in 1971/2 to modify the plant to permit the reprocessing of fuel from the 250 MW(e) prototype fast reactor (PFR). This is mixed plutonium oxide–uranium oxide fuel, the reprocessing of which required a more sophisticated and extensive plant than the FR/IFR. It was felt that modification of the old plant would show considerable savings over construction of a new facility. It would also provide valuable operational experience and information on the feasibility of decommissoning a fuel reprocessing plant, which would be of value to the designers of the larger-scale fuel reprocessing plant which would be required to service a commercial fast breeder reactor programme.

The plant to be modified consisted of a series of interconnected caves and cells (see Figs 16.7-16.10) in which fuel elements were broken down using a chop/leach technique. This was followed by solvent extraction on the dissolver solution in mixer–settler equipment. The thickness of the biological shielding ranged from 1.2 m concrete to 230 mm brick. There were also a number of redundant glove-box facilities.

In the process of modification, some areas were completely demolished and rebuilt (for examples the DFR blanket fuel cave) or were completely gutted leaving only the cell fabric, into which new equipment was installed. Other areas were subjected to lower levels of strip-out, and to rearrangement of the residual and new equipment.

To achieve the degree of modification it was necessary to enter nearly all sections of the plant; this had been proved practicable by previous experience.

The caves and cells were all contaminated to varying extents with both fission-product and plutonium activity. The contaminants were widely distributed in areas such as the cave and conveyor tunnel. They were predominantly located within the primary vessel containment in some of the extraction cells. These two situations were approached in different ways as follows.

The active extraction cells had gamma-radiation levels of about 100 R/hour at 1 m from the process vessels. The equipment and cell internals were also contaminated with plutonium alpha activity and fission-product beta activity in excess of 1000 counts per second (counts/s) alpha and 100 mR/h beta–gamma.

Decontamination of the vessel and pipework internals was effected by repeated washings using nitric acid and steam followed by selective washing with more aggressive decontaminants such as sulphuric acid, caustic, and nitric acid–sodium fluoride solutions. The washing continued until remote radiation monitoring indicated that it was safe to open up access into the cell. More extensive monitoring was then carried out. The contamination levels were then reduced by hosing down the cell and internal vessels with high-pressure water blasting equipment. This generally reduced the cell radiation levels below the 1–20 R/h range permitting short-term entry in order to establish radiation source areas. These were then treated specifically. For example, contaminated redundant equipment was cut out and disposed of as (solid) waste. Other equipment scheduled for reuse was further decontaminated *in situ* or in the site decontamination centre.

Unlike the active cells, the cave facilities had no comprehensive built-in liquor-washing system. Initial radiation levels within the cave and conveyor tunnel exceeded 100 R/h gamma (500 R/h beta–gamma) in some areas. Most of this came from contamination (which also contained significant plutonium alpha activity) distributed amongst the operating equipment.

The technique used was to clear up, sweep up, and swab all accessible surfaces using the built-in manipulators. The liquid effluents were removed in flasks to the highly active waste stores. The equipment and work-bench were cut out by electric arc cutting, using the manipulators. After removal of the most active material, limited man entry became possible to complete the gutting and removal of debris. A dramatic reduction in the contamination/radiation levels was then achieved by use of the high-pressure water jet.

Where necessary, structural concrete shielding was removed usually by diamond drilling or pneumatic drilling. Large sections were then broken out by using hydraulic jacks against the weakened structure. The method was successful but slow, and gave rise to high dust levels and large amounts of slurry. In the case of the cave therefore, where demolition was extensive, the work was done by specialist thermal lancing techniques. The work was thus completed safely in a fraction of the time that would otherwise have been necessary and consequently at a fraction of the radiation exposure. Extra containment, ventilation, and personnel protection arrangements were needed to cope with the quantities of toxic gases and smoke generated by this technique.

By such means the fully stripped-out areas were reduced to accessible contamination levels below 5 counts/s and below 10 mR/h. The active cell areas were reduced to contamination levels below 1000 counts/s and radiation levels below 50 mR/h. These levels were compatible with the future use of the plant areas as active facilities. Clearly the residual levels fall short of those to be achieved in total decommissioning, but this was not the intention. Total decommissioning would in fact have been easier, since the components which were responsible for the residual activity would not have been left installed but cut out and disposed of.

During the five years of the exercise (during which time the work was interrupted by two reprocessing campaigns) the average annual dose per man amounted to 2.21 rem: on a year-by-year basis, the dose fell from an average of 3.77 rem per man in the first year—the highest value for any year over the period—to 0.86 rem per man in the final year, 1976. The deviation of individual doses from the mean was small. The total man-rem dose from decommissioning and refurbishing operations over the period to 1976 was 914 man-rem.

Overall the exercise demonstrated that it is practicable to decommission and demolish a plutonium-contaminated gamma-active facility. The exercise was not intended to revert the area to the 'green-field' condition, a fact which tended actually to complicate the exercise rather than the reverse.

Figures 16.7–16.10 show details of the plant before, during, and after the modifications described.

Reference to a reactor design concept viz the use of a two-phase biological shield structure is made by courtesy of G.W. Meyers and B.F. Ureda from Design Considerations for Facility Decommissioning (1979). Use of information on the Elk River Project is acknowledged with thanks to the USAEC, under contract to whom United Power Association of Elk River prepared the Elk Power Reactor Program Report, CO 0-651-93.

Fig. 16.7 Dounreay reprocessing plant. West wall of blanket cave early 1973 before demolition

Fig. 16.8 Dounreay reprocessing plant. West wall of blanket cave June 1973 during demolition

Fig. 16.9 Dounreay reprocessing plant. Reconstruction of blanket cave for use as PFR fuel disassembly cave 1974

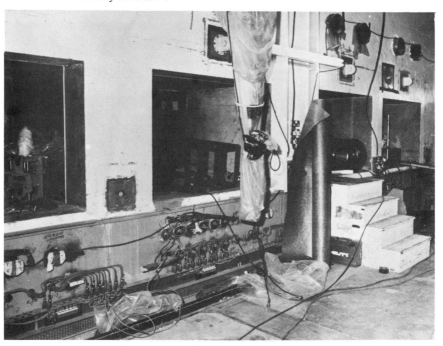

Fig. 16.10 Dounreay reprocessing plant. Completed PFR fuel disassembly cave as at January 1979

References

1. UNITED POWER ASSOCIATION. *Final Elk River Reactor Program report. COO–* 165–93. UPA, Elk River, Minnesota (1974).
2. INTERNATIONAL NUCLEAR FUEL CYCLE EVALUATION. *Reprocessing, plutonium handling, recycle, report of Working Group 4. INFCE/PC/2/4.* International Atomic Energy Agency, Vienna (1980).
3. WOOLLAM, P.B. *The radioactive inventory of a decommissioned Magnox power* station. *RD/B/N* 4231. Central Electricity Generating Board (1978).
4. SMITH, R.I. KONZEK, G.J., and KENNEDY, W.E. *Technology, safety and costs of decommissioning a reference pressurised Water reactor power station. NUREG/ CR–*0130. Battelle Pacific Northwest Laboratories (1978).
5. ESSMANN, J., BROSCHE, D., THALMANN, G., VOLLRADT, J., WATZEL, G.V.P. *Provision for decommissioning LWR power plants by the German utilities. IAEA–SM–*234/2. IAEA, Vienna (1978).
6. MANDAHL, B. *Technology and costs for dismantling a Swedish nuclear power plant, SKBF Project KBS. KBS TR*79–22. Kärn,–Bränsle–Säkerhet, Stockholm (1979).
7. UREDA, B.F. and MEYERS, G.W. *Design considerations for facility decommissioning.* Atomics International, Canoga Park, California (1979).
8. INTERNATIONAL ATOMIC ENERGY AGENCY. *Factors relevant to the decommissioning of land-based nuclear reactor plants.* IAEA Safety Series Report No. 52. IAEA, Vienna (1980).
9. MEYERS, G.W. and KITTINGER, W.D. *Progress report on dismantling of the sodium reactor experiment. IAEA–SM–*234/20. IAEA, Vienna (1978).
10. BARRETT, T.R. and THOM, D. *Decommissioning and reconstruction of the fast reactor fuel reprocessing plant, Dounreay. IAEA–SM–*234/9. IAEA, Vienna (1978).
11. BROOTHAERTS, J. DETILLEUX, E., GREENS, L., HILD, R. *Industrial experience gained in the decontamination and partial dismantling of a shut-down reprocessing plant. IAEA–SM–*234/40. IAEA, Vienna (1978).
12. SCHNEIDER, K.J., and JENKINS, C.E. *Technology, safety and costs of decommissioning a reference nuclear fuel reprocessing plant. NUREG–*0278. Battelle Pacific Northwest Laboratories (1977).
13. INTERNATIONAL ATOMIC ENERGY AGENCY. *Proc. Symp. on Decommissioning of Nuclear Facilities, Vienna, November 1978.* IAEA, Vienna (1979).

Suggestions for further reading

1. INTERNATIONAL ATOMIC ENERGY AGENCY. *Decommissioning of nuclear facilities.* IAEA Technical Committee Report No. 179. IAEA, Vienna (1975).
2. *Decommissioning of Nuclear Facilities.* IAEA Technical Committee Report No. 205. International Atomic Energy Agency, Vienna (1977).
3. AMERICAN NUCLEAR SOCIETY. *Proc. Conf. on Decontamination and Decommissioning of Nuclear Facilities, Sun Valley, Idaho, September 1979.* ANS, Sun Valley. (Ed. M. M. Ousterhont). Plenum Press, New York (1980).
4. FORE, C.S. and KNOX, N.P. Decommissioning of nuclear facilities. In *A selected bibliography. ORNL/EIS–*154/V1. Oak Ridge National Laboratory (1979).
5. MANION, W. J. and LaGUARDIA, T. S. *Decommissioning handbook.* US Department of Energy Contract EP–78–C–02–4775 (1980).

17

Thorium fuel cycles

J. D. THORN, C. T. JOHN, R. F. BURSTALL

Almost every type of reactor has been associated at one time or another with a proposal to utilize a thorium fuel cycle. Commercial-scale experience with the use of thorium fuel cycles has however been extremely limited to date. Thorium cycles offer the attraction of good fissile material utilization in thermal reactors, and if high priority is given to this aspect, thorium could in principle have commercial attractions in this application. Given, however, the relatively advanced state of development of the U–Pu fast reactor cycle, with its even better fissile material utilization, that attraction is no longer of much significance. Thorium is inferior to depleted uranium as a fertile material in fast reactors.

The article considers in detail the use of thorium fuel cycles both in thermal and fast reactors, and the parameters governing the choice between thorium-based and uranium-based cycles in these various applications.

All stages of a thorium fuel cycle, including the mining of ore, conversion, and the reprocessing and fabrication of ^{233}U fuels must be taken into account when assessing its merits. In the unlikely event that thorium reprocessing were available, while for some reason ^{238}U–plutonium reprocessing were not, then the HTR, HWR and LWR would all be possible candidates for thorium-based thermal reactor fuel cycles.

Contents

1 Introduction

The idea of using thorium as a nuclear fuel, instead of uranium, is one of long standing. It is important to appreciate from the outset however that thorium is not a substitute for natural uranium, and lacks the essential property of the latter which has made nuclear reactors possible. Representing 0.7 per cent of natural uranium is the isotope ^{235}U, which is termed *fissile*, that is, when its nucleus is in collision with neutrons of appropriate energies, it can split (*fission*) into smaller nuclei, giving off energy in the process, and also releasing more neutrons which enable further energy-producing reactions to continue (a *chain reaction*). There is no such fissile component in thorium, and it cannot therefore of itself fuel a reactor and produce energy.

The role which thorium can play is secondary and comparable to that of the remainder of the natural uranium, the isotope ^{238}U. These materials are termed *fertile*, because they capture neutrons and in the process produce new fissile material. In a typical reactor the fission process provides more neutrons than would be necessary to maintain the chain reaction if it were not for the fact that fertile material is also present. This presence makes the best possible use of the extra neutrons. In the case of ^{238}U, the new fissile material is ^{239}Pu, and with thorium, it is ^{233}U.

The purpose of developing nuclear power is to supplement the other available economic sources of energy, recognizing that those sources of proven capability to meet the bulk of our current requirements are being depleted, whilst the need for them may well continue to increase. Supplies of fissile material are also limited, so that the possibility of the ultimate depletion of this resource must also be guarded against. This provides an incentive to utilize the available material as effectively as possible, by reprocessing the irradiated fuel, in order to separate and recycle at least the fissile material it contains. This comprises the remainder of its original content and also new material formed during irradiation, which may in certain circumstances exceed the amount used and so constitute a new source.

It is pertinent to ask whether that fissile material could be better used in conjunction with some fertile material other than ^{238}U, i.e. thorium. The following are conceivable reasons for pursuing this course.

1. *Questions of neutronics*. The use of a Th–^{233}U cycle might lead to an improvement in the overall balance of neutronic events, making better use of the available fissile material, and perhaps increasing the quantity of new fissile material formed in the process.
2. *Physical and chemical characteristics*. The use of a thorium cycle could enable engineering plants, such as reactors or reprocessing plants, to be cheaper or more efficient.
3. *Other social benefits*. A thorium cycle could have advantages from the point of view of waste management, or from the viewpoint of protecting fissile materials from diversion at a national or sub-national level.

It will be clear from what has been said above, that reasonable abundance, though an indispensible condition for the realization of any advantages, is not in itself a particular advantage as long as the anticipated shortage is of fissile, not fertile, material.

Whether advantages can be identified in the use of thorium will depend on what comparison is made. Thus, a comparison can be confined to the fuelling of thermal reactors on the assumption that fast reactors are not available. This could apply if fast reactors are never exploited, or it might refer to a shorter timescale, in which case the limited period concerned would need to be recognized and taken into account. In this comparison, the advantage sought is likely to be (and that claimed usually is) an increase in the amount of power which can be produced from a limited quantity of natural uranium. This might result from an improved neutron economy, achieved by exploiting the fact that when ^{233}U fissions in thermal reactors more net neutrons are produced than with either ^{235}U or ^{239}Pu.

Alternatively, a comparison might be made with fast reactors, in which case the neutron economy would be inferior and the relative advantages claimed for thorium would have to be quite different.

No full-scale application of thorium has yet been made in nuclear power, though development work has been undertaken with various types of reactors, particularly high-temperature reactors. Very little has been done to establish the complete fuel cycle, including reprocessing and recycle, under realistic conditions.

This chapter presents information about the thorium fuel cycle in some detail and discusses its relative advantages and disadvantages compared with the conventional fuel cycle based on ^{238}U. Most of the technical material is to be found in the next four sections, which deal with the occurrence of thorium in nature and its extraction, with its nuclear characteristics, with its use in various types of reactors, and with the associated fuel cycle. Sections 6–8 relate to economic and environmental aspects of thorium use and to non-proliferation aspects. Each section is aimed at being as self-contained as possible, and a summary has been provided at the end of each. The reader whose interest centres in a limited number of sections may find the summaries in the other sections helpful.

2 Occurrence and extraction

2.1 Resources

Thorium is extracted from mineral sources which are widely distributed over the surface of the earth. There are a number of thorium isotopes, as described further in §3, and most of the longer-lived ones are found in nature. By far the most abundant of these, and the most important for present purposes, is ^{232}Th, the fertile material which on capturing a neutron becomes the fissile ^{233}U.

Theories of cosmic abundance suggest that there should be about three times as much thorium as uranium in the earth's crust. The estimates by the Organization for Economic Cooperation and Development / Nuclear Energy Agency for 'reasonably assured resources' amount, however, to only half a million tonnes, with 'estimated additional resources' bringing the total to about one-and-a-half million tonnes; this is only about one-third of the estimated uranium resources quoted in the same study. This lower figure for thorium probably reflects the fact that thorium resources have not been surveyed as intensively as uranium. A further reason may lie in the fact that, on geochemical grounds, thorium deposits can be expected to be well dispersed and relatively dilute. If this is the case, however, then it raises doubts about the economic recoverability of thorium deposits, and leads one to ask how much of the available thorium should properly be categorized as a 'resource'.

Table 17.1 shows how identified resources are distributed geographically. Unassessed quantities also occur in the sands of Bangladesh, Korea, Madagascar, South Africa, Sri Lanka, Thailand, and Uruguay.

Such high concentrations of thorium as occur are thought to result from the low solubility of the thorium salts in natural waters leading to their accumulation in residues from the weathering of rocks. The main deposit from which thorium is industrially extracted is in fact in monazite sands. It is also associated with uranium ores, and recovered from uranium mills as a by-product.

Table 17.1 Thorium resources (excluding USSR, Eastern Europe and China)

Country	Thorium (tonnes) Reasonably assured resources	Estimated additional resources
Australia	18 500	0
Brazil	58 200	3 000
Canada	0	250 000
Denmark	15 000	0
Egypt	14 700	280 000
India	320 000	0
Iran	0	30 000
Liberia	500	0
South Africa	11 000	0
Turkey	500	0
United States	52 000	270 000
Total	490 400	833 000

Monazite is one of the components resulting from the crystallization of once-molten mineral solutions. It is a mixture of isomorphous rare-earth phosphates, and is found to contain between 1 and 20 per cent of thorium phosphate and thorium silicate. Deposits were formed by weathering of the parent rock, followed by the gravity concentration of heavy minerals in sand-beds through the actions of wind and water. As such weathering most typically occurs in the tropics, it is in the coastal areas of these regions that most deposits of monazite sand are found. In India, for example, there are extensive deposits, some sands containing up to 40 per cent monazite, though the average is less than 1 per cent. Present production of thorium (about 200 tonnes per annum) is almost entirely as a by-product of rare-earth extraction from monazite sands, and a major expansion would be needed to support a nuclear power programme using a thorium fuel cycle (see Table 17.6).

2.2 Exploration

An understanding of the relationship between the geological properties of the mineral being sought and the terrain being explored is essential in mineral exploration. The natural radioactivity of thorium minerals forms the basis of most prospecting instruments, through what is measured is not the activity of the thorium, but of the natural isotopes formed by its decay. Most instruments measure beta and gamma activities, though some are designed to measure alpha activity.

Geological base maps of a proposed prospecting area are essential to a systematic exploration survey, and these can be enhanced by aerial photography and aeromagnetic surveys. Airborne gamma-ray detection and gamma spectrometry are also good mapping aids. The gamma-radiation surveys are made with Geiger and/or scintillation counters, which are carried by hand, ground vehicles, or aircraft, or are lowered into bore holes, or trailed on lake or sea beds. Gamma-spectrometer surveys are particularly advantageous for thorium prospecting as they can discriminate between the three principal radioactive elements, potassium, uranium, and thorium. Interpretation of spectrometer data needs special knowledge, but combined with photography, the method can provide very good initial evaluation of a chosen area. An emanometer or radon monitor is useful for detecting in rocks or sub-soils irregular concentrations of the noble gas isotopes, radon (^{220}Rn) and thoron (^{222}Rn), and their immediate decay products. This technique, which can distinguish between radon and thoron, provides a measurement independent of background gamma or beta radiation. It is a potentially powerful tool for exploration.

2.3 Mining and extraction

Deep and shallow mining are both used and also the open-cast mining of rich, shallow seams. The primary concentration of heavy minerals found in combination with monazite is carried out using pinched sluices and trays, cone concentrations, spirals, wet tables, or similar plants. A feed containing 1–2 per cent of heavy minerals can be concentrated in this way to 90 per cent heavy minerals, with an overall recovery of 85–90 per cent of the heavy minerals present in the deposit.

The individual heavy minerals—ilmenite, rutile, monazite, zircon, sillimanite, and garnet—are separated from one another by methods depending upon differences in physical properties, i.e. specific gravity, magnetic susceptibility, electrical conductivity, and surface properties. The wet concentrate is passed through rotary driers at up to 150 °C, the dried feed then being treated electrostatically or electromagnetically. The electrically conductive ilmenite and rutile constituents are thus first separated by this method. In the second stage of this separation process, the non-conducting monazite, which is heavy and moderately magnetic, is isolated from other minerals by the use of high-intensity magnetic separators, and air or wet tablés. The resulting concentrate contains 98 per cent monazite.

A flowsheet for concentrating monazite from beach sand is given in Fig. 17.1

2.3.1 *Chemical treatment*

The chemical properties of thorium and the rare-earth elements associated with it are closely similar, and so their separation is difficult and time-consuming. Concentrates of the rare earths must be chemically processed to separate them from the other components of the mineral, and from impurities. Monazite, the chief thorium ore, is chemically inert, and any chemical treatment for extracting thorium must intially be very severe to achieve the complete dissolution necessary for the separation of the rare-earth elements (a valuable by-product), uranium (which is frequently present), and phosphates. Many processes are available having their own variations and techniques and are often proprietary.

The most common dissolution processes are:

(1) the acid process—using highly concentrated sulphuric acid—and

(2) the alkaline process—using highly concentrated sodium hydroxide.

Other processes, such as sodium carbonate roasting, chlorination, and carburization, have occasionally been adopted.

The acid process. There are several possible acid processes of which just one will be described here. Monazite, ground to below 100 mesh, is reacted at above 200 °C in highly concentrated sulphuric acid in agitated heated pots for about 24 hours. The slurry is then cooled and added to cold water in leach tanks fitted with agitators. The rare-earth and thorium sulphates dissolve, leaving silica, rutile, ilmenite, and zircon residues, which are filtered off. Sodium pyrophosphate is added to the solution, and thorium pyrophosphate is precipitated and recovered by filtering, leaving rare-earth sulphate in solution.

The alkaline process. The alkaline process involves reacting finely-ground monazite with 50–70 per cent sodium hydroxide for about four hours at about 140 °C in a cast-iron pot, equipped with an agitator. The slurry is settled for several hours, and the solution, containing water-soluble trisodium phosphate and excess sodium hydroxide, is decanted and filtered, leaving the rare-earth elements and thorium as insoluble hydroxides. Treating these hydroxides with hydrochloric acid (pH 3.5) dissolves the rare-earth hydroxide, and leaves solid thorium hydrate with only slight impurities of rare-earth hydroxides. The thorium is recovered by a stronger acid solution.

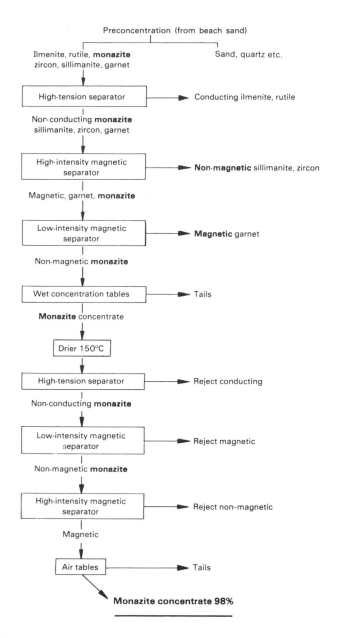

Fig. 17.1 Typical flow sheet for concentrating monazite sands or ores

2.4 Refining

For use as a nuclear fuel, thorium production is required to achieve a high level of purity, particularly with respect to neutron-absorbing elements. This purity is obtained industrially by (1) solvent extraction, (2) ion exchage, or (3) direct chemical precipitation.

1. *Solvent extraction* involves the preferential transfer of the desired material from one solution (the carrier) to another (the solvent). The carrier and solvent liquids must be immiscible with one another, so that they can subsequently be separated. The required material can then be recovered in purer form from the solvent by a subsequent extraction.

In purifying thorium by this process, the crude thorium, produced by a chemical treatment as discussed in §2.3, is converted to a solution of nitrate in water. When the impure solution is contacted with a suitable organic solvent (typically tributyl phosphate dissolved in kerosene or hexane), thorium selectively passes into the organic solvent as a complex salt, $Th(NO_3)4.2TBP$. The organic solution is then separated off, and the thorium is *stripped* from the solvent using dilute nitric acid. Nuclear-grade thorium nitrate is obtained with more than 99 per cent thorium recovery by successive operations.

2. *Ion exchange* is an effective way of separating individual lanthanides (rare-earth elements) in a pure state. Basically, it is a chromatographic technique for separating ions by their sorption from solutions on a suitable *ion exchange* medium, followed by differential elution (washing off) of the ions with a suitable solution.

3. *Direct-precipitation* processes involve either the addition of foreign ions to precipitate insoluble salts, or else exploit variations in solubility with pH of complex salts. Repeated precipitations are needed to ensure adequate purity.

2.5 Reduction to metal

Purified thorium nitrate can be used as feed material for producing thorium. The reduction of thorium compounds to the pure metal thorium is not easy, because at its high melting point of about 1700 °C thorium reacts readily with hydrogen, oxygen, nitrogen, carbon, and many oxides. The metal is usually produced as a sponge or powder by one of the following methods.

1. Thorium oxalate is precipitated from thorium nitrate with oxalic acid. The oxalate is converted to oxide by heating at 650 °C, and the oxide is then reduced to metal with calcium at about 1000 °C, using calcium chloride as a flux.

2. Thorium tetrafluoride is prepared by reacting thorium oxide at 325 °C with anhydrous hydrofluoric acid gas. The thorium tetrafluoride is then reduced in an exothermic reaction with calcium at 800 °C, using zinc chloride as a booster, and the zinc is subsequently removed by pyrovacuum treatment at 1360 °C.

3. Thorium tetrachloride is obtained by chlorinating a mixture of thorium oxide and carbon at 600 °C, and purifying by redistillation. The molten salt electrolysis of thorium tetrachloride in a graphite crucible which acts as the anode, results in the deposition of thorium metal at a molybdenum cathode. A similar electrolytic method can be used for the production of thorium metal from thorium tetrafluoride.

2.6 Summary

On theoretical grounds one would expect thorium to be about three times as abundant as uranium. Resources that have been identified are however a factor of three lower than those reported for uranium. This may be due to a lack of incentives for exploration, or it may be that much of the thorium occurs in dilute form. The principal deposits are in monazite sand, though thorium is also recovered as a by-product with uranium. The countries known to be rich in thorium are India, Egypt, and the USA.

Commercial extraction processes exist. Current production rates, however, are trivial compared with the requirements of even a modest nuclear power programme using this fuel.

3 Nuclear reactions

3.1 General review

Thorium is a useful material in nuclear reactors because neutron capture in ^{232}Th leads to ^{233}U, a fissile isotope of uranium with similar nuclear properties to the better-known isotope ^{235}U. Figure 17.2 shows the isotopes of most

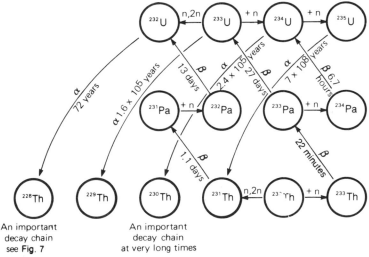

Fig. 17.2 Main isotopes in ^{232}Th – ^{233}U fuel

importance in the ^{232}Th–^{233}U fuel cycle, with their half-lives, decay modes, and neutron reactions. All the isotopes shown can undergo fission if sufficiently energetic neutrons are available. The probability of fission, however, varies considerably from isotope to isotope. Of the other possible neutron reactions, neutron capture is by far more probable than (n, 2n) reactions, in which neutron capture is followed by the emission of two neutrons. The two (n, 2n) reactions shown in Fig. 17.2 are included only because they are involved in the production of the isotope ^{232}U, which is of particular importance for fuel fabrication and reprocessing, as explained in §5.3.

Both decay processes and neutron reactions are accompanied by radioactive emissions. Neutron capture is invariably followed by gamma-ray emission. Decay involves the emission of either beta particles accompanied by gamma rays or alpha particles. The isotopes which undergo beta decay are characteristically of a relatively short half-life, whereas the isotopes which suffer alpha decay have relatively long half-lives.

A number of the isotopes shown in Fig. 17.2, namely ^{231}Th, ^{233}Th, ^{232}Pa, and ^{234}Pa, are so short lived that, apart from the radioactive emissions associated with their decay, they can be ignored for most practical purposes. (For example, ^{233}Pa can be regarded as being produced directly from ^{232}Th.) Some isotopes, namely ^{232}Th, ^{231}Pa, ^{233}U, ^{234}U, and ^{235}U, are by contrast so long-lived that they may by regarded as being stable for most purposes. The decay of these long-lived isotopes will be significant only in the context of long-term waste disposal studies.

The remaining two isotopes, ^{233}Pa and ^{232}U, have half-lives which are in an intermediate range, and for this reason both of them play a significant role in the assessment of the ^{232}Th–^{233}U fuel cycle.

3.2 Consideration of individual isotopes

3.2.1 *Thorium-232*
The nuclear property of ^{232}Th which is of most importance in determining its behaviour in a reactor is its neutron capture cross-section. Plots of the variation of this cross-section with neutron energy are shown in Figs 17.3 and 17.4. Figure 17.3 shows the variation up to an energy of 1 keV, and Fig. 17.4 the variation above 1 keV. Figure 17.4 also shows the variation of other cross-sections of interest.

It is a general rule that neutron capture cross-sections vary inversely with neutron velocity (or with the square root of neutron energy). This rule applies fairly well over much of the lower-energy range of Fig. 17.3. At higher energies (the *resonance region*) the cross-section rises to peaks at certain energies. Depending on the neutron flux in the resonance region, many neutrons may be absorbed at these resonance energies.

Figure 17.4 plots for comparison the capture cross-section of ^{238}U. It will be seen that the ^{238}U-value is lower than the ^{232}Th-value throughout the energy range, typically by a factor of about 1.5.

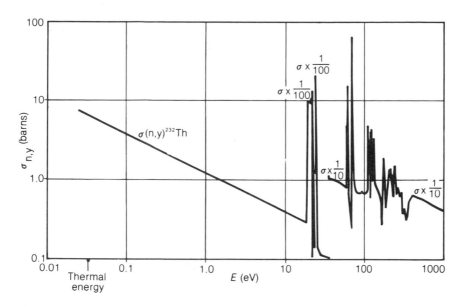

Fig. 17.3 Capture cross-sections of ^{232}Th

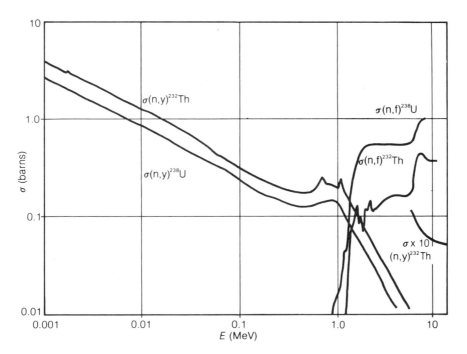

Fig. 17.4 Capture and fission cross-sections of ^{231}Th and ^{238}U

Figure 17.4 also plots the fission cross-section of ^{232}Th. Below 1 MeV this cross-section is zero, and only above 1.4 MeV does it rise rapidly. Even in a fast reactor, only 10–15 per cent of the neutron flux is typically in the energy range above 1 MeV, and this is why fission of ^{232}Th is of minor importance. In contrast, the fission cross-section of ^{238}U, which is also shown in Fig. 17.4, is already significant at energies below 1 MeV, and since there are many more neutrons at these slightly lower energies, fast fission of ^{238}U is of significance in all reactors, and is particularly important in fast reactors.

For the purposes of reactor calculations, it is necessary to divide the energy range of Figs 17.3 and 17.4 into discrete intervals and to represent the cross-section within each interval by an average value applicable for that group of neutrons. These energy intervals are chosen to reflect the cross-section variations of the different neutron reactions in a manner adequate to the purpose for which the calculation is being carried out. For accurate analysis it is almost always necessary to divide the range into many intervals, but for the purposes of qualitative comparison it is convenient to use an approximately equivalent cross-section applicable over a wider range of energies. In certain cases it may be sufficient to use a single, average cross-section value for the whole range of interest, this being termed a *one-group cross-section*.

Different types of thermal reactors have different neutron energy spectra, so that for each reactor type there will be a different one-group cross-section. It is therefore customary in the case of thermal reactor systems to quote two parameters, one being the cross-section at a neutron velocity of 2200 m/s (which is an energy of 0.025 eV), the most probable velocity at room temperature. The other parameter is the *resonance integral*, the integral cross-section value over all the energies within the resonance region, weighted by the reciprocal of the energy. In the case of fast reactors, where the variation between reactor types is not significant, it is possible to quote one-group cross-sections as indicative parameters. Table 17.2 shows typical parameters for the fission and capture cross-sections for the fertile isotopes, ^{232}Th and ^{238}U.

From the thermal reactor data, it will be seen that ^{232}Th has a higher

Table 17.2 Condensed nuclear data for fertile isotopes

	Thermal reactor				Fast reactor	
	2200 m/s value (0.025 eV)		Resonance integral		One-group	
	^{232}Th	^{238}U	^{232}Th	^{238}U	^{232}Th	^{238}U
Cross-section (barns):						
fission	0	0	0	0	0.01	0.05
capture	7.6	2.7	85	275	0.35	0.3
Neutrons/fission (on average)	—	—	—	—	2.3	2.75

capture cross-section than ^{238}U at a thermal energy of 2200 m/s, but the resonance integral for the capture cross-section of ^{232}Th is lower than that for ^{238}U. The importance in any particular reactor of the resonances which occur at neutron energies above thermal will depend on the degree of neutron moderation in the reactor core. In well-moderated reactors, such as current heavy-water-moderated reactors, which have a soft spectrum (i.e. relatively few neutrons at the high-energy end), these higher resonance energies will be of least importance. The capture cross-section at higher resonance energies is of most importance in reactors with a hard spectrum such as the high-temperature reactor (HTR), which depend on some, but less, moderation. The effective degree of moderation depends not only on the overall ratio of moderator to fuel, but also on the extent to which the fuel is segregated. since this allows more neutrons to slow down through the resonance energy region before meeting a fuel nucleus, and so reduces, the probability of capture. The more thoroughly the fuel is dispersed in the moderator, as in early HTR designs, the more resonance capture takes place, and the higher the concentration of fissile material (*enrichment*) necessary to maintain the chain reaction. In such a case a low resonance integral becomes increasingly desirable, and in this respect thorium shows an advantage.

There are two main differences in the one-group parameters for fast reactors. Firstly, the fission cross-section of ^{232}Th is much lower than that of ^{238}U. Secondly, the neutron capture cross-section of ^{232}Th is higher by about 15 per cent than that for ^{238}U.

The importance of the first effect, that of the fission cross-section, is that in uranium–plutonium-fuelled fast reactors, a significant proportion (about 15 per cent) of the fissions take place in ^{238}U. This effect is a double bonus, because each fission which takes place in a fertile material means a corresponding reduction in the destruction of the valuable fissile material, and the extra fission neutrons produced contribute towards the formation of fresh fissile atoms. In a ^{232}Th–^{233}U-fuelled fast reactor, only about 2 per cent of the fissions take place in ^{232}Th, and the breeding gain is accordingly lower. This result is further amplified by the number of neutrons produced per fission also being lower. In thermal reactors, the fission of fertile material is less important and is very small in ^{232}Th, but is still quite significant in ^{238}U.

The importance of the second effect, the higher capture cross-section of ^{232}Th, is the impact it has on the fuel enrichment requirement. The fuel enrichment may be defined as the required ratio of fissile atoms to total fissile plus fertile atoms in the reactor fuel. A high capture rate in the fertile material implies a need for a high feed enrichment, which increases the fissile material doubling time resulting from a given breeding gain.

3.2.2 *Uranium-233*

The cross-section of ^{233}U of most importance is usually that of fission, though the capture cross-section is by no means negligible. Fig. 17.5 gives the variation of the fission cross-section with energy up to 1 keV and shows

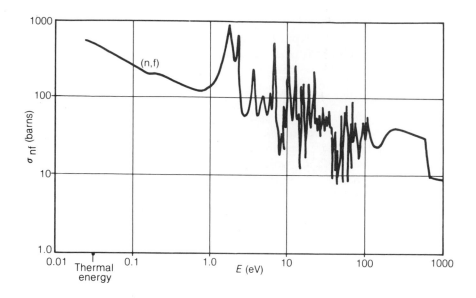

Fig. 17.5 Fission cross-sections of ^{233}U

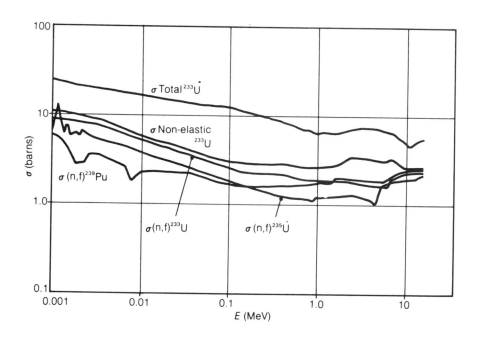

Fig. 17.6 Cross-sections of fissile isotopes

that there are many resolved resonances in the ²³³U fission cross-section, in the range 1–200 eV.

Fig. 17.6 shows the variations of ²³³U cross-sections above 1 keV and includes the fission cross-sections of the main alternatives to ²³³U, i.e. ²³⁵U and ²³⁹Pu, showing that overall the fission cross-sections of these last two isotopes are lower than that of ²³³U.

3.2.3 *Protactinium-233*

The importance of ²³³Pa lies in its half-live of 27 days. This time is long enough to create a significant delay in the production of ²³³U from ²³²Th, which has the following adverse effects.
1. Capture in ²³³Pa robs the cycle of a neutron.
2. It also robs the cycle of a fissile atom which would have been formed by decay of the ²³³Pa.
3. The delay in the production of a fresh fissile atom has important effects on reactivity.

When a ²³³Pa atom captures a neutron, the effective result is to prevent the production of a potential ²³³U atom. As may be seen from Fig. 17.2, neutron capture in ²³³Pa leads to ²³⁴Pa, which decays to ²³⁴U. The losses to fissile isotope production that are incurred in this way depend firstly on the rate at which ²³³Pa decays (to form ²³³U), and secondly on the rate of neutron absorption in the ²³³Pa. While the former process is determined solely by the natural decay constant of the isotope, the latter is directly dependent on the neutron flux level of the reactor. At typical thermal reactor (PWR) flux levels, the natural radioactive decay rate is generally much higher than the rate of neutron capture, so that only a few per cent of the potential ²³³U production is lost. At higher flux levels however, this problem becomes more significant, and the higher losses detract from and ultimately limit the economic advantages which might be expected from designing for a compact core, to the extent which otherwise might be possible, for example, in an HTR.

Another feature of ²³³Pa is its decay after the reactor is shut down. The resulting emission of beta particles and gamma rays contributes a significant porportion to the total radioactivity of the fuel. The decay of ²³³Pa also leads to an increase in the ²³³U content of the core, and to a decrease in neutron capture probability. On both counts the fuel becomes more reactive which may have implications for the control of the reactor.

3.2.4 *Uranium-232*

The main importance of the isotope ²³²U is that it is at the head of a decay chain, some of whose members emit gamma rays of high energy. The decay chain is shown in Fig. 17.7. All the members of the chain have relatively short half-lives except for the first two, and ²²⁸Th has a much shorter half-life than ²³²U. The most important of the gamma-ray emitters is ²⁰⁸Tl, with a

Decay type		Half life	Gamma emmission
	^{232}U		
α	↓	72 y	low energy
	^{228}Th		
α	↓	1.9 y	low energy
	^{224}Ra		
α	↓	3.6 d	0.24 MeV
	^{220}Rn		
α	↓	55 s	0.54 MeV
	^{216}Po		
α	↓	0.15 s	none
	^{212}Pb		
β	↓	10.6 h	0.3 MeV
	^{212}Bi		
α	↓	60 min	0.78
	^{208}Ti		
β	↓	3 min	2.6 MeV
	^{208}Pb	stable	

Fig. 17.7 Decay of ^{232}U

2.6 MeV gamma energy. Gamma rays are hazardous owing to their high penetration of most materials, and this energy is in a range at which it is particularly difficult to reduce the dose to acceptable levels. This delayed gamma-ray emission from ^{232}U, which is chemically inseparable from ^{233}U, constitutes a serious drawback to the use of the ^{232}Th–^{233}U fuel cycle. It is fortunate that the production of ^{232}U depends on (n, 2n) reactions which are of low probability. Nevertheless, because of its presence, it is necessary to carry out the fabrication of fuel elements by remote handling behind adequate shields. As the irradiation of the fuel is extended, either in a single residence in the reactor, or by recycle, increased shielding is required.

3.3 Thermal reactors

As already mentioned, the neutron flux spectra differ significantly for the various thermal reactors. However, in comparing the various thorium-based fuel cycles with their uranium–plutonium-fuelled counterparts, one finds that the physics parameters differ in basically the same way though to a varying degree, for all the reactors. Table 17.3 gives the 2200 m/s cross-sections and the resonance integrals of the main fissionable isotopes for thermal reactors.

From the nuclear data in Table 17.3, it will be seen that the advantage of ^{233}U over other fissile materials lies in its low neutron capture cross-section.

Table 17.3 Condensed nuclear data for fissile isotopes in thermal reactors

	^{233}U	^{235}U	^{239}Pu
Cross-section (barns)			
Fission:			
2200 m/s value	527	579	741
resonance integral	764	275	301
Capture:			
2200 m/s value	54	100	267
resonance integral	140	144	200
Neutrons/fission (on average)	2.5	2.4	2.9

Thus for a given number of fissions (that is to say, for a given energy production) there is a much smaller total destruction of primary fissile material for this isotope than for either of the other two. Hence, even though ^{239}Pu produces a higher number of neutrons per fission than ^{233}U, the lower capture cross-section of ^{233}U more than compensates for this, giving it the highest net contribution of neutrons to the chain reaction of any of these three materials.

The presence of ^{233}Pa as an intermediate product between the fertile ^{232}Th and the fissile ^{233}U detracts from this advantage by an amount depending on the compactness of the core design adopted, as discussed in 3.2.3. This is of particular importance in thermal reactor conditions, in which the economics and logistics of the thorium fuel cycle push reactor optimization towards high ratings where the loss of neutrons and the subsequent loss of new fissile material begin to increase rapidly. This same feature leads to marked transients in core reactivity during sustained shutdown or part-load operation, and thus to problems of fissile and control rod inventory at the high ratings that are desirable.

The advantages of ^{239}Pu are its high fission cross-section and its high release of neutrons per fission. The high fission cross-section means that fewer neutrons are lost in unproductive capture in structural materials and other non-fissionable materials. The high neutron release of ^{239}Pu is important because it means that, as far as this effect is concerned, fewer atoms of ^{239}Pu are needed to sustain the chain reaction.

A further factor which is important in comparing ^{232}Th–^{233}U systems with uranium-plutonium systems is that the higher plutonium isotopes build up to a greater extent than do the higher uranium isotopes from ^{233}U. The cause is the higher capture cross-section of ^{239}Pu as may be seen from Table 17.3. Neutron capture in ^{239}Pu leads to the new isotope ^{240}Pu, which itself has a high capture cross-section in thermal reactors. Thus as irradiation proceeds the probability of neutron capture increases rapidly and the amount of fissile material necessary to provide sufficient reactivity to sustain long irradiations is correspondingly increased.

Table 17.3 shows that ^{235}U is in an intermediate position between the other fissile isotopes since, although it has a higher fission cross-section than ^{233}U, it also has a higher capture cross-section. ^{235}U differs from either ^{233}U or

^{239}Pu, inasmuch as successive neutron captures do not result in the production of further fissile isotopes. Thus neutrons and fissile atoms lost by capture in ^{235}U can be regarded as permanently lost.

A proper comparison between ^{233}U and ^{239}Pu needs of course to take into account the circumstances of their original production and the properties of their parent materials, ^{232}Th and ^{238}U, discussed in §3.2.1. There is no simple and unique way of using the fuel parameters in Tables 17.2 and 17.3 to derive a universal figure of merit. Even considering the data in Table 17.3 alone, there are at least two separate criteria of economic significance, against which they need to be evaluated: the inventory of fissile material in the reactor, and the net consumption or production rate of the fuel. Both of these, furthermore, are dependent on the reactor design as well as on the properties of the fuel itself. In summarizing the situation for thermal reactors, it is nevertheless broadly true to say that the thorium–^{233}U cycle may well require a higher fissile inventory, but is likely to consume less fissile material, net per fission.

3.4 Fast reactors

Table 17.4 gives one-group data for a typical fast reactor. The fission cross-section of ^{233}U is considerably higher than for ^{235}U or ^{239}Pu, and the capture cross-section is lower. ^{239}Pu has the highest capture cross-section, though the difference is less marked here than it is in the case of thermal reactors. This change makes ^{239}Pu relatively more attractive as a fuel for use in fast reactors than in thermal reactors. From the data in Table 17.4, ^{233}U again appears the most attractive fissile material for fast reactor use. However, the overall production of new fuel compared with fuel consumption depends on contributions to the neutron balance from other sources also. Unfortunately, the apparent attractiveness of the ^{233}U isotope is more than offset by the fission contribution from the fertile ^{232}Th isotope being much lower than that from ^{238}U in the uranium–plutonium fuel cycle (see Table 17.2 and §3.2.1). This also tends to swing the balance of advantage away from the thorium cycle back in favour of the ^{238}U–^{239}Pu cycle. As a result, the ^{232}Th–^{233}U fuel cycle proves to be relatively unattractive and, as will be seen in §4, fast reactor studies suggest that a thorium–^{233}U cycle would give about 20 per cent less breeding of fissile material than a comparable ^{238}U–plutonium cycle.

Table 17.4 Fast reactor one-group nuclear data for fissile isotopes

	^{233}U	^{235}U	^{239}Pu
Cross-section (barns):			
fission	2.8	2.0	1.9
capture	0.3	0.5	0.6
Neutrons/fission (on average)	2.5	2.5	2.9

3.5 Summary

Comparing the fissile material utilization of the ^{232}Th–^{233}U and the ^{238}U–plutonium fuel cycles in thermal reactors, a balance has to be struck between a number of advantages and disadvantages. The important advantageous feature of the thorium cycle is the low capture cross-section of ^{233}U compared with ^{235}U or Pu. The beneficial effect of this is counterbalanced by:

(1) losses of neutrons in ^{233}Pa, which is present as an intermediate product between the fertile ^{232}Th and the fissile ^{233}U;

(2) the high neutron capture cross-section of ^{232}Th (the source material of ^{233}U), as compared with ^{238}U; and

(3) the fact that the build-up of higher isotopes resulting from neutron captures starting from ^{233}U is less useful than the equivalent process in the case of plutonium.

The individual significance of these various factors will depend on the reactor type and on even more specific design features. Thus the ^{233}Pa effect is intensified by the use of compact core designs with high power density and could preponderate if this design objective is greatly emphasized. On the other hand, the use of a low degree of neutron moderation, whilst increasing the neutron capture in either ^{232}Th or ^{238}U, has relatively less effect with the former.

It can be concluded that if good fissile material utilization is given top priority in selecting a reactor type and in determining its design, and that if that selection is confined to the range of thermal reactors, a thorium cycle is likely to provide the lowest consumption of fissile material, though it may well require a relatively high inventory.

If fast reactors are brought into the comparison, however, the disadvantages of ^{232}Th outweigh all that might be gained from the use of ^{233}U, one consequence being that the achievable breeding gain is at best very small and may even be negative. In this case a primary factor is that in the uranium-plutonium cycle direct fission in ^{238}U makes a significant contribution (15–20 per cent) to the total number of fissions, virtually equivalent to the whole breeding gain, whereas in the thorium cycle, the contribution from direct fission in ^{232}Th is negligible.

Finally it should be noted that all ^{232}Th–^{233}U systems lead to some production of ^{232}U. Although the quantities are not large and make no significant impact on neutron and fissile material balances, there is a long decay chain leading from ^{232}U to ^{208}Tl, which then decays to ^{208}Pb with the emission of a very high-energy gamma ray. This particular nuclear reaction leads to complications in reprocessing and recycling operations.

4 Utilization of fissile material

4.1 **Introduction**

Virtually every type of reactor has been associated at one time or another with some proposal to utilize a thorium fuel cycle. The purposes behind such proposals have varied, and have influenced the related details to some extent. Few have been the subject of major development work. Perhaps the earliest and certainly the most thoroughly studied application is to high-temperature reactors,[1] and the history of that development reflects changes in circumstances and in objectives over the period of thirty years or so since its first inception. More recently, concern about limited supplies of ^{235}U has led to some interest in the possibilities of thorium cycles for light water and heavy water reactors.

It is self-evident that the best utilization of fissile material, expressed as the energy capable of being produced, starting with a given quantity of that material, will be provided by reactors which breed; that is to say, in which the amount of fissile material destroyed in producing power is less than the amount newly created by the conversion of fertile materials. In a breeder system, it is possible, by repeated reprocessing and recycling of fuel, to make virtually the whole of the fertile material available as a source of energy. It is also possible to expand the system of breeder reactors without calling on any further supplies of fissile material from outside the system, at a rate which depends on the breeding ratio and on the fissile inventory required by the fuel cycle. This leads to the concept of *doubling time*, the time over which sufficient new fissile material is accumulated so that the breeder system can be doubled. A short doubling time is thus a most desirable breeder reactor property, at least until sufficient fissile material is being produced to meet the requirements for power production.

Some attempts to achieve breeding in thermal reactors will be mentioned later in this section, but a simple consideration of fissile material utilization leads to a preference for fast reactors, of which the paramount characteristic is their ability to provide shorter doubling times. From this point of view, the only interest in thorium would lie in comparing it with ^{238}U as the fertile material in fast reactors.

The fundamental physical reasons why thorium is relatively unattractive in fast reactors have been outlined in §3 and the point is further illustrated in terms of doubling times in §4.9. As far as immediate practical interest in thorium-fuelled reactors is concerned, that might appear to be the end of the matter. However, various other reasons have been put forward for taking an interest in the possible use of thorium in thermal reactors. A view might be taken that fast reactors or their essential concomitant, reprocessing, either will not or should not be exploited on an effective scale. Some of these arguments such as those based on fears of the possible proliferation of nuclear weapons, are discussed in later sections. A different argument stems from the fact that fast reactors will certainly not be available as the dominant source of nuclear power for many years to come, and it may be considered that meanwhile quite radical departures from current thermal reactor practice should receive attention.

The interest in the use of thorium is thus primarily related to systems consisting solely of thermal reactors and it is in these systems that the possibility of some benefit over the use of the uranium–plutonium cycle requires closer examination. This therefore becomes the main subject of this section. In discussing it, whatever reason has been adopted for discounting fast reactors has to be borne in mind.

4.2 Comparisons

In selecting a reactor and a fuel cycle, many matters have to be taken into consideration. The principal case for the thorium fuel cycle postulates a future shortage of fissile material, such that it becomes of great importance to achieve a high utilization of such material as is available. Even in confining one's attention to fuel utilization however, the selection and optimization of design and performance parameters should remain within reasonable bounds of technical feasibility and cost. Full assurance on these matters is almost impossible, and personal judgement cannot be eliminted. The attempt has been made here, in assessing fissile material utilization, to keep these considerations in mind. Even so, judgements can legitimately differ, and may result in different design optimizations and hence in different fissile material utilizations being quoted, even for ostensibly the same reactor type. For the HTR, for example, where there is a fairly good background of information, defensible assessments of fissile material requirements can vary by up to, say, 20 per cent. With some other types of reactor, the plausible range is broader. Nevertheless, in spite of these uncertainties, analysis can provide informative comparisons and, with due caution, useful conclusions can be drawn.

Before discussing particular types of reactor in turn, it may be helpful to have an approximate ranking order in mind. A broadly indicative selection of estimates is presented in Table 17.5. The method adopted is to quote for each reactor type and fuel cycle the quantity of natural uranium which will need to be provided over a representative service life. This requires some explanation.

All reactors require an initial inventory of fissile material. All reactors which recycle fuel (other than breeders) also require a make-up of fissile material in periodic refuelling. If this fissile material is ^{235}U, it has to be derived from natural uranium, by a separation (enrichment) process. A direct relationship therefore exists between the amounts of ^{235}U required, and of its natural uranium source, which depends only on the performance of the enrichment plant.

Alternatively, either ^{233}U or plutonium could be used instead of ^{235}U for

Table 17.5 Uranium requirements of various thermal reactor systems

	Tonnes nat. U*
PWR	
U once-through	4700
Improved U once-through	4000
Thorium–^{233}U recycle (denatured)§	3300
Uranium–plutonium recycle	3000
HWR	
Nat. U	4000
1.2% ^{235}U	2800
Thorium–^{233}U recycle	1800
Uranium–plutonium recycle	1900
HTR‡	
Once-through	3400
Thorium once-through	3200
Thorium–^{233}U recycle (denatured)§	2500
Thorium–^{233}U recycle	1800

* tonnes natural U per GW(e) capacity operated for 30 years at 75% load factor, assuming 0.2% tails from enrichment process
‡ batch refuelling; continuous refuelling is better by a few hundred tonnes
§ for an explanation of the term *denatured* and of its relevance, see §8

the fissile material make-up. Both of these are, however, *secondary* fissile materials, and their production from the fertile materials, ^{232}Th and ^{238}U, depends ultimately upon ^{235}U having been used to provide the required neutron flux. Using ^{233}U or ^{239}Pu therefore equates, albeit indirectly, with a use of ^{235}U elsewhere in the system, and thus to a requirement for natural uranium.

If a scheme does not include reprocessing and recycle of material, that is to say, if it is a once-through cycle, then the make-up required is, of course, the full amount of fissile material to be loaded into the reactor. In all cases, it is necessary to take account of the likely quantities of material which would be lost from the useful cycle in the course of operations such as fuel fabrication. Similarly, the hold-up of material in fuel cycle operations has to be included in the total inventory requirement.

This method is of course a simplification. For example, the distinction between inventory and make-up affects the timing of requirements, which could be an important consideration. Nevertheless, the total forward commitment, when ordering a reactor, to a future need for natural uranium is perhaps of greatest interest, and certainly constitutes a convenient single indicator which enables direct comparisons to be made between different reactors or fuel cycles.

In the recycle cases shown in Table 17.5, it is assumed that plutonium or ^{233}U is returned to the same type of reactor that produced it. Thus the possibility of using the fissile material to provide the first charges of fast reactors has been ignored. Such a course would result in some additional preference for plutonium-producing cycles over thorium cycles producing ^{233}U.

4.3 High-temperature reactors (HTR)

In discussing particular reactor types, the high-temperature reactor must have pride of place. This reactor system was originally intended to use a thorium cycle, and the development of the reactor part of the cycle has reached a stage of near-commercialization. The choice of thorium for the system was related to the objectives of achieving a high outlet temperature and a compact reactor core.

The HTR is a graphite-moderated, helium-cooled system, in which the fuel is relatively dispersed in the moderator in the form of separate particles, rather than, as in the case of most reactor types, being segregated as stacks of fuel pellets in metallic cans. One result is that the core is composed entirely of ceramic materials capable of withstanding relatively high temperatures. In early designs, as an additional aid to achieving a high coolant outlet temperature, virtually the whole of the graphite moderator was used so as to present a very large heat-transfer surface to the coolant. This in turn required a heat-transfer path from the fuel to the coolant passing through the graphite, and the temperature difference required was kept low by a notionally complete dispersion of the fuel in the moderator. As explained in §3.2, this led to nuclear characteristics which in relative terms favour the use of thorium. Thorium also permits the use of relatively low neutron moderation and is again compatible with the concept of a relatively compact core design. From simple theoretical studies, it seemed that conversion factors approaching unity might be achieved.

As designs were developed, more detailed attention was given to fission-product behaviour and to fuel-manufacturing costs. This led to the use of coated fuel and some segregation of the fuel in the moderator, with adverse effects on neutron economy, and supplemented other reasons for considering an HTR concept using the ^{238}U–plutonium fuel cycle.[2] These other reasons included the lack of an established reprocessing route for thorium, the limitations to power density set by neutron capture in protactinium (^{233}Pa), and the improving availability of ^{235}U at moderate enrichment levels. The plutonium produced could be recycled in the HTR, or, more effectively, in a fast reactor. However, although the HTR is no longer associated exclusively with the thorium–^{233}U cycle, it is closer than any other reactor system to being applied in that role. Small power reactors have been or are being built at Fort St. Vrain, Colorado,[3] and at Schmehausen, North-Rhine–Westphalia.[4] The latter embodies a distinctive *pebble-bed* core, consisting of graphite spheres of about 60 mm in diameter, which contain fuel particles within their structure. This design facilitates continuous fuel charge and discharge, which offers improved fissile material economy, since it minimizes the amount of control poison inserted into the core.

Because there is no established thorium fuel reprocessing route, commercial attention has so far been directed to a once-through mode of operation, at least as a temporary regime, though this requires nearly double the amount of ^{235}U over a reactor lifetime compared with the ^{233}U recycle case. Even so, the ^{235}U requirement is still much less than that of the current light water reactors operating in a once-through mode (see Table 17.5).

4.4 Heavy-water-moderated reactors (HWM)

Heavy water is an excellent moderator material, particularly where economy is sought in the use of fissile material, because of its relatively low capture of neutrons. In the best known form of heavy water reactor, CANDU, the heavy water is contained in a vessel (*calandria*), through which pass a number of tubes providing channels to contain the fuel and the flowing coolant. The latter is also heavy water but under pressure. The neutron economy with this type of reactor is so good that in its present commercial application the uranium fuel supplied has only its natural ^{235}U content.

Since the objective with thorium is to obtain good neutron economy, it might be expected that reactors such as CANDU, which have inherently low neutron losses, would be best suited for the purpose. It should be noted, however, that in the standard CANDU cycle, no reprocessing is carried out, and burn-up of the fuel is low, being limited by reactivity requirements. For the best neutron economy, the use of thorium would involve the introduction of reprocessing, in order to recycle the ^{233}U, and would also involve more expensive fuel fabrication (due to the activity of the recycled fuel), which must be offset against any other benefit. This would in turn lead to extended fuel irradiation in order to reduce the impact of those costs. As a result, some of the neutronic benefit would be lost due to increased neutron capture in fission products. Nevertheless, if attention is confined to reactor types already established in commerce, and without many or extreme modifications being made to them, CANDU appears as one of the more promising candidates for the application of the thorium cycle.

In considering the optimization of design to meet some assumed future situation of high uranium price, a range of alternative possibilities becomes apparent. Table 17.6 illustrates this point, with reference to CANDU.[5] (All the cases benefit, like the HTR pebble bed, from 'continuous' refuelling which is the standard practice in CANDU reactors.)

A fuel cycle can be conceived which is self-supporting to the extent that enough ^{233}U is created to replace the amount that is fissioned, though it remains necessary to provide fissile material from elsewhere to provide inventories for new construction. In order to make the system self-supporting in this way, fuel burn-up has to be limited, probably to about 10 MW d/kg h.a. as shown in the last column of Table 17.6. Depending on the various costs involved, however, it may be more economic to aim for higher fuel burn-up and to accept the need for a continuing supply of fissile material for topping up. In fact, the main doubt is not on whether this economic pressure will exist, but rather on what level of burn-up can be satisfactorily reached in commercial operation, within the limits of fuel element capabilities, licensing

Table 17.6 Fissile isotope requirements of various CANDU fuel cycles

	Natural U	Thorium high burn-up	Thorium lower burn-up
Burn-up (MW d/kg of heavy atoms)	7.5	37	10
Feed fuel (as all oxides)	Nat. U	(thorium + recycle U+ Pu)	
Equilibrium feed rates (per GW(e) at 80% Load Factor)			
Fissile Pu (kg/y)	—	135	0
Nat. uranium (t/y)	133	—	0
Thorium* (t/y)	—	27	100
Inventories (t)‡			
Fissile Pu	—	3.7	4.9
Nat. uranium	140	—	—
Thorium*	—	79	115

* assumes no thorium recycle
‡ includes out-of-reactor component for one year of recycle time

considerations, etc. Taking the 'high burn-up' case in Table 17.6, and assuming its plutonium requirements to be supplied by natural uranium CANDUs, the reactors using thorium would make up about two-thirds of the combined system. In addition to the start-up requirements of the plutonium-producing reactors, an equilibrium feed of natural uranium would be required of about 45 t/y per GW(e) of total system output.

Because of uncertainties about many of the items of cost involved, it is difficult to relate optimum design very closely to assumptions about future uranium prices, and much more practical experience is required before this can be done with confidence. It is clear, however, that if uranium prices rise sufficiently, relative to the other costs involved, then a thorium cycle becomes preferable to the existing natural uranium once-through cycle.

4.5 Light water reactors (LWR)

In light water reactors, the moderator is 'ordinary' (though highly purified) water, which is contained under pressure and serves at the same time as the coolant. Hydrogen has a relatively high neutron capture but it has the advantage that, because the hydrogen nucleus has the same mass as a neutron, the energy of the latter is reduced, per collision, by the greatest possible amount. This leads to a compact core arrangement, at the cost of neutron economy, and a need for fuel enrichment. Most reactors currently operating or under construction are of this type, there being two sub-varieties, the pressurized-water reactor (PWR), in which the heat is transferred from the coolant in a steam generator to a separate steam–water circuit driving the turbine, and the boiling-water reactor (BWR) in which the coolant is made to boil as it passes through the reactor core and the steam formed is taken directly to the turbine. More attention has been given to the use of thorium in the PWR than in the BWR, and the PWR will be discussed here.

PWRs at Indian Point and Elk River used thorium for a time, initially with ^{235}U enrichment, but the practice was rather quickly abandoned on economic grounds. Although uranium requirements are higher in this type of reactor than in an HWR or HTR, there could be interest in back-fitting a thorium cycle[6] in the event of a sharp change in uranium prices. However, over the period required for the development and introduction of the thorium cycle, improvements can in any case be expected in the performance of LWRs on the ^{238}U–plutonium once-through cycle, and these will have to be taken into account in a comparison. The adoption of thorium with ^{233}U recycle might further reduce the ^{235}U requirements by 20–25 per cent. However, with the increase in fuel fabrication costs due to gamma activity from the associated ^{232}U and with the expected cost of thorium–^{233}U reprocessing, it seems unlikely that such a cycle could become economic, particularly since similar savings in ^{235}U are possible in the ^{238}U–Pu case by adopting recycle.

There is also some interest in what might be achieved with a radically new design aiming to use thorium with great emphasis on neutron economy. An example of this approach is the *light water breeder reactor* under development in the USA since 1965, a core of this type having now been installed in the Shippingport reactor for practical testing.[7] In this concept, the aim is to achieve high ^{233}U production, perhaps sufficient to give a small breeding gain. To do this, the design adopts a lower moderator ratio (i.e. a lower ratio of water to fuel volume), this being about half that used in conventional PWRs. There is therefore a reduced area available for coolant flow, requiring either greater pumping power or a longer core with a lower power density. A lower power density also contributes to improving the conversion ratio, but requires a higher initial fissile loading. The Shippingport core has two other special features to improve the breeding characteristics. The fuel charge is of the *seed-blanket* type. That is to say, there are two types of fuel pin in a module. One, the *seed*, has a higher ^{233}U content, and is the primary producer of neutrons. The *blanket* is the primary fertile component. The other special feature is that reactivity control is provided by moving the seed fuel, so avoiding the loss of neutrons that would occur in conventional absorber control rods. It might theoretically be possible to back-fit a complete core of this type to existing reactors, but this would require derating, in addition to the rejection of the previous fuel charge and an appreciable outage time.

The level of conversion ratio aimed for in this development is similar to that of the 'self-sufficient', CANDU version already mentioned. It also requires the initial fissile charge to be ^{233}U which must be supplied from other reactors (*pre-breeders*), using thorium and either ^{235}U or plutonium. Whether, once initiated, a self-sustaining system can be achieved with an economic fuel burn-up remains to be seen. The 'special tricks' necessary with the LWR are symptomatic of the inferior neutron economy compared with CANDU.

4.6 Once-through cycles

All the thermal reactor types discussed above (HTR, HWR, LWR) are capable of operating on once-through cycles with uranium fuelling. There is an inevitable loss of fissile material in such a scheme but reprocessing costs are saved and if the fissile material is cheap enough, such cycles can be economic. Rising uranium prices put pressure on the neutron economy of such reactors and unless resort is made to reprocessing and recycling, the only major answer is to utilize more of the fissile material in the one pass through the reactor. In an idealized situation, a reactor which was self-sustaining in reactivity could operate indefinitely, and the loss of fissile material on discharge would be confined to the last charge on decommissioning the reactor.

Technology is still a long way from achieving a fuel life comparable with the life of the reactor plant and, indeed, it is hardly to be expected that this will be possible, in view of the relatively intense irradiation damage experienced by fuel element materials. However, if fuel could be developed to withstand a sufficiently high irradiation dose, then benefit could be taken from a fuel cycle which bred sufficiently to compensate for the loss of reactivity caused by neutron absorption in fission products, so that the reactivity of the core was maintained. In thermal reactors, this situation can be most closely approached with a thorium–^{233}U cycle.

Theoretical studies of such cycles have been carried out, but alternative ways of meeting fissile material supply limitations by adopting reprocessing and recycle appear much more practicable. The reactor system that comes nearest to being a possible exception to this is the HTR, but even in this case, because of the higher inventory of fissile material required for such a thorium–^{233}U cycle, with current technology it proves possible to do only slightly better than with the once-through uranium–plutonium cycle.

4.7 Fast reactors and combined systems

As already mentioned in §4.1, fast reactors not only breed but also offer the best doubling time. Thorium–^{233}U cycles prove to be inappropriate to this purpose, and the comparison with the ^{238}U–plutonium cycle is unfavourable to thorium by a considerable margin. This follows from the neutronic properties of the materials, discussed in §3, where it is explained why the breeding gain with the thorium cycle is inferior to that on other cycles. Briefly, although ^{233}U has some advantages over ^{235}U and ^{239}Pu, they are insufficient to compensate for the very low fission probability in ^{232}Th. A loss of over 0.2 in breeding ratio is fairly representative. Indeed, using thorium and ^{233}U alone, the *breeding gain* (i.e. the breeding ratio minus unity) in a study of representative reactor designs[8] was only 0.04, compared with 0.28 when using the uranium–plutonium cycle. Furthermore, because of the negative reactivity effect of thorium in the core, a higher enrichment and a larger fissile inventory are required, so that the doubling time became 112 years, compared with 16 years for the U–Pu cycle.

It is perhaps interesting to note in passing that the all-thorium–^{233}U fast reactor has a lower sodium-void coefficient than the classic ^{238}U–plutonium case. Sodium-void coefficient is, however, no longer seen as a serious impediment to the exploitation of plutonium-fuelled fast reactors and the sacrifice in the neutronic parameters mentioned above seems so crucial as to make this point of only academic interest.

4.8 The long term

Taking a long-term view, theoretical possibilities exist of manufacturing fissile from fertile materials by irradiation not in nuclear reactors as we know them, but by the use of neutrons derived either from particle accelerators or fusion devices. One possibility might be to use thorium in this way to produce ^{233}U and this has been examined in several studies. For the accelerator-breeder to have any significant impact on the world's energy supply, it appears that accelerators would be required which are very much larger than any that have so far been built. For example, with a linear accelerator breeder as discussed by Steinberg of Brookhaven,[9] it has been suggested that a 1 GeV proton accelerator delivering 300 MW of beam power would be required in order to produce between 1 and 2 tonnes of fissile material per annum. This may be compared by way of example with the inventory requirements shown in Table 17.6. There would also be a need to dissipate heat generated in the target with consequent design and cooling problems of a magnitude quite novel to accelerator targets. If this line were followed, the engineering outcome could well be more like a reactor, with an accelerator as a neutron booster, than an accelerator-target assembly.

The case for fusion breeding might arise rather differently. The general aim in fusion appears to be to generate energy and extract heat from the fusion device itself. The engineering problems of doing this are formidable, and whilst it would be unreasonable to expect much progress in this direction, even with ideas, before the production of fusion energy has yet been achieved, it is conceivable that the best solution could be to utilize the neutrons produced to manufacture fissile material for subsequent use in more conventional reactors rather than to produce power directly in the single plant.

4.9 Summary

In selecting a reactor and fuel cycle, many matters have to be taken into consideration. The principal case that has been advanced for the thorium fuel cycle postulates a future shortage of fissile material such that achieving a high utilization of that material becomes of greatly enhanced importance. It would be necessary to quantify that importance, if the significance of good fissile material utilization were to be placed properly in perspective. Nevertheless, the requirement for fissile material is inevitably a feature of interest which can be separately assessed, though some restraint must be exercised in discussing designs which incorporate features—for instance, very low burn-up — which might favour fissile utilization to a modest extent, but which would in some other way seriously impair the economic attraction.

Thorium fuel cycles could conceivably be applied in all types of reactor currently regarded as being of prospective commercial interest. In no case however does the thorium cycle approach the ^{238}U–plutonium fast reactor cycle in terms of its utilizaton of fissile material. If fast reactors are left out of account and attention is confined to comparisons within the range of thermal reactors, then a thorium cycle with ^{233}U recycle could in principle result in some improvement over the fissile material economy of the same type of reactor operating on the ^{238}U cycle with plutonium recycle. Analyses suggest that this is clearly true beyond the margins for uncertainty, only in the case of the HTR.

In the event that thorium reprocessing was available, while for some reason ^{238}U–plutonium reprocessing was not, then the HTR, HWR and LWR would all be possible candidates for thorium. Back-fitting to existing LWRs could then be of practical interest, though otherwise the prospects for the HTR and HWR seem more promising. Of these, the HTR may meet a wider range of applications, but it is behind the HWR in development and shows little sign at present of being exploited on a wide scale.

5 Reprocessing and fabrication of thorium and waste management

5.1 Status of development

The purpose of reprocessing irradiated fuel is to recover material which can be refabricated into new fuel for reuse. At the same time radioactive nuclear wastes are separated from the recovered material ready for storage and disposal. This section describes the processes required and outlines chemical principles and schematic processing routes which might form the basis of future plant design. The thorium fuels envisaged are of two main types: oxide with metal cladding and ceramic-coated particles usually in a matrix with graphite.

Processes were under development in the USA, during the 1950s and 1960s, in anticipation of HTR and LWBR requirements (see §4). Much of this work was carried out with lightly irradiated materials having relatively low concentrations of ^{233}U and ^{232}U, and reprocessing of fuel with a high burn-up remains to be demonstrated. Little progress has been reported in the last ten years towards a full-scale active demonstrations plant.

A pilot plant was constructed in Italy in which thorium oxide (ThO_2) fuel elements from a the American Elk River LWR were reprocessed in the

1960s. The plant included a refabrication facility, but there appears to have been no incentive to develop the thorium cycle further in Italy, and only a minimal research and development programme is continuing.

Research and development on the reprocessing of HTR fuels in several countries has given some emphasis to head-end operations though other aspects of reprocessing have also been the subject of laboratory and pilot plant work, mostly with materials of low radioactivity, and mainly in the USA and West Germany.

The lack of reprocessing under realistic conditions has also limited the experience available of refabrication, and of waste management where again knowledge is based on laboratory studies.

It appears that the commercial availability of a closed thorium fuel cycle can only result from a major effort and cannot be expected before the year 2000.

5.2 Reprocessing

For the uranium–plutonium cycle, the technology for the recovery and separation of uranium and plutonium from irradiated fuels is well established. The process (*Purex*) is based on one particular version of the solvent extraction technique which has already been described in general terms in §2. A process suitable for thorium-based fuels is likely to have a broad similarity to the Purex process, but with important differences. For example, the solvent extraction process in Purex depends on the fact that tetravalent plutonium is more highly soluble in tributyl phosphate (TBP) than is hexavalent uranium. For the separation of uranium and thorium, the process on which most work has been done is the *Thorex* process, which depends on the way that the relative solubilities of the two materials in nitric acid and in TBP vary with the acid concentration.

Practical development of thorium reprocessing has not yet reached the stage where process details can be treated as definitive, but the following illustrates some of the features involved. Figure 17.8 shows schematically the operations in reprocessing thorium fuels using Thorex. It involves four basic steps:

    (1) separation of the irradiated fuel from associated materials such as fuel cladding;

    (2) separation of fission products from heavy metals;

    (3) separation of uranium from thorium;

    (4) waste treatment and disposal.

Step 1 begins with the mechanical separation, as far as is practicable, of the irradiated fuel from associated materials such as the fuel element structure. This 'head-end' process is carried out in order to reduce the amount of extraneous material going on into the dissolver for final separation. It depends on the physical features of the reactor fuel element design, and is related to the particular type of reactor, not to the fuel cycle used. It will not be further discussed here.

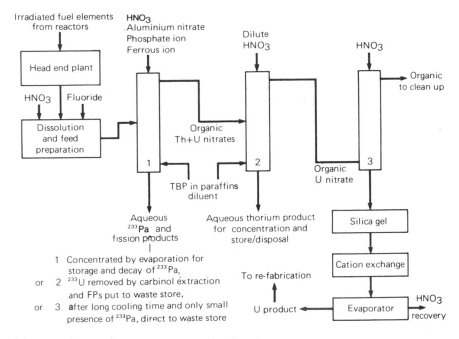

Fig. 17.8 Schematic Thorex reprocessing flowsheet

The second part of this first step, the chemical dissolution of the fuel in nitric acid, also has close analogies to the corresponding phase in the uranium–plutonium cycle, the main difference being that dissolution of thorium in nitric acid is so slow that a fluoride-ion catalyst is required. The presence of fluoride poses corrosion problems throughout the stages of the plant in which it is present. The resulting corrosion rate of stainless steel is decreased by the presence of aluminium ions, and aluminium nitrate is added either as feed to the dissolver, or at the subsequent stage as shown in Fig. 9.8 The aluminium added must be kept to a minimum, because it passes through the plant with the fission products and adds to the bulk of active material requiring storage.

In step 2, a solvent extraction process is used to extract the thorium and uranium nitrates in diluted TBP, leaving the fission products in an aqueous raffinate stream, (Fig. 17.8, column 1). The scrub solution in this column includes phospate to assist in the removal of protactiniuim from the thorium–uranium, and a ferrous ion for the removal of trace chromium compounds produced by the corrosion of the dissolver vessel. The TBP solution is finally centrifuged, to remove insoluble solids, and the resultant liquor is then ready for step 3 (Fig. 17.8, column 2).

The separation of uranium from thorium is carried out by a further solvent extraction process. The thorium is preferentially stripped from the solvent by dilute nitric acid, while the uranium remains preferentially in the organic solvent. The valuable product, the uranium with its ^{233}U content, is then stripped from the organic solvent by an even weaker nitric acid scrub (Fig. 17.8, column 3), the TBP going on to a clean-up plant before reuse. Finally, the aqueous uranium product is passed over a bed of silica gel to remove remaining trace impurities, such as zirconium and niobium, and then through a cation exchange column to remove traces of thorium and some corrosion products.

Returning now to consider the fission-product stream in the aqueous phase from column 1, this will contain ^{233}Pa, another valuable product because it decays to ^{233}U (with a half-life of 27 days). The relative amount of ^{233}Pa will depend on the length of time which has passed between fuel irradiation and reprocessing (*cooling time*). If the cooling time has been long, little ^{233}Pa will be present, and the fission-product stream will pass straight on to step 4, the waste management and disposal stage, discussed later in this section. If the cooling time has been short, and the ^{233}Pa content relatively high, the fission products may be concentrated by evaporation for storage and further ^{233}Pa decay, followed by another extraction of ^{233}U. An intermediate possibility is to extract the ^{233}Pa immediately with dibutyl carbitol (Butex), the fission products then going on to step 4.

Current proposals of a process on the above lines are based on the assumption that thorium and fission products are better separated, if only for the purposes of subsequent storage and final disposal. If this assumption should prove incorrect, the reprocessing flow-sheet could be simplified. One step could replace steps 2 and 3, and by a suitable adjustment of the nitric acid concentration in the solvent extraction process, the thorium would be allowed to remain with the fission products, only the uranium being separated off.

5.3 Refabrication

The processes for refabricating fuel containing recycled material would be expected to follow those methods used initially for new fuel elements, but there is an important difference. Recovered thorium contains ^{232}Th and ^{228}Th which decays with a half-life of 1.9 years through a series of hard gamma-emitting daughter products. These would cause severe handling problems in the refabrication of thorium fuel elements, making the reuse of recovered thorium unattractive. The recovered uranium isotopes, however, after removal of fission products will at first be essentially free from gamma activity but, because of the decay of the associated ^{232}U to ^{228}Th, unless refabrication can be undertaken within a few days, remote handling is essential. It is anticipated that refabrication plants will need extensive development to adapt for remote handling as they will be heavily shielded and maintenance will be complicated.

It is to be noted that the requirement for shielding continues throughout the handling and transport route from the fabrication plant up to loading in the reactor.

The recycled ^{233}U is also alpha active, creating a dust hazard. Because of this, consideration is given, as with plutonium-fuelled fast reactors, to processes which minimize the generation of dust. There is therefore an interest in the development of gel precipitation processes to produce oxide spheres of a suitable range of sizes. These could form the basic fuel kernel for HTR coated particles, or might be loaded into metal-clad fuel pins for vibro-compaction.

In general, the fuel element assembly processes are essentially identical to those required for fresh fuel, but with the added complication of remote handling which adds considerably to the cost.

5.4 Waste management

Once again, there are many similarities between the uranium–plutonium and the thorium–uranium fuel cycles in the general requirements for the storage and disposal of wastes, but at the same time there are some notable differences.

The radioactivities of the 'normal' wastes—fission products and actinides— are for practical purposes similar in the two cases. If thorium is not to be recycled because of its radioactivity, its disposal becomes an additional problem which appears to have received little consideration.

Some extra problems do arise, however, with the normal wastes in the thorium cycle, owing to the different chemistry of the separation process. The generally favoured plan for the disposal of highly active reprocessing wastes, such as fission products, is to incorporate them into the structure of a glass. In the case of the thorium cycle, the presence of sulphates, phosphates, and fluorides in the raffinate of the reprocessing plant would have a major impact on the vitrification process, both on the design of the plant and on the chemistry of the glass produced. One effect would be a considerable corrosion problem at the high temperatures used in vitrification, which could be somewhat pervasive as about 20 per cent of the fluoride volatilizes during the solidification stage of the glass manufacture. This might be reduced by the addition of further material such as calcium. A recent estimate suggests that, on a comparable basis, the glass volume required would be about 70 per cent greater than in the case of the uranium–plutonium cycle and the Purex process.

5.5 Summary

Experience in reprocessing thorium-based fuels falls short of commercial demonstration in various respects, including the scale of operation, the irradiation level of fuel which has been reprocessed, and some unresolved technical problems. Only by a major effort can commercial reprocessing be expected to become available by the end of the century. A likely basis for such development is the Thorex process which has been briefly described.

Inevitably, little experience exists of the fabrication of fuel elements with recycled material. The operations need to be well shielded because of the high-energy gamma radiation emitted, and dust must also be controlled, because of an alpha hazard. The remote operations required are therefore relatively expensive and more akin to the fabrication of fuel with recycled plutonium than to currently conventional thermal reactor fuel manufacture.

There are extra problems in vitrifying the highly active wastes over and above those for the uranium cycle, because of the salts, including fluorides, likely to be present in the reprocessing raffinate.

6 Economics

6.1 Costs

There is insufficient experience with the various steps in the thorium-based fuel cycles, for any reliable cost estimate to be made. However, conclusions can be drawn on the basis of more qualitative arguments, which are likely to be as sound as some attempts at quantitative economic predictions.

Leaving aside some rather special circumstances, it appears that the only clear economic benefit of the thorium cycle would lie in the superior neutron economy it offers in thermal reactors. There appears to be little ground for expecting net saving in areas other than this one of fissile material utilization. Let us first examine the fissile material aspects.

The cost of providing the fertile material for the U–Pu cycle (^{238}U) is difficult to estimate, because ^{238}U arises only as a by-product from the extraction or use of fissile uranium (^{235}U). On any reckoning however it must remain for a long time only a trivial part of the overall cost of nuclear power. If, by breeding, fissile fuel was made abundant, the supply of fertile material could conceivably become relatively important but resources are such that a shortage of ^{238}U could hardly arise for some centuries. Meanwhile, it is sufficient to note that the information available on the abundance and extraction costs of thorium does not rule out the possibility of its economic use.

The use of thorium does result in some differences in the design of thermal reactors, but any benefit in reactor plant costs can only be marginal. The most significant effect might be expected to be due to difference in the power density of the reactor core. ^{233}Pa, whose equilibrium level increases with power density, is not only a neutron poison, but also, on capturing neutrons, removes material from the direct, fissile production chain. It therefore has a double detrimental influence on neutron economy, which sets an upper limit to the power density economically achievable. The design of high power density reactors, such as HTRs, for the thorium cycle is already influenced by this consideration, whereas higher power densities and fuel ratings can be achieved with a uranium–plutonium fuel cycle without encountering such a constraint.

In closing the fuel cycle it seems that cost differentials are bound to be adverse to thorium. In the short term, the non-existence of the necessary plants reduces theoretical cost considerations to an irrelevance. It is a fairly general opinion that even if pursued with some vigour, it would not be until the end of this century that reprocessing of thorium to permit the recycle of ^{233}U could begin to be introduced on a commercial scale.

Current concepts for reprocessing thorium are unlikely to prove cheaper than for uranium–plutonium if only because of the more corrosive nature of liquor and the use of fluorination (see §5). The fabrication of recycled fuel also appears likely to be as expensive as that for plutonium-containing fuel, because of the activity stemming from the ^{232}U ^{228}Th decay chain and the near comparability of ^{233}U and ^{239}Pu as dust hazards.

6.2 Prospects

By these general arguments, one arrives at the conclusion that for a thorium cycle to be economically attractive in the general case, the advantages of better neutron economy must outweigh other considerations. On the other hand, if this factor becomes sufficiently important, fast breeder reactors would be even more advantageous, to a degree sufficient to outweigh any adverse capital cost differential associated with their use. Thus, the general case depends on an assumption that for some reason plutonium breeders are not proceeded with. Reasons for that which have been put forward derive from a range of political viewpoints, the discussion of which are beyond the scope of this chapter, though it is to be noted that independence of national energy supplies, based on indigenous deposits of thorium could be one such reason.

If the introduction of fast reactors were delayed simply because supplies of uranium remained relatively abundant, then the incentive for the use of thorium would also be lacking. If for some other reason and in spite of constraints upon uranium availability, the use of thorium in thermal reactors became the preferred strategy, this could only apply for a limited period of time, or it would limit the scope for utilizing nuclear power in general and thorium in particular.

It remains to consider the possibility of more particular circumstances that might favour the use of thorium. A conceivable shortage of ^{238}U in the remote future has already been mentioned as one reason why thorium might ultimately become a necessary supplement. At a somewhat earlier date, if the problem of fissile material supply had been resolved by the deployment of breeder reactors, there might be some benefit in a partial return to the use of non-breeding reactors. Such benefit might include the use of reactors for hydrogen production, coal gasification or liquefaction, and steel manufacture, all of which require high temperatures for which breeder reactors might not be suitable, whereas a future development of the HTR might. Other examples which can be visualized for which a limited deployment of thorium fuel might be justified include the possibility that some reduction in capital costs might be achieved compared with an 'all-fast' system. The scale of such deployment of non-breeding reactors would still depend on the degree of neutron economy achievable in them and thorium could as a result be preferred. In all such cases, the prospect is of ^{233}U being bred, in, say, the radial blanket of a fast reactor, and then being supplied for the thermal reactor in a symbiotic arrangement.

6.3 Breeding

A case is sometimes put that thermal reactors using thorium could 'just breed', i.e. produce more fissile material than is lost, and that any ability to breed, however small, makes it possible to utilize virtually all the available fertile material through eventual conversion into fissile material. This calls for some comment.

Firstly, the feasibility of achieving such thermal breeding in conditions of normal commercial practice remains to be demonstrated, and requires a close approach to theoretical ideals in many respects. However, there is another, philosophically more important, aspect of the argument. This is the implication that 'just breeding' achieves all that is required. All nuclear systems require an investment of fissile material in the first charge to start them off. In an expanding nuclear power supply, there is a continuing demand for such inventories, in addition to whatever regular make-up might be required in steady operation. Even if that make-up is reduced to zero (by 'just breeding'), this still leaves the necessity for further inventories to be found from elsewhere. Thus, 'just breeding' allows the nuclear system to be self-supporting only on an assumption of zero growth, and if the system is to be expanded, a higher breeding gain becomes necessary.

6.4 Summary

As long as fissile material (^{235}U) remains available at a reasonable price, no economic reason is seen for the general adoption of a thorium–^{233}U cycle, because in other respects it is unlikely to be competitive with the use of the ^{238}U–Pu cycle in thermal reactors.

If, on the other hand, fissile material becomes expensive and good neutron economy has an enhanced value, then fast breeder reactors become relatively the more economic.

In special circumstances, the use of thorium might be adopted. The possibilities vary with the timescale considered, and may become more widespread in the latter half of the next century.

7 Environmental impact and safety

7.1 **Mining**

For a nuclear programme based on thorium, a continuing supply of fissile uranium would be required. Used in thermal reactors, thorium would permit a possible reduction in the relative scale of uranium mining and extraction of about a factor of two. This reduction would of course be associated with a new requirement to mine and extract thorium.

It may be noted that thorium deposits, like uranium, are not free from the emanation of radioactive gas. Often thorium and uranium are in fact associated (see §2). It is to be presumed that similar biological standards would apply to either uranium or thorium mining, and underlie regulations such as those regarding ventilation requirements and the treatment of tailings. Disposal of tailings has a considerable impact on land use, additional to that of the mine itself or the open-cast workings. Questions of water use also arise. The balance between these various considerations is clearly specific to local circumstances.

If the comparison is to be made with the use of uranium–plutonium fuel in fast reactors, the thorium cycle requires much more, not less, uranium to be mined in addition to the thorium. On this basis, it is clear that the thorium cycle would have a disadvantage.

7.2 **Reactors and fuel cycle operations**

In the fabrication of recycled fuel, the alpha and gamma activities quoted in Figs. 17.2 and 17.7 are relevant. The dust hazard (alpha) with ^{233}U, although lower than with Pu, requires similar handling precautions. With regard to gamma activitity, the balance goes the other way, though the provision of shielding would be such as to give similar working conditions. No real difference is seen in the safety and environmental effects, beyond the cost implications of achieving the same safety levels.

Secondary differences in, for example, the control characteristics of the various fuel cycles, could influence the detailed design of thermal reactors, but no major safety distinction is seen either way, and other environmental aspects would also be largely unaffected. A comparison between thermal and fast reactors is outside the scope of this article, though clearly similar overall safety standards would be applied through regulations, inspections, operating rules, etc. One point that would need to be taken into account in comparing thermal and fast systems, is the need for an enrichment plant for most thermal reactor cycles, with a consequent increase in demand on the whole energy supply system.

In reprocessing, the principal concern is the treatment and disposal of the wastes arising. Here, some differences do become apparent though it is doubtful whether, taken together, they need be of much significance. If, as seems probable, a process akin to Thorex is used, then as mentioned in §5, there would be extra problems in the vitrification of wastes for ultimate disposal. More research and development would be needed to achieve comparable results, and the volume of glass for disposal could well be greater, though environmentally this appears to be rather a detail since volumes are in any case small, the main impact being on costs. However, the general proposal is that the thorium should not be recycled. This adds further to the waste problem.

The radioactive-decay characteristics of the fission products and actinides also differ, as discussed in §3. Activity levels in the wastes from the two cycles are initially similar. Subsequently, first the uranium cycle and then the thorium cycle are higher than the other by factors of about ten. It is difficult however to identify any net difference of particular significance.

7.3 Summary

While there are differences of detail, the most significant distinction between the environmental and safety of the thorium and uranium cycles aspects is likely to be in the associated mining operations.

In mining, the comparison between the thorium thermal reactor cycle and the fast reactor cycle must be adverse to thorium because of the relative quantities involved. In comparing thermal reactor cycles, a judgement would depend on the local circumstances of any particular mine sites.

8 Proliferation resistance

8.1 Comparison with plutonium

The knowledge that plutonium can be used to manufacture nuclear weapons and that it is chemically separable from the uranium in which it is formed by irradiation, can lead to a simple view that therefore the ^{238}U–plutonium fuel cycle should be abandoned, and the thorium–^{233}U cycle adopted in its place. Although this is not the place for an exhaustive discussion of proliferation risks and preventive measures to control them, nevertheless comparative statements like the foregoing require some examination.

First of all, it must be understood that any fissile material might conceivably be used to make a weapon. ^{233}U is no exception and indeed, depending on his particular capabilities, might be considered by the weapon maker to be his first preference. It is of less general concern than alternatives at the present time merely because of its relative rarity.

Secondly, as will be clear by now to the reader, thorium is only a fertile material. To establish or expand a thorium cycle, fissile material is required, either in the form of plutonium or of ^{235}U, so that thorium cycles still require the use of these potential weapons materials, and of the enrichment and reprocessing plant required to provide them.

Table 17.7 Comparison of U/Pu data relevant to proliferation resistance

	Commercial Pu	Commercial ^{233}U	Denatured ^{233}U
Critical mass	1	1	very large
α –activity	30	1	0.1
γ –activity	medium but soft	high and hard	medium and hard
n–activity	high	low	low

The difference in the proliferation sensitivity of these two fuel cycles is therefore not one of kind, but at most one of degree. Any comparison between them will need to take into account the form and the composition of the fissile materials as they would arise in the civil nuclear fuel cycle. For example, the plutonium produced by irradiation is not purely the isotope ^{239}Pu, but contains other isotopes which complicate and could even prevent the preparation of weapons-usable material by simple chemical separation. The use of ^{235}U in weapons is again rendered more difficult by the natural presence of significant quantities of ^{238}U, so that only highly enriched uranium (i.e. much more nearly pure ^{235}U) is of direct proliferation concern, unless there is access to an isotope separation plant of the required capability. It will be clear that the sensitivity of these materials depends in the first instance on the form in which they enter or leave the reactor, and secondly on the control which can be exercised over the plants required for their further treatment.

In the Th–^{233}U cycle, the ^{233}U is chemically separable from its parent material thorium. As formed in irradiation, it will however be associated with other uranium isotopes which, as in the case of plutonium, make it more difficult to use. Table 17.7 presents some comparative data. It will be seen that ^{233}U, if available, may well be the easier material for the manufacture of weapons in all the respects shown except for gamma activity. This arises from the ^{232}U–^{228}Th decay chain previously discussed in §3. ^{228}Th would, however, be removed in reprocessing as would also its daughter products, leaving a brief period of low gamma activity before the decay of ^{232}U has produced much further ^{238}U, though it may be difficult in practice for a weapons fabricator to take full advantage of this fact.

On balance, it is judged that commercial ^{233}U is easier to turn into weapons than is commercial plutonium, and that the adoption of the straightforward Th–^{233}U cycle would, from this particular point of view, be a backward step.

8.2 'Denatured' fuel cucle

The possibility exists of reducing the concentration of ^{233}U in the uranium component of the fuel by the admixture of some ^{238}U: the resultant fuel is often referred to as *denatured thorium* or more properly as *denatured* ^{233}U. To extract purer ^{233}U from fresh fuel of this type, it would be necessary not only to separate the uranium from the thorium, which can be done chemically, but also to enrich the ^{233}U content. It should be noted that the enrichment of ^{233}U in ^{238}U requires relatively little separative work. It is often suggested that the ^{235}U enrichment for common use is desirably kept below 20 per cent. It is arguably just as easy to re-enrich ^{233}U from a concentration of 5 per cent or possibly lower. The adoption of denatured ^{233}U with the uranium component having 12 per cent ^{233}U content (the composition adopted in most such proposals) is equivalent to using 20 per cent ^{235}U only if it is assumed that the relationship depends entirely on critical masses, and that possibilities of re-enrichment can be ignored.

It should be noted that, even if it is assumed that re-enrichment is not available, the denatured ^{233}U cycle is nevertheless not free of proliferation concerns. In such a case, the recycled fuel would require a make-up quantity of highly enriched uranium in order to top up its fissile content, and this material would have to be present as a feature of enrichment and fabrication processes. The irradiated denatured fuel would moreover contain chemically separable plutonium as well as ^{233}U, though in a smaller quantity per unit of electricity generated than with the simple ^{238}U–Pu cycle. It would also contain a comparable amount of ^{233}Pa, which could be separated and become a source, through decay with a 27-day half-life, of ^{233}U, free not only from ^{238}U but also from ^{232}U. Even the proliferation route based solely on chemical separation is therefore still present for the irradiated fuel.

The adoption of this denatured fuel cycle involves an economic penalty in the increased difficulty of reprocessing, since the three elements thorium, uranium, and plutonium will all be present in the irradiated fuel, and both the last two of these have a fissile value. If these are recovered and recycled, the utilization of fissile material represents a compromise between the uranium and thorium cycles, and similarly some of any proliferation advantage which thorium may have had, is lost. Assuming, however, that the plutonium is rejected, then not only is there a loss of valuable fissile material, which has to be made good by additional supplies of natural uranium (compare the last two entries in Table 17.5), but also, as with 'once-through' uranium cycles, plutonium is, most undesirably, added to the waste for disposal.

8.3 **Summary**

To reduce risks of weapon proliferation, substitution of ^{233}U for ^{239}Pu as a product of civil nuclear fuel irradiation could be a mistaken policy. Whilst a would-be weapons manufacturer would face rather different problems in trying to obtain and use ^{233}U, it is by no means clear that he would find the overall task more, rather than less, difficult.

The adoption of a denatured ^{233}U cycle would add some extra complication to the process of extracting weapons material from fresh thorium-cycle fuel. As an aid to the safeguarding of the fuel cycle as a whole, this appears to be too marginal to justify, on its own account, the increased complexity of legitimate reprocessing, the inferior fissile economy, and the partial loss of any other advantage of the thorium cycle.

References

Section 4

1. SHEPHERD, L. R., HUDDLE R A U, HUSAIN, L. A., LOCKETT, G. E., WORDSWORTH, Dr., STERRY, F. The possibilities of achieving high temperature in a gas-cooled reactor. In *Proc. 2nd Int. Conf. on the Peaceful Uses of Atomic Energy, Geneva, September 1958*, vol. 9, pp. 289–305. United Nations (1958).
2. SCHRODER, E., BLOMSTRAND, J. H., and BRUNEDER, H. Fuel cycles for power reactors. *J. Br. Nucl. Energy Sec.*, **5**, no. 3 (1966).
3. DAHLBERG, R. C., TURNER, R. F., and GOEDDEL, W. V. Fort St. Vrain—core design characteristics. *Nucl. Engng. Int.*, **14**, 1073–7 (1969).
4. MULLER, H. W. Design features of the 300 MW THTR power station. In *Proc. Symp. on Advanced and High-Temperature Gas-Cooled Reactors, Julich, October 1968*, pp. 135–53. International Atomic Energy Agency (1969).
5. SLATER, J. B. *An overview of the potential of the CANDU reactor as a thermal breeder*. AECL–5679. Atomic Energy of Canada Ltd. Chalk River, Ontario (1977).
6. SHAPIRO, N. L., REC, J. R., MATZIE, R. A., *Assessment of thorium fuel cycles in pressurized water reactors*. EPRI NP–359. Electric Power Research Institute Palo Alto, California (1977).
7. ENERGY RESEARCH AND DEVELOPMENT ADMINISTRATION (US). *Light water breeder reactor program—final environmental statement*, vol. 1. ERDA–1541 (1976).
8. CHANG, Y. I., TILL, C. E., RUDOLPH, R. R., DEEN, J. R., KING, M. J. *Alternative fuel cycle options*. ANL–77–70. Argonne National Laboratory, Argonne, Illinois (1977).
9. STEINBERG, M. *Linear accelerator breeder for energy security*. ERDA–CONF–770107 (1977).

Suggestions for further reading

Section 2

CUMMINS, A. B. and GIVEN, I. A. *SME mining engineering handbook*, vols. 1 and 2. Society of Mining Engineers of the American Institute of Mining, Metallurgical and Petroleum Engineers (1973).

DAVIDSON, C. F. and BOWIE, S. H. U. Methods of prospecting for uranium and thorium. In *Proc. 1st Int. Conf. on Peaceful Uses of Atomic Energy, Geneva, 8–20 August 1955*, vol. 6, pp. 659–72. United Nations (1955).

ORGANIZATION FOR ECONOMIC CO-OPERATION AND DEVELOPMENT/NUCLEAR ENERGY AGENCY. *Uranium resources, production and demand*. OECD/NEA (1977).

PARKER, J. G. and BAROCH, C. T. *The rare-earth elements yttrium and thorium—a materials survey*. Inf. Circ. US Bureau of Mines 8476 (1977).

PILKINGTON, E. S. and WYLIE, A. W. Production of rare-earth and thorium compounds from monazite (Part 1). *J. Soc. Chem. Ind.*, **66**, pp. 387–394 (1947).

TAGGART, A. F. *Handbook of mineral dressing*. Chapman and Hall (1945).

WOLLENBURG, H. A. Fission-track radiography of uranium and thorium in radio-active minerals. In: M. J. Jones (ed.). In *Proc. 4th Int. Geochem. Exploration Symp., London, 17–20 April 1972*. The Institution of Mining and Metallurgy (1972).

INSTITUTION OF MINING AND METALLURGY AND COMMISSION OF EUROPEAN COMMUNITIES. *Proc. Int. Symp. on Geology Mining and Extractive Processing of Uranium, London, 17–19 January 1977*. Institution of Mining and Metallurgy, London (1977).

Section 3

BURSTALL, R. F. Importance of transactinium nuclear data for fuel handling. In *Proc. IAEA Advisory Group Meeting on Transactinium in Isotope Nuclear Data, Karlsruhe (November 1975), vol. 1, paper A5.* International Atomic Energy Agency, Vienna (1975).

CRITOPH, E. MILGRAM, M. S. VEEDER, J. I. BANERJEE, S. BARDAY, F. W. HAMEL, D. *Prospects for self-sufficient thorium cycles in CANDU reactors. AECL 5501.* Atomic Energy of Canada Ltd., Chalk River, Ontario (1975).

HUGHES, D. J. Brookhaven National Laboratory, New York *Neutron cross-section. BNL* report 325 (edns. 1, 2 and 3 1957, 1958 and 1973). National Technical Information Service, Springfield, Virginia.

INTERNATIONAL ATOMIC ENERGY AGENCY. *An index to the literature on microscopic neutron data (IAEA). CINDA*-75 vol. 2. IAEA (1975).

LEDERER, C. M. *Table of isotopes* (6th edn). John Wiley, New York (1967).

LEWIS, W. B. *Large-scale nuclear energy from the thorium cycle.* In *Proc. 4th Int. Conf. on the Peaceful Uses of Atomic Energy, Geneva, 1971*, vol. 9, pp. 239 ff. IAEA (1972).

MASSIMO, L. *Physics of high-temperature reactors.* Pergamon, Oxford (1976).

MERZ, E. R. The thorium fuel cycle. In *Proc. Int. Conf. on Nuclear Power and its Fuel Cycle, Salzburg, May 1977.* vol. 2, pp. 37–53. IAEA (1977).

NUNN, R. M. The requirements for transactinium nuclear data for the design and operation of nuclear power plants. In *Proc IAEA Advisory Group Meeting on Transactinium, Isotope Nuclear Data, Karlsruhe, November 1975*, vol. 1, paper A4. IAEA (1975).

REICH, C. W. Status of beta- and gamma-decay and spontaneous-fission data from transactinium isotopes. In *Proc IAEA Advisory Group Meeting on Transactinium Isotope Nuclear Data, Karlsruhe, (November 1975)*, vol. 3, paper B7. IAEA (1975).

TEUCHERT, E. AND RUTTEN, H. J. Core physics and fuel cycles of the pebble bed reactors. *Nucl. Engng. and Design, 34,* no. 1 (1975).

WYMER, R. G. (ed.). *Proc. of 2nd Int. Symp. on Thorium Fuel Cycle, Gatlinburg, 1966.* Washington, United States Atomic Energy Commission (1968).

Section 5

BURCH, W. D. AND LOTTS, A. L. Developments in reprocessing technology for high temperature and fast breeder fuels. In *Proc Int. Conf. on Nuclear Power and its Fuel Cycle, Salzburg, Austria, May 1977*, vol. 3, pp. 673–91. IAEA, Vienna (1977).

CANDELIERI, T., MOCCIA, A., AND SIMONETTA. *Operating Experience on the first Elk River fuel element reprocessing campaign* [in Italian]. *DOC CNEN (RT1) COMB-RITR* (76) 203. Rome, Comitato Nazionale per L'Energia Nucleare (1976).

ENERGY RESEARCH AND DEVELOPMENT ADMINISTRATION. *Light water breeder reactor program. Final environmental statement*, vol. 4 *ERDA* 1541 (1976).

KAISER, G. Status of reprocessing technology in the HTGR fuel cycle. In *Proc. Int. Conf. on Nucler Power and its Fuel Cycle, Salzburg, May 1977*, vol. 3, pp. 661-71. IAEA (1977).

Section 7

CLARKE, R. H. MACDONALD, H. F. FITZPATRICK, J, GODDARD, A. J. H. Waste disposal aspects of the long-term cooling characteristics of irradiated nuclear fuel. *Ann. Nucl. Energy, 2,* 451–466 (1975).

Energy Research and Development Administration. *Light water breeder reactor program—final environmental statement. ERDA* 1541 (1976).

International Atomic Energy Agency/International Labour Oganization. *Manual on radiological safety in uranium and thorium mines.* Safety series, No. 43, p. 82. IAEA/ILO (1976).

Meyer, H. R. Till, J. E. Bomar, E. S. Radiological impact of thorium mining and milling. *Nucl. Safety*, **20**, pp. 319–330 (1979).

Section 8

Carter, W. L. Rainey, R. H. Johnson, D. R. Thorium fuel cycles for LWRs. In *Proc. National Topical Meeting on the backend of the LWR Fuel Cycle, Savannah, Georgia, 19–22 March 1978.* Report No. CONF 780304 American Nuclear Society (1978). (Describes fuel diversion assessments and recyle requirements.)

Glossary

Absolute filter: an efficient filter for removing particulate matter from gases.

Absorbed dose: the energy deposited by ionizing radiation per unit mass of material (such as tissue).

Absorber: material which reduces radiation by removing energy from it; the magnitude of the reduction depends on the type of radiation, the type of material, its density, and its thickness. A sheet or body of such material.

Absorption process: a process where a component in a gaseous or liquid mixture is removed by contacting the gaseous or liquid mixture with a liquid or solid absorber. Silica gel is a common solid absorber.

Accelerator: a machine for producing high-energy charged particles by electrically accelerating them to very high speeds. Types include betatron, cyclotron, synchrotron, Cockcroft-Walton, Van de Graaff, tandem generator, and linac (linear accelerator).

Accident transient: the variation with time of the neutron population, the power, and the temperature of a reactor following some postulated accident.

Actinides: actinium and the elements following it in the Periodic Table; the most important are actinium, thorium, uranium, neptunium, plutonium, americium, and curium. Many of them are long-lived alpha-emitters.

Activation: the process of inducing radioactivity by irradiation, usually with neutrons, charged particles, or photons

Activation cross-section: effective cross-sectional area of target nucleus undergoing bombardment by neutrons, etc. Measured in barns.

Active: often used to mean radioactive.

Active area: part of a laboratory where radioactivity may be present, and where exposure to individuals is constantly controlled.

Activity: the number of nuclear disintegrations occurring per unit of time in a quantity of a radioactive substance. Activity is measured in curies. Often used loosely to mean radioactivity.

Activity, specific: the activity per gram of material.

Additive compounds: compounds formed by additive reactions, in which a double bond is converted into a single bond by the addition of two more atoms or radicals.

Adiabatic: without loss or gain of heat.

Adsorption: the retention of dissolved substances on the surface of a substance (adsorbent).

Advanced gas-cooled reactor (AGR): the successor to the Magnox reactors in the UK nuclear power programmes. It uses slightly enriched oxide fuel canned in stainless steel with graphite moderator and carbon dioxide coolant.

Aerosol: a colloidal system such as a mist or fog, in which the dispersion medium is a gas.

Aggressive salts: these are chemicals like sodium and magnesium hydroxides or chlorides which are very corrosive if they exist in high concentrations at high temperature (200–300 °C).

Alara: as low as reasonably achievable.

Alpha particle: a charged particle having a charge of 2 and a mass of 4 atomic mass units. It is emitted in the decay of many heavy nuclei and is identical with the nucleus of a helium atom, consisting of two protons and two neutrons.

Alpha (radiation): helium nucleus emitted by some radioactive substances, e.g. plutonium-239.

Americium: artificially made transuranic element.

Anharmonic: said of any oscillation system in which the restoring force is non-linear with displacement, so that the motion is not simple harmonic.

Anhydrous: a term applied to oxides, salts, etc. to emphasize that they do not contain water of crystallization or water of combination.

Anisotropic: said of crystalline material for which physical properties depend upon direction relative to crystal axes. These properties normally include elasticity, conductivity, permittivity, permeability, etc.

Annihilation: spontaneous conversion of a particle and corresponding antiparticle into radiation, e.g. positron and electron, which yield two gamma-ray photons each of 0.511 MeV.

Annihilation radiation: the radiation produced by the annihilation of a particle with its corresponding antiparticle.

Antibody: a body or substance invoked by the stimulus provided by the introduction of an antigen, which reacts specifically with the antigen in some demonstrable way.

Antigen: material which sensitizes tissues in an animal body by contact with them and then reacts in some way with tissues of the sensitized subject *in vivo*, or with his serum *in vitro*.

Aqueous phase: the 'watery' solution in a solvent-extraction process.

Argillaceous rocks: sedimentary rocks having a very small mineral grain size, as in clay.

Aspect ratio (torus): the ratio R/r of the major to minor radii.

Atom: the smallest particle of an element, which has the chemical properties of that element. An atom consists of a comparatively massive central nucleus of protons and neutrons carrying a positive electric charge, around which electrons move in orbits at relatively great distances away.

Atomic: strictly, relating to the behaviour and properties of entire atoms—nuclei and orbital electrons; it is more usually a synonym for *nuclear*, it is as in 'atomic energy'.

Atomic absorption spectrometry: a method of physical analysis. A small aliquot of the sample to be analysed is introduced into a flame, the heat from which excites the outer electrons of the atoms causing the emission of light characteristics of the constituent elements. This light preferentially absorbs light of identical wavelengths when this, emitted from a standard comparison lamp, is passed through the flame. This provides a means of identifying elements of a given material as well as an estimate of the amount of each present from the extent of absorption.

Atomic displacement cross-section: a measure of the probability of a neutron displacing an atom from its normal position in the crystal lattice structure of a material. The probability is expressed as a target area or cross-section.

Atomic mass unit (amu): one-twelfth of the mass of an atom of carbon-12. Approximately the mass of an isolated proton or neutron.

Atomic weight: the average mass of the atoms of an element at its natural isotopic abundance, relative to that of other atoms, taking carbon-12 as the basis. Roughly equal to the number of protons and neutrons in its nucleus.

Atomic number (Z): the number of protons in an atomic nucleus. Nuclei with the same atomic number but different mass numbers are isotopes of the same chemical element.

Attenuation: reduction in intensity of radiation in passing through matter.

Autoclave: a vessel, constructed of thick-walled steel for carrying out chemical reactions under pressure and at high temperatures.

Average effective dose-equivalent: the measure of the risk from exposure to radiation, which takes account of the different sensitivity of various organs of the body and allows for the effects of different types of radiation.

Azimuthal power instability: eccentric neutron behaviour which results in uneven nuclear conditions in the reactor.

Backfitting: making changes to plants already designed or built.

Background: (1) in discussing radiation levels and effects, it refers to the general level of natural and man-made radiations against which a particular added radiation component has to be considered; (2) in discussing radiation measurement techniques, it may also include spurious readings due to the *noise* characteristics of the instrument and its power supplies, and to the presence of local radioactive contamination, etc.

Bare sphere critical mass: the mass of pure fissile material which if formed into spherical shape, with no outer layer of neutron reflecting material, will just sustain a chain fission reaction.

Barn: a unit of area (10^{-24} cm^2) used for expressing nuclear cross-sections.

Barytes: barium sulphate, a common mineral in association with lead ores, occurring also as nodules in limestone and locally as a cement of sandstones.

Base load: in electricity generation, the minimum steady power demand on the system over a period.

Batch (process): a process not operated continuously.

Bearing pads: pads attached to the outer faces of fuel rod wrappers which contact with similar pads on neighbouring elements either initially (restrained core) or as a result of distortion (free-standing core).

Bearing resonances: low-speed synchronous whirling frequencies, which are determined by the inertia of the rotor and the support stiffness of the bearings.

Bearing systems: the supports to hold a rotating shaft in its correct position.

Becquerel (Bq): the new unit of activity in the SI System; it is equivalent to 1 disintegration per second or roughly 2.7×10^{-11} Ci.

Belt grinding: an abrasive belt process for removal of a thin surface layer from a tube outer surface.

Benchmark: a name of American origin to describe a well-defined problem or experiment which then provides a reference standard for inter-comparison of various methods of solution or prediction.

Beta particle: an electron or positron emitted from a nucleus in certain types of radioactive disintegration (beta-decay).

Beta quenching: rapid cooling of uranium from the β-phase region.

Beta radiation: nuclear radiation consisting of β-particles.

Bifurcate: twice forked, forked.

Binder/binderless routes: pelleting methods employing or not employing a binder to assist the powder compaction.

Binding energy: the energy theoretically needed to separate a nucleus into its constituent protons and neutrons; it gives a measure of the stability of the nucleus.

Biological shielding: heavy concrete shielding erected around certain sections of plant containing radioactive materials in order to protect the operators from nuclear radiations.

Biosphere: that part of the earth and the atmosphere surrounding it, which is able to support life.

Blanket: *fertile* material (usually depleted uranium) arranged round a fast reactor core to capture neutrons and create more fissile material (usually plutonium); in a fusion reactor the blanket may be of lithium to capture neutrons and create more tritium.

Bled-steam feed-heating train: a series of heat exchangers in which steam is *bled* or extracted from the main expansion path through a steam turbine and is used to raise the temperature of feed water being returned to a steam generator from the condenser at the exhaust of the turbine.

Blowdown: rejection of liquid from a vessel under pressure to reduce dissolved solids.

BNFL: British Nuclear Fuels Limited.

Boiling-water reactor (BWR): a light water reactor in which the water is allowed to boil into steam which drives the turbines directly.

Boltzmann equation: the fundamental particle conservation diffusion equation based on the description of individual collisions, and expressing the fact that the time rate of change of the density of particles in the medium is equal to the rate of production less the rate of leakage and the rate of absorption.

Boral sheeting: a composite formed of boron carbide crystals in aluminium with a cladding of commercially pure aluminium.

Bore grinding: a grid abrasion process to remove a thin surface layer from a tube bore.

Boron: element important in reactors, because of large cross-section (absorption) for neutrons; thus, boron steel is used for control rods. The isotope ^{10}B on absorbing neutrons breaks into two charged particles ^{7}Li and ^{4}He, which are easily detected, and is therefore most useful for detecting and measuring neutrons.

Boron counter: an ionization chamber or proportional counter for detecting thermal neutrons by their interaction with boron-10 nuclei.

Brachy-therapy: treatment of tumours by radiation from sources placed in or near to the tumour.

Branching: alternative modes of radioactive decay which may be followed by a particular nuclide.

Brazing: the process of joining two pieces of metal by fusing a layer of brass or spelter between the adjoining surfaces.

Breeder: short for fast breeder reactor.

Breeding: the process of converting a fertile isotope, e.g. ^{238}U into a fissile isotope, e.g. ^{239}Pu. Fast reactors can be designed to produce more fissile atoms by breeding that are lost by fission. The process is also referred to as conversion.

Bremsstrahlung: X-rays produced when rapidly moving charged particles, e.g. electrons, interact with matter (from German 'braking radiation').

Broad-group library: a set of nuclear cross-sections tabulated as average values over a few (about 40) relatively broad energy groups.

Buffer tank: a vessel usually charged with a gas connected to a system containing a liquid, allowing the liquid to be expelled from the system and the out-flow brought to rest in a controlled manner by the cushioning effect of the gas in the tank as it is compressed above the liquid.

Bundle: see *fuel assembly*.

Burn-up: (1) in nuclear fuel, the amount of fissile material burned up as a percentage of the total fissile material originally present in the fuel; (2) of fuel element performance, the amount of heat released from a given amount of fuel, expressed in megawatt-days per tonne.

Burst: a fuel cladding defect which allows fission products to escape into the coolant; it need not be more than a very small crack or pin-hole.

Busbar: an electric conductor of large current capacity connecting a number of circuits.

Butex: name given to dibutyl ether of diethylene glycol, an organic liquid used in solvent-extraction processes.

Calandria: a closed tank penetrated by pipes so that liquids in each do not mix.

Calorimetry: the measurement of thermal constants, such as specific heat, latent heat, or calorific value; such measurements usually necessitate the determination of a quantity of heat, by observing the rise of temperature it produces in a known quantity of water or other liquid.

Campaign: the period from plant start-up to plant shutdown in a nuclear fuel reprocessing operation—usually a few months long.

Can: the container (usually of metal) in which nuclear fuel is sealed to prevent contact with the coolant or escape of radioactive fission products etc., sometimes also to add structural strength and to improve heat transfer. It may be made of Magnox, Zircaloy or stainless steel and may carry fins to increase the rate of heat tranfer.

CANDU: a type of thermal nuclear power reactor developed in Canada and widely used there; it uses natural (unenriched) uranium oxide fuel canned in Zircaloy, and heavy water as moderator and coolant.

Canyon concept: a reprocessing plant layout favoured in the USA, where the plant is constructed partly below ground in concrete-lined vaults or canyons.

Capture: the process in which a particle (e.g. a neutron) collides with a nucleus and is absorbed in it.

Carbon-dating: a means of dating by measuring the proportion of radioactive carbon. Atmospheric carbon dioxide contains a constant proportion of radioactive ^{14}C, formed by cosmic radiation. Living organisms absorb this isotope in the same proportion. After death it decays with a half-life 5.57×10^3 years. The proportion of ^{12}C to the residual ^{14}C indicates the period elapsed since death.

Carbon-14: see *carbon-dating*.

Carcinogenesis: the production and development of cancer.

Carcinoma: a disorderly growth of epithelial cells which invade adjacent tissue and spread via lymphatics and blood vessles to other parts of the body.

Carrier-free: a carrier-free preparation of an isotope consists of atoms of that isotope alone.

Cartridge: a unit of nuclear fuel in a single can.

Cascade: (1) in nuclear fuel processing etc., a progressive sequence of operations in which the process material flows from one stage of the plant to the next, for example, in solvent-extraction processes (in gaseous diffusion or gas centrifuge separation processes there may be several hundred almost identical stages—flow may be in both directions); (2) the emission of gamma rays by a radioactive nucleus in sequence separated by a very short time interval.

Cascade (ideal): a cascade in which the flow is graded along it to avoid mixing losses, yielding maximum separative power.

Cascade (jumped): a cascade in which the product or waste from a stage is not neccssarily connected to the stage above or stage below, respectively. If the cut is small, the waste may be connected to a point several stages below in the cascade.

Cascade (squared off): a cascade built in sections whose outlines follow those of an ideal cascade—each section of a squared-off cascade is a square cascade.

Catalytic hydrogenation: chemical reactions in which molecular hydrogen is added to the organic compound in the presence of a secondary element or compound, the catalyst. The catalyst effectively takes no part in the reaction but is essential for its completion. Typical hydrogenation reactions are the hydrogenation of olefins to paraffins, aromatics to napthenes, and the reduction of aldehydes and ketones to alcohols. Typical catalysts are Ni, Pd, V_2O_5, etc.

Cation exchange: the process by which suitable solid agents such as zeolites, artificial resins or clays can remove cations (i.e. positively charged atoms or molecules) from solution by exchanging them with another cation. Commonly used in water softening whereby calcium and magnesium ions are replaced by sodium ions; the removed ions are held by solid cation exchanger. In the context of radioactive waste disposal the process would be typically the removal of strontium and plutonium ions by replacement by sodium or potassium ions from a mineral.

Cave: a heavily shielded compartment in which highly radioctive materials can be safely kept, handled or examined by remote manipulation; sometimes called a *hot cell*.

Centrifuge: apparatus rotating at very high speed, designed to separate solids from liquids, or liquids from other liquids dispersed therein.

Ceramic: hard pottery-like materials having high resistance to heat, e.g. oxides and carbides of metal; it is used for nuclear fuels operating at high temperatures.

Cermet: an intimate mixture of metallic and ceramic particles which combine some of the desirable qualities of both, e.g. for reactor fuels.

Chain reaction: a process which, once started, provides the conditions for its own continuance. In the chain reaction of nuclear fission, neutrons cause nuclear fission in uranium or plutonium, producing more neutrons, which cause further fissions, and so on.

Channelling: the escape of radiation through flaws in the moderator or shielding of a reactor, etc. leading to high levels of radiation in the regions affected.

Charcoal delay bed: beds of charcoal to which gases can be admitted providing hold-up or delay by absorption and desorption. In a nuclear plant such delays are used to allow the decay of activity.

Charge-face: of a nuclear reactor, that face of the biological shield through which the fuel is inserted.

Charge/discharge machine: a mechanical device for inserting or removing fuel in a nuclear reactor without allowing the escape of radiation and, in some reactors, without shutting the reactor down.

Charged particles: nuclear or atomic particles which have a net positive or negative electric charge; they include electrons (and beta particles), positrons, protons, deuterons, alpha particles, and positive or negative ions of any of the chemical elements, but not neutrons.

Chemisorption: irreversible adsorption in which the adsorbed substance is held on the surface by chemical forces.

Chopping: process of cutting nuclear fuel into small lengths.

Cisternography: visualisation of body spaces.

Cladding: the protective layer, usually of metal, covering the fuel in a nuclear fuel element.

Clean critical assembly: a reactor in its initial stage before irradiation has caused changes in the fuel composition; the material composition is usually very well known.

Closed cycle, closed-cycle cooling: a completely enclosed path, e.g. a Magnox coolant circuit.

Cluster: see *fuel assembly*.

Coastdown: the process of slowing down of a pump or turbine once the drive mechanism has stopped or been disengaged.

Coated fuel particle: a compact of nuclear fuel coated with a refractory material which restricts release of fission products.

Codecontamination: the decontamination of U and Pu together from fission products.

Cold criticality: the establishment of a low-power, nuclear fission chain reaction under conditions of essentially zero heat generation.

Cold drawing: a continuous cold metal working process in which a tube is pulled through a die, reducing its wall thickness and/or diameter.

Cold pilgering-type operation: an intermittent metal-working process to reduce tube wall thickness and/or diameter.

Cold work: the plastic deformation of a material at temperatures far below the melting-point, carried out usually in order to increase its strength.

Collimator: a device to confine radiation to a narrow beam by preferentially shielding against radiation in other directions.

Column: vertical, cylindrical apparatus for carrying out a chemical operation, e.g. solvent extraction, absorption, etc.

Commissioning: running a machine etc. (e.g. a nuclear reactor) up to power and checking that it complies with the specifications before the supplier hands it over to the customer.

Committed dose-equivalent: the total integrated dose-equivalent over 50 years to a given organ or tissue from a single intake of radioactive material into the body.

Common-mode failure: the failure of two or more supposedly independent parts of a system, e.g. a reactor, from a common external cause or from interaction between the two parts.

Compound nucleus: a highly excited nucleus, of short lifetime, formed as an intermediate stage in a nuclear reaction, e.g. ^{236}U prior to fission.

Concentration limits: a technique of criticality control using concentration limits as the control.

Concentration pulse: a term coined to describe an enhancement, which exists for a particular period only, in the concentration of one of the constituents of a general medium. In the case of a radioactivity concentration pulse, a graphical plot of specific activity against time demonstrates a profile rising to a maximum and, thereafter, falling towards the usual background level.

Condenser off-gas: incondensable gases isolated in the steam condenser at the exhaust of a turbine that have to be drawn off if the condensing process is to continue efficiently.

Conditioning: the addition of chemicals to a solution in order to adjust the chemical composition; it is usually carried out in a 'conditioner'.

Confinement time: in nuclear fusion research, the average life-time of a particle in a plasma containment system.

Constant-volume feeder: see *CVF*.

Contactor: (reprocessing) generic term for solvent-extraction apparatus.

Containment: physical boundaries constructed to confine radioactive material from a reactor or plant used in reprocessing.

Contamination: the presence of unwanted radioactive matter, deposited on solid surfaces, or introduced into solids, liquids or gases.

Continuous operation: a method of operation of a process or plant in an unbroken sequence (see *batch*).

Continuous refuelling: replacing fuel channels one at a time at the required interval rather than in a batch at longer intervals. Refuelling with the reactor on-load is a particular case.

Control rod: a rod of neutron-absorbing material (e.g. cadmium, boron, hafnium) moved in or out of the reactor core to control the reactivity of the reactor.

Control rod worth: the reactivity change resulting from the complete insertion of a fully withdrawn control rod into a critical reactor under specified conditions.

Conversion: a term used for breeding (see separate entry) when the main fissile element consumed is different from the main fissile element bred. Thus the term applies to reactors fuelled with $^{235}U/^{238}U$ as opposed to $^{239}Pu/^{238}U$.

Conversion factor: the ratio of the number of fissile nuclei produced by conversion to the number of fissile nuclei used up as fuel.

Converter reactor: a reactor in which the conversion process takes place, but in which breeding, with a net gain of fissile material, does not.

Coolant: the gas, water or liquid metal circulated through a reactor core to carry the heat generated in it by fission (and radioactive decay) to boilers or heat exchangers.

Cooling, radioactive: progressive diminution of radioactivity, especially of nuclear fuel after removal from a reactor. This is accompanied by a diminution of heat output, so the word may be used in either sense.

Cooling pond: a water-filled tank in which used fuel elements are placed while cooling (in the radioactive and the thermal senses) is allowed to proceed; the water provides both radiation shielding (conveniently transparent) and means of removing the heat of radioactive decay.

Core-catcher: commonly used to describe the device designed to retain the products of the melted reactor core after a postulated accident. It may be within the reactor vessel (internal) or below the vessel (external).

Core follow: a mathematical technique for constraining the theoretical prediction of the behaviour of a reactor to follow the measured behaviour during operation.

Core, homogenous: core materials so distributed that the neutron characteristics of the reactor are homogenous.

Core power: the rate of production or use of energy in the core.

Core, reactor: the central region of a nuclear reactor containing the fuel elements, where the chain reaction of nuclear fission proceeds.

Coriolis coupling: the coupling between vibrations and rotations in a molecule.

Coulomb barrier: the potential barrier between charged particles due to mutual electrostatic repulsion.

Counter: an instrument for counting pulses of radiation, or the electric pulses that these cause, and displaying or recording them in digital form; also used loosely for any form of radiation detection or measuring instrument.

Counter-current: opposing flows, as for example, where the organic phase carrying U + Pu flows in one direction, while the aqueous phase containaing fission products flows in the other direction.

Counting rate: the rate at which radioactive events, for example, the emission of beta particles, are registered by the measuring device; it is usually expressed in counts per minute, counts per second etc. The counting rate is less than the total radioactive disintegration rate by a factor which expresses the overall counting efficiency of the particular measuring device for the radiation in question.

Coupled control system: a form of power station control which is inherently load following; in the nuclear application, the system is sometimes referred to as 'core follows turbine'.

Coupled hydrodynamic–neutronic instability: in a BWR thermohydraulic instability is complicated by a feedback through the link between the amount of steam in the core (voidage) and the power generated in the fuel. This feedback effect can be dominant when the time constant of a hydraulic oscillation is close to the same magnitude as the time constant of the fuel element. Strong nuclear-coupled thermohydrodynamic instabilities occured in early experiments at Idaho where a metallic fuel with a low time constant was operated in a low-pressure boiling-water flow.

Creep (radiation): the time-dependent, non-reversible dimensional change in a material subject to both a mechanical load and a neutron flux. Radiation creep is caused by the stress-induced preferential segregation of the atoms displaced by irradiation damage. It is observed at temperatures too low for thermomechanical creep to occur. At higher temperatures, e.g. in a reactor core, both types of creep may occur simultaneously.

Creep (thermal or thermomechanical): the low, time-dependent, non-reversible dimensional change in a material when subject to a mechanical load less than that required to produce plastic deformation. Creep is very temperature-dependent, significant only at temperatures above about 0.4 of the melting-point on the absolute scale.

Critical, criticality: a nuclear reactor or other assembly of fissile material is said to have 'gone critical' when its chain reaction has just become self-sustaining.

Critical mass: the amount of fissile material needed to maintain a nuclear chain reaction.

Critical material: the material in which the concentration of radioactivity resulting from a given discharge is highest, when expressed as a fraction of the appropriate derived working limit.

Critical pathway: the pathway by which most radioactivity reaches the critical material.

Critical population group: the group of persons whose radiation doses, resulting from a given practice, are highest.

Criticality incident or excursion: inadvertent accumulation of fissile material into a critical assembly, leading to criticality and the sudden and dangerous emission of neutrons, gamma rays and heat.

cross-section, nuclear: the target area presented by a nucleus to an approaching particle relating to a specified nuclear interaction, e.g. capture, elastic scattering, fission. The cross-section varies with the type of nucleus, the type of energy of the incident particle and the specified interaction. Cross-sections are measured in barns, and give a measure of the probability of the particular reaction.

Cryostat: low-temperature thermostat.

Curie (Ci): the unit of radioactivity, being the quantity of radioactive material in which 3.7×10^{10} nuclei disintegrate every second. Originally it was the activity of 1 gram of radium-226. The curie has now been superseded in the SI system by the becquerel (Bq), equal to 1 disintegration per second.

Curium: artificially made transuranic element.

Cut: the fraction of the feed to a separation stage which emerges in the product stream.

CVF (constant-volume feeder): a rotating device with 'buckets' on the ends of radial arms, which scoops up liquid and delivers it at a rate proportional to the rotational speed.

Cycle: in solvent extraction, used to denote one complete sequence of extraction, scrubbing and stripping.

Cyclotron: an accelerator in which charged particles follow a spiral path in a magnetic field and are accelerated by an oscillating electric field.

Dating, radioisotope: determination of the age of an archaeological or geological specimen by measuring its content of a radioactive isotope in relation to that of its precursor or decay product, or of its stable isotope; applied particularly to radio-carbon dating of archaeological specimens. (See *carbon dating*).

Daughter product: the nuclide immediately resulting from the radioactive decay of a parent or precursor nuclide. If it is radioactive, it will in due course become a parent itself.

D–D: symbol for reaction between two nuclei of deuterium atoms.

Decade: any ratio of 10:1

Decay chain: a series of radionuclides each of which disintegrates into the next, until a stable nuclide is reached.

Decay constant (decay, law of radioactive): the probability per unit time that a nucleus will decay spontaneously. If the number of nuclei is N, the rate of decay dN/dt and the decay constant λ, then the law of radioactive decay is: dN/d$t = -\lambda N$.

Decay heat: the heat produced by radioactive decay, especially of the fission products in irradiated fuel elements. This continues to be produced even after the reactor is shut down.

Decay product: synonym for *daughter product*.

Decay, radioactive: the disintegration of a nucleus through emission of radioactivity. The decrease of activity due to such disintegration.

Decommissioning: the permanent retirement from service of a nuclear facility and the subsequent work required to bring it to a safe and stable condition.

Decommissioning, stage of: the sequence of stable stages of partial or total decommissioning.

Decontamination: the process of removing radioactivity or any other unwanted impurity.

Decontamination factor: the ratio of the proportion of contaminent to product before treatment to the proportion after treatment.

Delayed neutrons: neutrons resulting from fission but emitted a measurable time after fission has taken place. They play an essential part in nuclear reactor control.

Depleted uranium: uranium with less than the natural content (0.71 per cent) of ^{235}U, e.g. the residue from an isotope enrichment plant or from a nuclear fuel reprocessing plant.

Derived working limit (DWL): a limiting value for the amount of radioactive material which may be present continuously in a given situation without risk that the basic international dose limitations will be exceeded.

Detector, radiation: a device for detecting and counting individual radiation pulses or for measuring radiation intensity. The variety of radiations and the kinds of measurements that need to be made require many forms of detector, which include: Geiger-Müller counters, proportional counters, solid and liquid scintillation counters, fission and ionization chambers, and semi-conductor detectors.

Deuterium: the hydrogen isotope of mass 2, 'heavy hydrogen'. (See *heavy water*.)

Deuteron: the nucleus of a deuterium atom, comprising one neutron and one proton.

DFR: the Dounreay fast reactor.

DFR fuel: fuel used in DFR, an alloy of enriched uranium and molybdenum.

Diagrid: the structure supporting the core, blanket and radial shield rods, which also distributes the coolant flow amongst these items.

DIDO: nuclear reactor situated at Harwell.

Die-filling: filling of a die (container) with powder prior to compressing into a pellet.

Diffractometer: an instrument used in the examination of the atomic structure of matter by the diffraction of X-rays, electrons, or neutrons.

Diffusion: in general, the random movements of particles through matter. Specifically used for: (1) diffusion of a gas through a porous membrane, notably in the enrichment of uranium by the diffusion of uranium hexafluoride gas; (2) the movement of fission neutrons through a moderator.

Diffusion plant: a plant for the enrichment of uranium in the ^{235}U isotope by gaseous diffusion of uranium hexafluoride through a porous membrane.

Direct cycle: where the turbine is driven by coolant directly received from the reactor, i.e. one primary circuit.

Disequilibrium: the converse of the stable condition described under *secular equilibrium* in the particular case where a transient preferential separation of one or more of the members in the isotopic decay chain upsets a previously established equilibrium state which then takes a time, dependent upon the longest half-life members disturbed in the chain, to re-establish.

Dishing: a shallow spherical or truncated conical depression in one or both end faces of a UO_2 fuel pellet.

Disintegration: any transformation of a nucleus, either spontaneous or by interaction with radiation, in which particles or photons are emitted. It is used in particular to mean radioactive decay.

Disposal: the removal from man's environment of unwanted or dangerous material, notably nuclear waste, to a place of safety, without the intention of retrieving it later.

Distillation: purification of a liquid by boiling it and condensing and collecting the vapours.

Distribution coefficient: the ratio of the total concentration of a substance in the organic phase (regardless of its chemical form) to its total analytical concentration in the aqueous phase.

Divergence: a nuclear chain reaction is said to be divergent when the rate of production of neutrons exceeds the rate at which they are lost, so that the fission reaction increases in intensity or spreads through a larger volume of material.

DNA: deoxyribonucleic acid (see *nucleic acid*).

Doppler broadening; Doppler coefficient; Doppler constant; Doppler effect: when the velocity of an atom (derived from its thermal energy) is comparable to that of interacting neutrons, the proportion of neutrons affected by resonances in the neutron cross-sections changes with temperature. This is because the resonances are effectively broadened as the velocity of the atom increases with temperature (Doppler broadening). The effect is important for a few isotopes present in quantity in a typical LMFBR, for example ^{238}U. As temperature increases, capture in ^{238}U increases and reactivity decreases; this is called the Doppler effect and is a very useful safety feature. The Doppler temperature coefficient is dR/dT where R is reactivity and T temperature. Because this coefficient is found to be approximately inversely proportional to the absolute temperature T, a Doppler constant D is also defined, where $D = T\,dR/dT$

Dose commitment: future radiation doses inevitably to be received by a person or group, e.g. from a radioactive material already incorporated in the body.

Dose-equivalent: the absorbed dose multiplied by a quality factor to measure the biological effectiveness of radiation irrespective of its type in rems or sieverts.

Dose-equivalent (effective): the *dose-equivalent* to the whole body having the same risk of causing biological harm as an exposure of part of the body.

Dose-radiation: generally, the quantity of radiation energy absorbed by a body.

Dose-rate: the dose absorbed in unit time, e.g. rems per year.

Dosemeter, Dosimeter, Dose-rate meter: an instrument which measures radiation doses or dose-rates

Doubling time: (1) in a divergent reactor, the time taken for the neutron flux density, and therefore the power, to double; (2) in a breeder reactor, the time taken to produce new fuel equivalent to a full replacement charge in addition to the fuel consumed during this time.

Down time: the period during which a reactor is shut down for routine maintenance.

Dragon: a high-temperature gas-cooled reactor experiment operated at Winfrith, UK, by an OECD project team from 1964 to 1975.

Dry cooling tower: a cooling system which uses the atmosphere as a heat sink by a combination of a jet condenser, closed water circuit, heat exchangers cooled by air, and a cooling tower.

Dry well: the region around the reactor vessel which is kept 'dry' during normal operation, but through which steam and water would discharge in the event of a loss of coolant accident.

Dryout margin: the factor relating a heat flux employed in a boiling system with the critical heat flux which would cause the heating surface to be blanketed by the vapour phase thereby raising the temperature of this surface.

D–T: symbol for reaction between nuclei of deuterium and tritium.

Ductility: the maximum dimensional change per unit length of a mechanically loaded material, as measured just before the point of failure, excluding any region of gross deformation.

EBR: experimental breeder reactor.

Eddy diffusion: the mixing of isotopes due to turbulent motion.

Effective dose-equivalent: the dose-equivalent to individual organs multiplied to give the 'effective whole-body dose-equivalent'.

Effluent: a waste stream from a chemical process, usually gaseous or liquid.

Egg boxes: a constructional feature of fast reactor fuel sub-assemblies.

Eigenfunction: the solution of an equation compatible with the boundary conditions associated with possible values of a parameter of the equation (the eigenvalue).

Elastic scattering: the outcome of collisions between particles, in which the total kinetic energy of the system is unchanged, but the directions of motion of the particles are altered (i.e. the particles simply bounce off one another with no net energy loss).

Electromagnetic radiation: radiation having the nature of electromagnetic waves. In the nuclear context it includes gamma rays and X-rays.

Electromagnetic separation: the separation of ions of different masses by deflection in a magnetic field.

Electron: one of the stable elementary particles of which all matter consists. It carries a single unit of negative electric charge equal to 1.6×10^{-19} coulombs and has a mass of 9×10^{-31} kilograms.

Electronegative residuals: a term that refers to the particular impurities that might form electronegative ions left in the gas after attempts have been made to remove them. Successful operation of a proportional counter depends upon the gas it contains being as pure as possible especially in regard to certain molecular species (e.g. Cl_2, O_2, NH_3, H_2O, etc.) which readily form electronegative ions. Formation of electronegative ions limits, or may even completely prevent, the operation of the counter in its intended mode.

Electronvolt: a unit in which energy is measured in the study of nuclear particles and their interactions. It is equal to the change in energy of an electron crossing a potential difference of 1 volt. Abbreviation eV and multiples keV (10^3 eV) and MeV (10^6 eV).

Electrophoresis: motion of colloidal particles under an electric field in a fluid, positive groups to the cathode and negative groups to the anode.

Element, chemical: a simple substance which cannot (by normal chemical means) be broken down into simpler components. All its atoms have the same number of protons in their nuclei and therefore occupy the same place in the Periodic Table.

Eluant: liquid which is added to an ion-exchange column and passes through it carrying the desired product which it has removed from the column.

Emanometer: radon monitor.

Energy, atomic: popular, though not strictly accurate, synonym for nuclear energy.

Energy containment time: the total energy of a confined plasma divided by the rate of energy loss from it.

Energy fluence: the energy intensity integrated over time of a short pulse of radiation.

Energy loss discrimination: a technique by which charged particles that may be identical in certain respects, e.g. in mass and momentum-to-charge ratio, are separately distinguished by means of the different rates at which they lose energy in passing through a thin detector (thin meaning the path length within the counter, over which the energy loss is measured, is small compared with the total range of the particle).

Energy, nuclear: the energy released when the particles constituting the nuclei of atoms undergo rearrangement, especially through neutron-induced fission in uranium or plutonium.

Engineered storage: storage of spent fuel or high-activity wastes in facilities specially constructed to ensure safe keeping until such time as processing or disposal is undertaken.

Enrichment: the process of increasing the abundance of fissionable atoms in natural uranium (which contains 0.7 per cent of fissle isotope ^{235}U). This is usually done in either the *centrifuge* process (where isotopes are separated by centrifugal force) or the *diffusion* process (where a series of screens retards the heavier isotopes).

Environmental pathway: the route by which a radionuclide in the environment can reach man, e.g. by progressive biological concentration in foodstuffs.

EURATOM: European Atomic Energy Community.

Event tree: a diagram which, starting from some initiating event, identifies the possible courses of an accident by a series of branches expressing the respone 'YES' or 'NO' to the question 'does this engineered safety feature work?'

Eversafe: a description given to plant whose dimensions are limited so that a critical quantity of plutonium (or highly enriched uranium) cannot be accommodated.

Excitation: the addition of energy to a system, e.g. a nucleus, transferring it from its 'ground' state to an 'excited' state.

Excursion: a rapid increase of reactor power above the set levels of operation. This increase may be deliberately caused for experimental purposes or it may be accidental.

Exothermic: accompanied by the evolution of heat.

Expansion coefficient: the fractional expansion (i.e. the expansion of unit length, area, or volume) per degree rise of temperature.

Exponential assembly: an experimental sub-critical assembly into which thermal neutrons are introduced at one face. The neutron flux density in the assembly decreases exponentially with distance from this face. Used in studies of reactor physics, etc.

Extraction (in solvent extraction): the transfer of a dissolved substance from an aqueous phase to an organic phase.

Extraction column: solvent-extraction apparatus where the contacting is done in a vertical column with aqueous inlet at top and organic inlet at bottom.

Fall-out: deposition of radioactive dust etc. from the atmosphere, resulting from the explosion of nuclear weapons or from accidental release.

Fast neutrons: neutons travelling with a speed close to that with which they were ejected from the fissioning nucleus, typically about 20 000 km s^{-1}.

Fast reactor: a nuclear reactor in which most of the fissions in the chain reaction are caused by fast neutrons, travelling with a speed close to which they were ejected from the fissioning nucleus. It contains no moderator, and is capable of generating more fissile material than it consumes.

Fault conditions: any condition in the reactor or plant which is a departure from the designed normal operating condition and which could lead directly or indirectly to an automatic shutdown, to damage or to an accident.

Fault tree: a diagram representing possible initiating events and sequences of successive failures that could lead to an accident.

Fecundity: capacity of a species to undergo multiplication.

Feed: a solution introduced into an extraction system.

Feed train: the series of components which exist between the power station condenser and the steam generator through which the feedwater must pass. Typically these components will be condensate booster pumps, de-aerators, feed pumps, and feed heaters.

Feedwater: the water, previously treated to remove air and impurities, which is supplied to a boiler for evaporation.

Fertile: material such as uranium-238 and thorium-232 which can be transformed by neutron absorption into fissile ^{239}Pu or ^{233}U, respectively.

Fine-group library: a set of nuclear cross-sections tabulated as average values over many energy groups, usually about 2000.

Fine-structure experiment: one in which the detailed variation of the neutron reaction rates from the moderating material, through the coolant, the structural materials, and into the fuel is measured.

FINGAL: fixation in glass of active liquid.

Fissile: a material readily capable of undergoing fission when struck by a neutron, notably ^{235}U, ^{233}U, ^{239}Pu, and ^{241}Pu.

Fission-counter: a detector consisting of a tube lined with fissile material or filled with a fissile gas which detects neutrons by the ionization produced in it by the fission products.

Fission, fast: fission induced by fast neutrons.

Fission gas bubble swelling: a swelling of UO_2 resulting from the accumulation of fission gases into bubbles, mainly on grain boundaries.

Fission neutrons: neutrons produced at the time of fission.

Fission, nuclear: the splitting of a heavy nucleus usually into two nearly equal fast-moving fragments, accompanied by fast neutrons and gamma rays. Fission may be either spontaneous or induced by the absorption of a particle or a high-energy photon.

Fission product(s): the nuclide(s) formed when a fissile material undergoes nuclear fission.

Fission spectrum: (1) the energy distribution of prompt neutrons in fission of a specified nuclide (2) the energy distribution of prompt gamma radiation arising from the fission; (3) the range and abundance of nuclides formed from the fission.

Fission, spontaneous: a mode of radioactive decay in which a heavy nucleus undergoes fission without being excited by any external cause. It occurs in ^{238}U as a mode of decay having a very long half-life.

Fission, thermal: fission induced by thermal neutrons.

Fission track dating: a physical method of dating applicable to glassy materials such as obsidian. Such materials frequently contain the isotope ^{238}U, which in undergoing spontaneous fission, forms two, energetic, heavy nuclei, which lose energy in travelling through the crystal lattice and cause visible damage along their paths (fission tracks). On formation, at high temperatures, the material is fully annealed, meaning all previous crystal lattice damage is reformed and no tracks are present. Thus, from a physical count of the number per unit area in a section of a sample and knowledge of the specific U concentration producing tracks, it is possible to estimate the elapsed time since the material's formation.

Fission yield: the number of nuclei of a particular mass resulting from 100 fissions (strictly the chain fission yield).

Fissium: collective term to describe all the fission products formed by nuclear fission of a fissile material.

Flask: a heavily shielded container used to store or transport radioactive material, especially used nuclear fuel.

Flame photometry: as for *atomic absorption* a method of physical analysis, but in this case the light emitted from the samples introduced into a flame is directly examined. The presence of a particular element is demonstrated by the identification of that element's characteristic light output seen as discrete lines when viewed through a suitable spectroscope, and the intensity depending upon the quantity present.

Flashing-off: the process whereby steam is formed from hot water by a reduction in the pressure of the system.

Flocculation: separation of radioactive waste products from water by coagulation of an insoluble precipitate.

Floc process: a process where material is precipitated as a mass of fine particles.

Flow path redundancy: a way of providing a series of alternative parallel routes for a water flow, so that if one route becomes blocked or fails, other parallel routes are still available.

Flowsheet: a schematic plan giving details of all the steps in the process including quantities of chemicals required.

Fluidized bed: if a fluid is passed upward through a bed of solids with a velocity high enough for the particles to separate from one another and become freely supported in the fluid, the bed is said to be fluidized.

Fluoroscopy: examination of objects by observing their X-ray shadow shown on a fluorescent screen.

Flux density: for a given point (especially in the core of a nuclear reactor) the number of neutrons or other particles incident per second on an imaginary sphere centred at that point, divided by the cross-sectional area (1 cm^2) of that sphere. It is identical with the product of the population density of the particles and their average speed.

Forced-draught cooling towers: a system used for cooling fluids, where the coolant—generally atmospheric air—is drawn or forced through a 'rain' of the fluid to be cooled under the action of motor-driven fans.

Form factor: the ratio of the effective value of an alternating quantity to its average value over a half-period.

Fossil fuel: fuel derived from fossilized organic matter; includes coal, crude oil, and natural gas.

Fractional electrolysis: concentration or separation of isotopes by application of electrolysis.

Fracture machanics: a method of analysis which can establish the response of a loaded structure to flaws or cracks postulated to be present at various locations in the material of the structure.

Free electron–hole pairs: the particle pairs formed in solid crystalline material when neutral impurity atoms contained in the material are dissociated by radiation or an electric field, and an electron is physically separated from the remaining positively charged atom (the whole).

Free energy: the capacity of a system to perform work, a change in free energy being measured by the maximum work obtainable from a given process.

Freon: a halogenated hydrocarbon. Used in refrigeration and as aerosol propellants.

Frequency following: the adjustment of the load of an electricity generator by attempting to restore the instantaneous frequency of the supply to a nominal value.

Fretting: a uniform, sometimes very rapid, removal of metal from one or both of two contacting surfaces between which there is relative movement or periodic impacting; sometimes called fretting corrosion.

Froth flotation: a method of separating particles of different densities by the use of frothing agents to vary the density of the liquid in which they are suspended.

Fuel assembly: a group of nuclear fuel elements forming a single unit for purposes of charging and discharging a reactor. The term includes bundles, clusters, stringers, etc.

Fuel channel: a channel in a reactor core designed to contain one or more fuel assemblies.

Fuel cycle: the sequence of steps involved in supplying and using fuel for nuclear power generation. The main steps are mining and milling, extraction, purification, enrichment (where required), fuel fabrication, irradiation (burning) in the reactor, cooling, reprocessing, recycling, and waste management and disposal.

Fuel cycle equilibrium: when the isotopic compositions of feed fuel, output level, and wastes become constant in a system where fuel is returned to the reactors.

Fuel element: a unit of nuclear fuel which may consist of a single cartridge, or a cluster of thinner cartridges (pins).

Fuel inventory: the total amount of nuclear fuel invested in a reactor, a group of reactors, or an entire fuel cycle.

Fuel pin threaded end plug: a modified fuel pin in which the end plug which seals the hollow Zircaloy-2 tube is extended somewhat, and the external surface is threaded to allow it to screw into the lower and upper tie plates and thus act as a structural member.

Fusion, nuclear: a reaction between two light nuclei resulting in the production of a nucleus heavier than either, usually with release of excess energy.

Gamma ray: very short wavelength electromagnetic radiation, emitted during many types of nuclear reaction.

Gamma-ray spectrometry: an analytical technique whereby radionuclides are identified and measured by determining the energies and intensities of the gamma rays they emit during radioactive decay.

Gangue: the portion of an ore which contains no metal.

Gas counter: Geiger counter into which radioactive gases can be introduced.

Gas counting: counting of radioactive materials in gaseous form. The natural radioactive gases (radon isotopes) and carbon dioxide (^{14}C) are common examples.

Gaseous diffusion: name given to the practical separation process based on the principle of molecular diffusion.

Gaseous wastes: generic term to denote gaseous fission products (e.g. iodine, krypton, etc.) or gaseous chemical wastes (e.g. steam, oxides of nitrogen, etc.).

Gas–graphite reactors: gas-cooled, graphite-moderated reactors such as Magnox, AGR, and high-temperature gas-cooled reactors.

Gas lift: technique of lifting liquor from one level to a higher level by entraining liquor in gas bubbles under pressure in a narrow tube.

Geiger-Müller counter: a simple and well-established form of radiation detector which produces electrical pulses at a rate related to the intensity of the radiation. Commonly called a 'Geiger counter'.

Gel precipitation (of fuel): a process for converting liquid metal nitrate into solid mixed-oxide spheres using a gelling agent. The plutonium is co-precipitated with uranium from the nitrate to produce the required enrichment or composition for fast reactor fuel. The process involves fairly simple fluid-handling procedures and can produce fuel with a very low dust content.

Geometric limitation: a method of criticality control which prevents neutron multiplication by appropriate design of the container of fissile material.

Germanium: a metalloid element; it occurs in a few minerals, including coal, and has exceptional properties as a semiconductor.

Glassification: see *vitrification*.

Glove box: a form of protection often used when working with alpha-emitting radioactive materials. Gloves fixed to ports in the walls of a transparent box allow manipulation of work within the box without the risk of inhalation or contact.

Granulocytes: a group of blood cells of the leucocyte division.

Graphite: a black crystalline form of carbon used as a moderator and/or reflector of neutrons in many nuclear reactors.

Gray (Gy): the unit of absorbed radiation dose (replacing the rad under the SI system).

Guide thimbles: another name for support tubes in PWR fuel elements.

Half-life: the characteristic time taken for the activity of a particular radioactive substance to decay to half of its original value—that is, for half the atoms present to disintegrate. Half-lives vary from less than a millionth of a second to thousands of millions of years, depending on the stability of the nuclide concerned.

Half-life, biological: the time required for the amount of a particular substance in the body to be halved by biological processes.

Halides: fluorides, chlorides, bromides, iodides, and astatides.

Handed: arranged as a mirror image of an adjacent section.

Hanger bar: the portion of a suspension section (q.v.) of a fuel assembly by which the fuel and scatter plug are hung from the seal plug.

Hard: of radiation, having a relatively high penetrating power, i.e. energy.

HARVEST: highly active residues vitrification engineering study.

Head end: that part of the reprocessing scheme before solvent extraction, i.e. fuel receipt, fuel breakdown, fuel dissolution, liquor clarification, and conditioning.

Health physics: the study of persons exposed to radiation from radioactive materials.

Heat exchanger: device for transferring heat from one body of fluid to another.

Heavy water (deuterium oxide, D₂O): water in which the hydrogen is replaced by 'heavy hydrogen' or deuterium. Because of the very low neutron absorption cross-section of deuterium, heavy water makes an excellent moderator and is used in, e.g. CANDU and SGHWR nuclear reactor. It is present in ordinary water at one part in about 5000.

Helical multipole coils: helical coils in which adjacent conductors carry opposing currents to produce a multipole magnetic field.

Heterogeneous core: one in which fertile or blanket sub-assemblies are loaded within the boundaries of the highly enriched core zone.

Hex: a colloquialism for uranium hexafluoride (UF₆), the gas used in isotope enrichment plants.

High-level: of radioactive wastes, those that require continuous cooling to dissipate the heat of radioactive decay.

Homogeneous diffusion reactor model: a mathematical description of a reactor in which the reactor is represented as a homogeneous medium of average material composition and the diffusion approximation of neutron transport is assumed to apply.

Honeycomb grid: description of a constructional feature of fast reactor fuel sub-assemblies

Hopping mechanism: molecules adsorbed on a surface can move by hopping from one adsorption site to another.

Hot: jargon for highly radioactive.

Hot cell: see *cave*.

Hot spot: (1) the point of highest temperature in a reactor fuel or its cladding; (2) a restricted area of comparatively intense radiation or radioactive contamination.

Hulls: small lengths of fuel pin cladding left after dissolution of the fuel.

Hydrocarbon: a compound of hydrogen and carbon.

IAEA: International Atomic Energy Agency

Immiscible: that cannot be mixed.

Immunology: the science dealing with the various phenomena of immunity, induced sensitivity, and allergy.

Imploding linear systems: fusion devices in which cylindrical plasmas are created by the implosion of material lining the reactor vessel.

In coincidence: two nuclear events occurring within a fixed (normally extremely short) space of time are said to be in coincidence. The fixed time is called the resolving time.

Indirect cycle: where the turbine is driven by steam produced from the heat of the reactor coolant, i.e. a primary and a secondary circuit.

Inertial confinement: short-term plasma confinement arising from inertial resistance to outward forces.

Individual risk: the probability of harm to an individual.

Induced activity: activity of the radionuclides produced within materials by neutron irradiation.

Inelastic collision: a collision in which kinetic energy is not conserved. With neutrons, a collision with a nucleus in which part of the initial energy is released as a gamma ray. A neutron is emitted from the nucleus which is of lower energy than the incident neutron.

Inelastic scattering: the outcome of collisions between particles, in which some energy is absorbed or emitted, i.e. they do not simply bounce off one another.

Inert gas: helium, neon, krypton, xenon, and radon are the so-called inert gases.

Infarct: death of tissue resulting from the arrest or sudden insufficiency of circulation in the artery supplying the part.

INFCE: International Nuclear Fuel Cycle Evaluation.

Infra-red radiation: invisible electromagnetic radiation with wave-lengths between those of visible light and those of radio waves, i.e. from approximately 0.8 m to 1 mm.

Integro-differential equation: a mathematical equation involving terms in which quantities are required to be integrated, and terms in which quantities require differentiation.

Intermediate-level waste: all those wastes not included in the categories 'high level' and 'low level'.

Invariant magnetic moment: the magnetic moment associated with a gyrating charged particle is an invariant of motion.

In vitro: literally 'in glass', term used to describe the experimental reproduction of biological processes in isolation from the living organism.

In vivo: term used to describe biological processes occurring within the living organism.

Ion: an electrically charged atom or group of atoms.

Ion exchange: interchange between ions of charge. The process is used to remove species from aqueous solution by exchange with other species held on an insoluble solid compound.

Ion pair: a positively charged ion together with the electron removed from it by ionizing radiation.

Ionization: the process of creating ions by dislodging or adding orbital electrons.

Ionization chamber: a device for measuring the intensity of ionizing radiation. The radiation ionizes the gas in the chamber and the rate at which ions are collected (on oppositely-charged electrodes) is measured as an electric current.

Ionization continuum: the energy region above the threshold for ionization of an atom.

Ionization radiation: radiation which removes orbital electrons from atoms, thus creating ion pairs. Alpha and beta particles are more densely ionizing than gamma rays or X-rays of equivalent energy. Neutrons do not cause ionization directly.

Irradiate: to expose to irradiation, particularly to penetrating forms such as gamma rays and neutrons. Often used to mean the 'burning' of fuel in a nuclear reactor. Also used as a measure of the extent of this burning, expressed in megawatt-days per tonne.

Irradiation swelling: changes in the density and volume of materials due to neutron irradiation.

Isomers (nuclear): nuclides having the same number of neutrons and of protons, but having different internal energy levels.

Isomorphism: the name given to the phenomenon whereby two or more minerals crystallize in the same class of the same system of symmetry and develop very similar forms.

Isothermal temperature coefficient: the rate of change in reactivity with temperature from one uniform temperature over the whole reactor to another uniform temperature.

Isotopes: forms of the same element having different atomic weights.

Isotopic abundance: in a specimen of an element, the percentage of atoms having a particular mass number (i.e. of a specified isotope).

Isotropic: said of a medium, the physical properties of which, e.g. magnetic suscepti-bility or elastic constants, do not vary with direction.

Jet nozzle process: process whereby isotope separation is obtained by the fast flow of uranium hexafluoride in a curved duct.

Joule: a unit of energy; 1 kW h = 3.6 million joules.

K-capture: a radioactive transformation whereby a nucleus captures one of its orbital electrons. Usually accompanied by emission of electromagnetic radiation.

Laminar flow: flow in which adjacent layers do not mix, except at molecular scale.

Laser compression: compression of matter induced by impinging laser beams

Laser fusion: fusion achieved by spherically symmetrical laser compression (q.v.)

Lattice: (1) the regular pattern of fuel arrangement within a reactor core; (2) the arrangement of atoms in the structure of a crystal.

Lattice constants: the simple factors, which characterize the neutron physics of the lattice, such as the fast fission factor, the resonance escape probability, the fuel absorption probability, and the number of neutrons produced per absorption in the fuel.

Leaching: the dissolution of a substance from a solid containing it, e.g. uranium from ore or fission products from vitrified radioactive waste.

Lead-time: the expected time required from the placing of an order for a plant to the commercial operation of the plant.

Leakage: the net loss of particles (e.g. neutrons, gamma rays) from a region or across a boundary of a region.

Leaning post: a structural item in the fast reactor core against which a group of six adjacent sub-assemblies are located or are sprung to provide radial restraint. Control rods are usually located and moved inside leaning posts. Leaning posts may be used also to contain instruments, experiments, and non-standard sub-assemblies.

Levitron: in fusion research, a toroidal magnetic trap formed by levitating a current-carrying ring in the plasma chamber.

Licensee: the holder of a licence issued by the regulatory body to perform specific activities related to the siting, construction, commissioning, operation, and decommissioning of a nuclear plant.

Light water: ordinary water, as distinct from heavy water.

Light water reactor (LWR): a reactor using ordinary water as both moderator and coolant. The term embraces boiling-water reactors and pressurized-water reactors.

Limiter: an aperture defining the boundary of a plasma.

Liquid scintillation counting: a method for measuring the rate of decay of a radioactive isotope, frequently a β-emitter, in which the material containing the isotope is converted to a liquid, which is then mixed with another liquid called a scintillant. The scintillant contains molecules of a solute (fluors), which fluoresce when excited by a transfer of energy, from, for example, an emitted β-particle. The resultant emission of light quanta (flashes of light), the number of which is proportional to the total energy of the initiating β-particle, are registered as a simple pulse by a closely placed photomultiplier tube (or tubes). The number of pulses counted in unit time (the counting rate) is thus related to the disintegration rate (activity) of the sample being measured.

LMFBR: liquid-metal-(cooled) fast breeder reactor.

load factor: the ratio of the average load during a year (or any other selected period) to the maximum load occurring in the same period. It is also used for the ratio of total output of a generating unit in a period to its designed or reference capacity.

Load throw-off: the rapid rejection of the load on an electric generator.

LOCA: loss of coolant accident. The conditions which might arise when the coolant level in the primary vessel, or in the secondary circuit, falls.

Loop-type reactor: one in which the primary coolant is piped outside the main vessel to external heat exchangers (c.f. *pool-type* reactors).

Loss-cone: in a *magnetic trap* (q.v.) the region of velocity space occupied by escaping particles.

Low-level waste: generally, those wastes which because of their low radioactive content do not require shielding during normal handling and transport. In the UK, it is usually interpreted as those solid, liquid or gaseous wastes that can be disposed of safely by dispersal into the environment.

LWR (light water reactor) box: the term used to describe the square-cross-section

sheath which encloses the fuel pins comprising the fuel element of an LWR.
LWR fuel: fuel used in LWRs, usually slightly enriched UO_2.

Magnetic confinement: in fusion research, the use of shaped magnetic fields to confine a plasma.

Magnetic field: effect produced in the region around a conductor carrying an electric current. It exerts a force on any moving electric charge, causing charged particles to travel in helical paths about magnetic field lines.

Magnetic field line: often called 'line of force', an imaginary line showing the direction of the magnetic field. The density of these lines is often used to denote field strength.

Magnetic mirror: when a particle gyrating round a line of force moves from weaker to a much stronger field, it can be reflected back. This arrangement is called a magnetic mirror.

Mangetohydrodynamics (MHD): the study of the motion of an electrically conducting fluid in the presence of a magnetic field.

Magnox: (1) an alloy of magnesium containing small amounts of aluminium and beryllium developed for cladding natural uranium metal fuel used in the Calder Hall reactors and the power stations of the first British nuclear power programme (the Magnox reactors): it absorbs few neutrons and does not react with the carbon dioxide gas used as reactor coolant (hence the name—'*magnesium, no oxidation*'); (2) the generic name given to the type of gas-cooled graphite-moderated reactor using Magnox-clad fuel, on which Britain's first nuclear power programme was based.

Mandrel: a rod inserted into a pellet die prior to filling with powder to form a central hole in the pressed pellet.

Mass balance area (MBA): section of plant or process area that can be isolated in order to determine the quantity of fissile material present.

Mass defect: the difference between the mass of a nucleus and the sum of the masses of its constituent nucleons.

Mass–energy equivalence: confirmed deduction from relativity theory, such that $E = mc^2$, where E = energy, m = mass, and c = velocity of light.

Mass limitation: technique for criticality control where the total mass of fissile material present is limited.

Mass number: the total number of neutrons and protons in an atomic nucleus.

Mass spectrometer: an analytical instrument in which accelerated positive ions of a material are separated electromagnetically according to their charge-to-mass ratios. Different species can be identified and accurate measurements made of their relative concentrations.

Mathematical modelling: the representation of a real physical process by a series of mathematical equations, whose solution thus aids the understanding of the process.

Maximum permissible level/body burden/concentration: the maximum permitted value for such quantities as radiation dose-rate, quantity or concentration of a radionuclide, as determined by health physics considerations. Usually based on the recommendations of the International Commission on Radiological Protection.

Meson: one of a series of unstable particles with masses intermediate between those of electrons and nucleons, and with positive, negative or zero charge.

Metastasis: a secondary tumour.

MeV: million electronvolts—a measure of energy.

Micrometre: one millionth of a metre.

Milling capacity: the quantity of uranium ore which can be handled by the fabricating plant. The throughput of uranium ore in a uranium processing mill will be restricted by the milling capacity.

Missile shield: a steel or concrete structure placed over the upper head of the reactor vessel to protect the containment building from missiles such as ejected control rods.

Mixed mean temperature, mixed outlet temperature: the most useful mean value of the temperature which, when multiplied by the mass flow of fluid and its specific heat, gives the transport of heat along a passage.

Mixed oxide fuel: fast reactor fuel consisting of intimately mixed dioxides of plutonium and uranium (which may be depleted). In a fast reactor the plutonium undergoes fission, while the uranium acts as a fertile material for breeding.

Mixer–breeder: the upper axial breeder zone of the fast reactor, which forms part of the core sub-assemblies and is a separate assembly of large-diameter wire-wrapped pins.

Mixer–settler: a solvent-extraction plant unit comprising two inter-connected tanks, in one of which two immiscible fluids of different densities are stirred together, then allowed to separate out in the other. The plant will comprise several such units arranged in a cascade with the two liquids flowing in opposite directions, one carrying the product and the other the impurities.

Moderating ratio: a figure of merit which accounts for both a moderator's ability to slow down neutrons and its propensity to capture them by absorption.

Moderator: the material in a reactor used to reduce the energy, and hence the speed, of fast neutrons to thermal levels, so far as possible without capturing them. They are then much more likely to cause fission in ^{235}U nuclei. The most important moderators are graphite, water, and heavy water.

Moderator coefficient: the rate at which the reactivity of the system increases with temperature.

Mole: the amount of substance that contains as many entities (atoms, molecules, ions, electrons, etc.) as there are atoms in 12 g of ^{12}C. It replaces in SI the older terms gram-atom, gram-molecule, etc. and for any chemical compound will correspond to a mass equal to the relative molecular mass in gram.

Molecular diffusion: the process by which molecules pass through very fine pores. The lighter molecules move faster and travel more easily down the holes.

Monazite sands: deposit from which thorium is mainly extracted.

Monte Carlo method: statistical procedure when mathematical operations are performed on random numbers.

MOX: mixed oxide fuel.

MTR: materials testing reactor (e.g. DIDO, PLUTO).

MTR fuel: enriched uranium–aluminium alloy clad in aluminium metal.

Multipass steam superheater: an arrangement for raising the temperature of steam above the boiling temperature associated with its pressure, wherein the steam flows successively through separate paths arranged through the heat source—for example—the core of a nuclear reactor.

Multiphoton dissociation: the process in which a molecule is dissociated by the absorption of many photons of the same energy.

Multiplication factor: the ratio of the rate of production of neutrons by fission in a nuclear reactor or assembly to the rate of their loss, symbolized by the letter k 1. When the reactor is operating at a steady level, $k = 1$, when divergent, $k > 1$, when shut down, $k < 1$.

Multi-start helical finning: a series of separate helical fins formed on the outer surface of a tube.

Muon: subatomic particle with rest mass equivalent to 106 MeV. It has unit negative charge, a half-life of about $2\,\mu s$, and decays into electron, neutrino, and antineutrino. It participates only in weak interactions without conservation of parity.

Myocardium: the middle layer of the heart, consisting of cardiac muscle.

Neutrino: a particle having no mass or charge which is emitted in radioactive beta decay along with an electron. Although of great interest as an elementary particle, it is of little concern in nuclear technology.

Neutron: an elementary particle with mass of 1 atomic mass unit (approximately 1.67×10^{-27} kg), approximately the same as that of the proton. Together with protons, neutrons form the nuclei of all atoms. Being neutral, a neutron can approach a nucleus without being deflected by the positive electric field, so it can take part in many types of nuclear interaction. In isolation neutrons are radioactive, decaying with a half-life of about 12 minutes by beta emission into a proton.

Neutron absorber: substance with the property of absorbing neutrons, e.g. boron, gadolinium, etc.

Neutron absorption: a nuclear interaction in which the incident neutron joins up with the target nucleus.

Neutron economy: for good neutron economy losses of neutrons must be kept to a minimum. In a reactor, some neutrons are unable to take part in the chain reaction because they are captured in the nuclei of the reactor material or they leak out of the system.

Neutron-induced voidage: the presence in a material of voids, which are small aggregates of vacancies stabilized by gas atoms; these give macroscopic increases in the volume of non-fissile components subjected to irradiation by neutrons.

Neutron poisons: see *neutron absorber*: absorbers dissolved in solutions of fissile materials or incorporated into plant equipment which holds or processes fissile materials are soluble or fixed poisons, respectively.

Neutron scatter plug: an arrangement in a fuel assembly which prevents neutrons streaming out from a reactor core along the coolant pipework by causing such neutrons to be deflected back into the core.

Neutron spectrum: the energy distribution of a neutron population.

Neutron yield: the average number of fission neutrons emitted per fission.

NII: Nuclear Installations Inspectorate.

Noble gases: another name for the inert gases, comprising the elements helium, neon, argon, krypton, xenon, and radon-222. Their outer (valence) electron orbits are complete, thus rendering them inert to most chemical reactions.

Nomogram: chart or diagram of scaled lines or curves for the facilitation of calculations.

Non-invasive: not interfering directly with bodily tissues.

Nuclear (energy etc.): resident in, derived from or relating to atomic nuclei (rather than to entire atoms—see *atomic*).

Nucleic acids: the non-protein constituents of nucleoproteins. Nucleic acids play a central role in protein synthesis and in the transmission of hereditary characteristics.

Octahedral symmetric anharmonicity: a term in the potential energy of a polyatomic molecule which is octahedrally symmetric.

Off-gas: gaseous effluents from a process vessel, a chemical process, or a nuclear reactor.

OK: odourless kerosene, a hydrocarbon mixture used as a diluent for the extractant in a solvent-extraction process.

Oklo: the uranium mine in Gabon, West Africa, where the first evidence of a natural nuclear reactor was discovered. Thousands of millions of years ago the isotopic abundance of ^{235}U was much higher than it is today. Local concentrations at Oklo were high enough, when moderated by incoming water, to form critical assemblies in which fission took place over periods of millions of years.

Once-through fuel cycle: where the fuel is used only once, and is not reprocessed for reuse.

Once-through type boiler: high-pressure boiler using superheated (dry) steam.

One-group cross-section: a single, average cross-section value for the whole range of neutron energies of interest.

Orbital: of electrons, revolving around the atomic nucleus at considerable distances (on the atomic scale) from it. The number of possible orbits is limited, and when an electron changes orbit, energy is given off (as light or X-rays), or absorbed.

Osteosarcoma: a malignant tumour derived from osteoblasts, composed of bone and sarcoma cells.

Oxide fuel: nuclear fuel manufactured from the oxide of the fissile material. Oxides will withstand much higher temperatures and are much less chemically reactive than metals.

Parameter: one of the measurable characteristics or limits of a given design, system or operation. For example, in a nuclear reactor the pressure vessel dimensions, coolant temperature limits, neutron doubling time, radiation levels, power output, etc.

Parent: the immediate precursor of a *daughter product* in radioactive decay processes.

Particles, elementary: particles which are held to be simple; in nuclear energy these are neutrons, protons, electrons, positrons, photons. In the study of high-energy physics there are a great many other 'elementary' particles such as mesons, pions, quarks, neutrinos, etc.

Partition coefficient: the ratio of the total concentration of a substance in the organic phase to its total concentration in the aqueous phase.

Passage grave: one of the main categories of megalithic or chamber tomb.

Pebble bed (core): a loose bulk of pebble-shaped fuel compacts made from graphite and uranium dioxide powder.

Perfusion: use of radioisotopes to measure the blood flow per unit volume of the organ under investigation.

Periodic Table, chart: a chart of the chemical elements laid out in order of their atomic number so as to bring out the relationship between atomic structure and chemical etc. properties.

PFR: prototype fast reactor, at Dounreay, Scotland.

PFR fuel: a mixture of uranium and plutonium dioxide used as fuel in the PFR.

Photochemical: the chemical effects of radiation, chiefly that due to visible and ultraviolet light.

Photodissociation: dissociation produced by the absorption of radiant energy.

Photomultiplier: a device in which electrons striking its first stage, owing to absorbtion of a photon of light, are multiplied in number so as to produce an easily detectable electrical pulse at the last stage.

Phytoplankton: planktonic plants.

Pickling: a chemical process for cleaning metallic surfaces by the dissolving of a thin layer of metal.

Pile: the former name for a nuclear reactor, particularly of the graphite-moderated type.

Pilot plant: small-scale plant used by chemical engineers to study behaviour of larger plants which they are designing.

Pin, fuel: a very slender fuel can, as used, for instance, in fast reactors.

Pinch effect: the constriction of an electric discharge due to the action of its own magnetic field.

Pitchblende: a particularly rich ore of uranium, historically important in the discoveries of radioactivity by Becquerel and the Curies. Radium was first discovered in this mineral.

Planchet: plain disc of metal.

Plankton: small plants and animals living in the surface waters of the sea.

Plasma: an electrically neutral gas of free ions and electrons (i.e. electrons that have been stripped from the original atoms).

Plasma temperature: temperature expressed in degrees K (the thermodynamic-temperature) or in electronvolts (the kinetic temperature); 1 keV = 10 000 K.

Plate crevice (tube to tube): the very narrow gap which is formed when a tube is located in the drilled hole in the thick tube plate.

Plateout: a general term used to encompass all processes by which material is removed from suspension to form a coating (or plating) on exposed surfaces.

Plenum (fuel pin): a space provided at one or both ends of a fuel pin for the fission gases released by the fuel during operation.

PLUTO: materials testing reactor at Harwell.

Plutonium (Pu): the important fissile isotope ^{239}Pu produced by neutron capture in all reactors containing ^{238}U. Higher isotopes, plutonium 240, 241, 242, and 243, are produced in lesser quantities by further captures.

Poison: see *neutron poison*.

Poison, burnable: a neutron absorber deliberately introduced into a reactor system to reduce initial reactivity, but becoming progressively less effective as burn-up progresses. It thus helps to counteract the fall in reactivity as the fuel is burned up.

Poisson distribution: statistical distribution characterized by a small probability of a specific event occurring during observations over a continuous interval (e.g. of time or distance).

Poloidal field: the magnetic field generated by an electric current flowing in a ring.

Polonium: radioactive element, symbol Po. Important as an alpha-ray source relatively . free from gamma emission.

Polyatomic: a molecule containing many atoms.

Polymerization: the combination of several molecules to form a more complex molecule having the same empirical formula as the simpler ones. It is often a reversible process.

Pool-type reactor: one in which the primary coolant circuits (i.e. including the intermediate heat exchangers and primary pumps) are within the primary vessel (cf. loop-type reactor).

Positron: the anti-particle of the electron, having a positive charge instead of the more usual negative charge. It is the only anti-particle of significance in the context of nuclear power.

Positron annihilation: a positron, or positive electron, can annihilate with a negative electron to produce electromagnetic radiation consisting of two gamma rays of 0.511 MeV energy emitted (if the positron and electron are at rest) in exactly opposite directions (180 ° to each other). If the electron is in a solid, the distribution of angle between the gamma rays provides information on the distribution of electron velocities in the solid, and hence about the physical structure of the material.

Posting: the name given to the transfer of radioactive or dangerous materials such as plutonium from store to sealed glove box or plant or from plant to plant etc. in such a manner that the materials are never exposed to personnel.

Power density: rate of production of energy per unit volume (of a reactor core).

Power ramping: an increase in reactor power from a pre-existing level to a higher level. The term is generally applied to a fairly rapid increase after a prolonged period at the lower level.

Precursor: in a radioactive decay chain any nuclide which has preceded another.

Pre-equlibration conditioning: preliminary treatment of an aqueous phase in order to convert the material to be solvent extracted into the most suitable chemical form.

Pressure-suppression containment: a form of reactor primary containment employing *pressure-suppression ponds* to reduce the pressure inside the containment, following an accident in which there was an escape of a condensable reactor coolant.

Pressure-suppression ponds: a large volume of water used to reduce the pressure in, say, the primary containment of a reactor cooled by boiling or pressurized water after a release of coolant, by condensing the steam fraction and the vapour flashed-off from the escaping liquid.

Pressure-tube reactor: a class of reactor in which the fuel elements are contained in a large number of separate tubes, through which the coolant water flows, rather than in a single pressure vessel. Examples include Britain's SGHWR, and Canada's CANDU.

Pressure-vessel: a reactor containment vessel, usually made from thick steel or prestressed concrete, capable of withstanding high internal pressures. It is used in gas-cooled reactors and light water reactors.

Pressurized-water reactor (PWR): a light water reactor in which ordinary water is used as moderator and coolant. The water is prevented from boiling by being kept under pressure and is circulated through a boiler in which steam is raised in a separate circuit for the turbo-alternators.

Primary circuit: the coolant circuit which removes heat directly from the core.

Primary separation plant: that part of a nuclear fuel reprocessing plant where the bulk of the fission product decontamination occurs, and plutonium and uranium are separated from each other.

Prismatic core: vertical columns of graphite prisms.

Proliferation: in the nuclear policy context, an increase in the size of nuclear weapons arsenals worldwide; this may involve the escalation of nuclear weapons capability to countries which currently are not nuclear weapons states.

Prompt: of neutrons or gamma rays, emitted immediately upon fission, or other interaction.

Prompt critical: the state of achieving criticality in a reactor by means of the prompt neutrons alone and therefore without the control effected through the delayed neutrons.

Proportional counter: a detector for ionizing radiation which uses the proportional region in a discharge tube characteristic, where the gas amplification in the tube exceeds unity but the output pulse remains proportional to initial ionization.

Protium: lightest isotope of hydrogen of mass unity (^{1}H) most prevalent naturally. The other isotopes are deuterium (^{2}H) and tritium (^{3}H).

Proton: the nucleus of hydrogen atoms of mass number 1 and a part of all other nuclides. The number of protons in a nucleus of any element is the atomic number, Z, of that element.

Pulsed inertial devices: fusion systems relying on *inertial confinement* (q.v.)

Purex: generic name for solvent-extraction processes using TBP as the extractant.

Pyrite: sulphide of iron crystallizing in the cubic system. It is also known as iron pyrite.

Pyrometallurgical process: process using heat to refine or purify metals.

Quality assurance (QA): a systematic plan of inspection necessary to provide adequate confidence that a product will perform satisfactorily in service.

Quality control: a statistically based procedure of operational checks and tests for the production of a uniform product within specified limits in accordance with design requirements.

Quality factor: of radiation, a factor used to express the biological effectiveness of different kinds of radiation.

Quartz fibre electrometer: an instrument which measures radiation exposures via the force between two charged quartz fibres; this charge is reduced as ionization occurs. It can be conveniently sized for monitoring personal doses, with an immediate visual display.

Rad: a unit of absorbed radiation dose, equivalent to 10^{-2} J/kg. The unit is being replaced by the SI unit, the gray (Gy), equal to 100 rads.

Radiation: electromagnetic waves, especially (in the context of nuclear energy) X-rays and gamma rays, or streams of fast-moving particles (electrons, alpha particles, neutrons, protons), i.e. all the ways in which an atom gives off energy.

Radiation area: an area to which access is controlled because of a local radiation hazard.

Radiation damage: undesired effects in a material arising from disturbance of the atomic lattice or from ionization caused by radiation. It is often deliberately incurred in the course of experimental work, especially on reactor materials.

Radiation dose: the quantity of radiation received by a substance.

Radiation risk: the risk to health from exposure to radiation.

Radiation source: a device which emits radiation, such as a quantity of radioactive material encapsulated as a sealed source, or a machine, e.g. for generating X-rays for medical purposes.

Radioactive source: any quantity of radioactive material intended for use as a source of radiation.

Radioactive waste: all materials arising from reactor operations which are deemed to be of no further value, but which, due to induced activity or contamination, or a combination of both, have a radionuclide content which exceeds a prescribed level.

Radioactivity: the property possessed by some atomic nuclei of disintegrating spontaneously, with loss of energy through emission of a charged particle and/or gamma radiation.

Radioactivity, induced: radioactivity that has been induced in an otherwise inactive material, usually by irradiation with neutrons.

Radioactivity, natural: the radioactivity of naturally occurring materials (e.g. uranium, thorium, radium, potassium-40).

Radiobiology: branch of science involving study of effect of radiation and radioactive materials on living matter.

Radiocarbon dating: (see *carbon dating*).

Radiochemistry: that part of chemistry which deals with radioactive materials, including the production of radionuclides etc. by processing irradiated or naturally occurring radioactive materials. The use of radioactivity in the investigation of chemical problems.

Radiography: a method of visually examining the interior of a specimen for defects etc. by passing a beam of penetrating radiation through it so that 'shadows' are cast by the denser or thicker parts. These can be examined on a fluorescent screen or a cathode ray tube, at the time, or recorded on photographic film. Medical diagnostic X-rays and industrial gamma-ray tests are the best known examples. A more recent development uses neutrons.

Radioiodines: radioactive isotopes of the element iodine.

Radioisotope: short for radioactive isotope.

Radiolysis: the chemical decomposition of material by radiation.

Radionuclide: radioactive nuclide.

Radiotherapy: treatment of disease by the use of ionizing radiation.

Radiotoxicity: a measure of the harmfulness of a radioactive substance to the body or to a specified organ following its uptake by a given process.

Radon: a zero-valent, radioactive element, the heaviest of the noble gases.

Raffinate: the waste stream remaining after the extraction of valuable materials from solution particularly in the reprocessing of fuel.

Rare-earth elements: a group of metallic elements possessing closely similar chemical properties. These are extracted from monazite, and separated by repeated fractional crystallization, liquid extraction, or ion exchange.

Rare-earth fission products: fission products which are rare earths.

Raster: a pattern of scanning lines arranged to provide complete coverage of an area.

Rating (linear, mass, of fuel): the rate at which heat is generated in fuel. Mass rating is expressed in watts per gram of fuel; linear rating of a fuel pin is usually expressed in watts per unit length of pin.

Reactant: a substance taking part in a chemical reaction.

Reactivity: a measure of the ability of an assembly of fuel to maintain a neutron chain reaction. It is equal to the proportional change in neutron population between one generation and the next. In terms of k_{eff} reactivity $= 1 - (1/k_{eff})$.

Reactivity worths: the reactivity change caused by a particular addition or removal of a material or sub-assembly to or from a reactor. The worth of an individual isotope is often expressed as the ratio of its reactivity at the core centre to that of a ^{239}Pu atom.

Reactor chemical: a vessel, or part of a plant, in which a chemical reaction is maintained and controlled, usually as part of a production process.

Reactor vessel: the container of a reactor in which the fuel, moderator (if any) coolant and control rods are situated.

Recuperator: an arrangement whereby hot fluid leaving a circuit, heats the incoming fluid.

Recycling: the reuse of the fissionable material in irradiated nuclear fuel after it has been recovered by reprocessing.

Reducing agent: a substance that will remove oxygen from or add hydrogen or electrons to a second substance, itself being oxidized in the process. Reducing agents are often used in the separation of Pu from U.

Reflector: an extra layer of moderator or other material outside the reactor core designed to scatter back ('reflect') into the reactor some of the neutrons which would otherwise escape.

Refractory elements: metals with a high melting point such as vanadium, niobium, tantalum, molybdenum, titanium, and zirconium.

Regulatory body: a national authority or a system of authorities designated by national government, assisted by technical and other advisory bodies, and having the legal authority for issuing of licenses.

Rem (roentgen-equivalent man): the unit of effective radiation dose absorbed by tissue, being the product of the dose in rads and the quality factor. The rem is being replaced by the SI unit, the sievert (Sv), equal to 100 rem.

Reprocessing: the processing of nuclear fuel after its use in a reactor to remove fission products etc. and to recover fissile and fertile materials for further use.

Reserves: of uranium, resources which can, with reasonable certainty, be recovered at a cost below a specified limit using currently proven technology.

Residence time: dwell time in a given section of a process.

Resin, useful lifetime: the period during which the resins arranged in a water purification plant of the ion-exchange type continue to be effective in service.

Resolved resonance region: region where the peaks in cross-section can be separately recognized.

Resonance integral: the integral cross-section value over all the energies within the resonance region, weighted by the reciprocal of the energy.

Resonance (reaction rate): the high, often narrow, peak in a reaction rate curve as a function of energy which is observed when the incident particle excites a specific energy level in a compound nucleus.

Resources: of uranium, 'reasonably assured resources' refer to ore known to exist and to be recoverable within a given production cost. 'Estimated additional resources' refer to ore surmised to occur around known deposits or in known uranium-bearing districts.

Reversed field pinch: a toroidal magnetic trap in which the toroidal field changes sign at an intermediate minor radius.

Reverse osmosis: a process essentially akin to filtration, involving the use of extremely high-quality membranes for the removal of dissolved salts from solution.

Rig: an experimental device, especially one designed to enable work to be carried out in, or in close association with, a nuclear reactor or chemical plant.

Roentgen: a unit of exposure to radiation based on the capacity to cause ionization. It is equal to 2.58×10^{-4} coulombs per kilogram in air. Generally an exposure of 1 roentgen will result in an absorbed dose in tissue of about 1 rad.

Roll and shock: motion imparted to a ship as a result of interaction with waves.

Rose Bengal: a red dye which has the property of accumulating in the liver.

Rotameter: instrument for measuring the flow of liquids or gases.

Rotating nut and translating screw device: a mechanical device consisting of a vertical threaded shaft together with a series of freely rotating roller nuts canted to match the lead angle of the threads on the shaft. As the shaft rotates the nuts turn within the threads of the shaft translating vertical motion to it, much as a turning nut would cause a bolt to raise or lower in a slot which prevents the bolt from turning.

Rotor precession: this is low-frequency orbiting motion superimposed on the high rotation frequency.

Safety rod: One of a set of additional reactor control rods used specifically for emergency shutdown and for keeping the reactor in a safe condition during maintenance etc.

Salting out: improving the extraction of a substance (in solvent extraction) by the addition of particular substances to the aqueous phase.

Saturated fluid: a liquid at its boiling-point corresponding to the imposed pressure.

Saturated steam: steam at the same temperature as the water from which it was formed, as distinct from steam subsequently heated.

Saturation temperature: the temperature at which the liquid and vapour phases are in equilibrium (at some given pressure). When the pressure is 1 atmosphere, the saturation temperature is called the boiling-point. Above the saturation temperature, the liquid phase cannot exist stably.

Scattering: general term for irregular reflection or dispersal of waves or particles. Particle scattering is termed elastic when no energy is surrendered during the scattering process—otherwise it is inelastic.

Scintillation counter: a radiation detector in which the radiations cause individual flashes of light in a solid (or liquid) 'scintillator' material. Their intensity is related to the energy of the radiation. The flashes are amplified and measured electrically and displayed or recorded digitally as individual 'counts'. (see *gamma-ray spectrometry*).

Scoop: a pipe which is used for extracting the product or waste from a centrifuge. It can also act as a braking mechanism and to stimulate friction flow in a machine.

Scram: the evacuation of a building or area in which it would be dangerous for the operators to remain; in US usage, emergency shutdown of a nuclear reactor or other potentially dangerous plant.

Scrub: the process of removing impurities from the separated organic phase containing the main extractable substances by treatment with fresh aqueous phase.

Sealed-face production lines: a design of production line for active fabrication in which all equipment is sited behind a single operating face, as distinct from the use of free-standing glove boxes.

Sealed source: a radiation source totally enclosed in a protective capsule or other container so that no radioactive material can leak from it.

Seal plug: specifically the pressure closure at the end of the channel of a tube reactor through which access may be gained for refuelling.

Secular equilibrium: the term given to the particular stable situation which exists in respect of the relative abundances of the isotopic members of a radioactive decay chain, e.g. A decays to B decays to C ... D etc., after a long decay period (long in terms of the decay constant of the longest half-life isotope at the head of the chain). Secular equilibrium is established when the quantity present of any given isotope within the chain is exactly balanced by, on the one hand, the rate of its production caused by the decay of the isotope one up in the chain, and, on the other hand, its own rate of decay to the isotope next down in the chain. At this point, the disintegration rates of all the members of the chain are equal.

Semi-conductor detector: detector for nuclear particles which makes use of a semi-conductor; this is a material whose resistivity is between that of insulators and conductors and can be changed by the application of an electric field.

Separation factor: this factor measures the difference in mole fraction of the desired isotope after passage through a separating element.

Separation nozzle process: in this process isotope separation is achieved by expanding gas through a nozzle.

Separative work: a measure, for costing purposes, of the work done in enriching a material such as uranium from the initial concentrations to the final desired enrichment.

Serum protein: protein in the blood serum; this is the fluid portion of the blood obtained after removal of the fibrin and blood cells.

SGHWR: steam-generating heavy water reactor.

Shear: a property of a twisted magnetic field whereby the amount of twist varies with depth; it is used in some plasma confinement experiments to reduce instabilities.

Shearing: cutting nuclear fuel into small lengths.

Shield, biological: a mass of absorbing material which reduces the level of ionizing radiation, e.g. from a reactor core, to an acceptably low level. It is usually made from high-density concrete, lead, or water.

Shield, thermal: in a reactor, a shield of thick metal plates placed inside the biological shield to protect it from damage by overheating.

Shim: a term used to describe a device which permits a small adjustment to be made. It may be used in relation to the small adjustment in reactivity control permitted by the use of soluble poison boric acid.

Shock heating: a method of heating plasma by a sudden increase in magnetic field.

Shroud tube: a vertical tube within the above-core structure, in which a control rod moves and is supported. Shroud tubes align with the control rod guide tubes; so that during operation a control rod may be partly within a shroud tube, partly within a guide tube, and partly within a leaning post.

Shutdown, emergency: the rapid shutdown of a reactor to remedy or forestall a dangerous condition.

Shutdown power: in a shutdown reactor, the continuing power output due to heat produced by fission product decay.

Shutdown, reactor: stopping the chain reaction of fission by inserting all control rods, making the reactor sub-critical (i.e. k less than 1).

Silanes: a term given to the silicon hydrides.

Single reheat: passing steam back to a superheater after it has been partially expanded in a turbine.

Sintering: a process of densifying and binding granules or particles. It is based on the increase in the contact area between granules resulting from the enhanced diffusion of molecules across the surfaces of granules when a material (such as mixed fuel oxide) is heated.

Site licence: a licence issued by the Nuclear Installations Inspectorate for the operation of a nuclear site. In the UK, under the 1965 Nuclear Installations Act, all nuclear sites (except those operated by government departments or the UKAEA—for which, however, similar conditions apply) must be so licensed.

Skimmer: this is used in the jet nozzle process to divide the flow into product and waste streams.

Slab geometry: a model of a reactor in which the fuel and moderating materials are represented as adjacent slabs. The slabs may be finite or of infinite extent.

Slumping (fuel): the movement of molten fuel under gravity.

Smear test: a method of estimating the loose, i.e. easily removed, radioactive contamination upon a surface. It is made by wiping the surface and monitoring the swab.

Societal risk: the probabiity of harm to numbers of people.

Sodium void coefficient: the reactivity change resulting from the loss of all coolant, or from the loss of coolant in a specified region of the reactor.

Sodium voiding: the removal of sodium from a specific location in the reactor core, or from the whole core, or from a section of the core.

Soft radiation: radiation having little penetrating power.

Sol: a colloidal solution, i.e. a suspension of solid particles of dimensions in the approximate range 10^{-4}–10^{-6} mm.

Solenoid: a cylindrical coil of wire in which an electric current through the wire sets up a magnetic field along the axis of the coil.

Sol-gel: a process in which the aqueous sol (colloidal dispersion) is converted into gel spheres by partial dehydration.

Solid waste: radioactively contaminated waste in solid form.

Solvent: the organic phase which may consist of a mixture of extractants and/or diluent and/or modifier.

Solvent degradation: deterioration of the solvent brought about by high irradiation levels.

Solvent recycle: the return of the (organic) solvent to the solvent-extraction process for reuse, usually after purification.

Somatic effects: the effects of radiation on the body of the person or animal exposed (as opposed to genetic effects).

Source term: the mathematical expression describing the source of neutrons of a particular energy at a point in space.

Spacer grids: skeletal structures which hold the fuel pins in the required spatial positions at intervals along the length of a fuel element.

Spatial modes: the components of the neutron flux distribution that can occur in a reactor analogous to the fundamental and higher harmonic standing waves in a vibrating string, or on a vibrating surface.

Spatial transient experiments: measurements on the variation of properties, such as the neutron population or the reactor temperature with time and position, following some initiating disturbance.

Sparger (sparge tube): a device consisting of a tube with holes in it through which water can be sprayed.

Specific energy loss: energy loss per unit path length.

Specific heat capacity: the quantity of heat which unit mass of a substance requires to raise its temperature by one degree.

Spectrometer: instrument used for measurements of wavelength or energy distribution in a heterogeneous beam of radiation.

Spent fuel: nuclear fuel which has reached the end of its useful life in a reactor.

Spider connecter: the device which links the control rod drive shaft to each individual absorber rod in the fuel assembly (see Fig. 4.8).

Spiked seating: fuel support allowing space for gases to circulate.

Spike (sub-assembly): the lowest part of a sub-assembly; it fits tightly into the coolant supply plenum (diagrid), and so maintains both the geometry of the coolant flow and the positional rigidity of the sub-assembly.

Spoil: waste.

Spot prices, spot market: although the majority of uranium procurement is carried out under long-term contracts, some small quantities of uranium are bought for immediate delivery on the 'spot market' at prices which tend to be considerably higher.

Sputter: the term 'sputtering' is used to denote the process whereby atoms, or clusters of atoms, charged or uncharged, are released from an electrode (generally a metal), held at a negative potential, under the impact of bombarding ions.

Square coupons: a term to describe thin square slabs of reactor material. When assembled in appropriate numbers, a desired average reactor composition is obtained.

Stage: a concept in solvent extraction where complete equilibrium between phases is attained. In mixer–settlers the concept of a stage is often synonymous with the physical unit of one mixer and one settler.

Standpipe: an open vertical pipe connected to a pipeline, to ensure that the pressure head at that point cannot exceed the length of the standpipe.

Steam driers: devices used to remove the last traces of water from steam. They usually consist of a series of passages in which the steam is made to change direction abruptly. Water droplets having inertia tend to travel straight on, impinge on the sides of the passage and the water is collected in vertical gullies or drains.

Steam-end pedestal: the structure containing the bearing and shaft-sealing glands of a turbine, specifically at the end of the cylinder.

Steam quality at TSV: the measure of the temperature and pressure of steam at the turbine stop valve. From a knowledge of the temperature and pressure of the steam, the amount of superheat can be calculated.

Stellarator: in fusion research, a toroidal magnetic trap with the magnetic fields generated entirely by conductors placed around the torus.

Step function: one which makes an instantaneous change in value from one constant value to another.

Stochastic, non-stochastic effects (of radiation): an effect is said to be stochastic if its probability is a function of the irradiation dose without threshold. If the severity of the effect is a function of the irradiation dose, with or without a threshold, then the effect is non-stochastic. Somatic effects can be stochastic or non-stochastic, while hereditary effects are usually regarded as stochastic.

Stoichiometry: exact proportion of elements to make pure chemical compounds.

Storage: emplacement in a facility, either engineered or natural, with the intention of taking further action at a later time, and in such a way and location that such action is expected to be feasible. The action may involve retrieval, treatment *in situ* or a declaration that further action is no longer needed, and that storage has thus become disposal.

Stratosphere: the earth's atmosphere above the troposphere.

Stress corrosion: crack propogation which depends on the combined action of chemical corrosion and tensile stress.

Stringer: see *fuel assembly*.

Stripping foil: a stripper, which is an essential feature of a tandem accelerator, in the form of a thin foil. Ions are accelerated towards the stripper which is maintained at the potential of the high-voltage terminal. In passing through it, molecular ions are totally disintegrated and atomic ions are stripped of the charge by which they were first accelerated and, in changing sign, are then freshly accelerated away from it. Sign changes can be either from +ve to −ve or, as in the case of carbon ion produced from a caesium *sputter* ion source (q.v.), from −ve to +ve.

Strip solution: the aqueous phase used for removing a particular solute from a loaded solvent or extract.

Sub-assembly, fuel, blanket: the basic, removable, unit of the fast reactor core. Each fuel sub-assembly has a steel, usually hexagonal, wrapper enclosing a bundle of fuel pins, usually with an inlet filter and a gag to regulate the coolant flow which removes heat from the fuel pin bundle. A blanket sub-assembly similarly contains blanket pins. The sub-assembly represents the least mass of fuel handled by the operator for movements to and from the core.

Superconductor: a material which exhibits zero resistance (at low temperature) to electric current flow.

Supercritical: exceeding the necessary conditions for the attainment of criticality.

Superheat: (1) the condition in which the temperature of a liquid is above the saturation corresponding to the pressure in the liquid, where the degree of supherheat, T, is the difference between the superheated liquid temperature and the saturation temperature; (2) the temperature rise given to steam passing through a superheater.

Superheat channel liner: a thin metal shield arranged inside the pressure tube (of a reactor) in which the coolant is being superheated, allowing secondary coolant to be introduced so that the pressure tube can be maintained at a temperature associated with the boiling process.

Surge tank: an appropriately elevated or pressurized tank connected to, say, the suction or inlet main of a group of feedwater pumps. It is able either to accept surplus condensate or to supply extra condensate when there is demand for a change in the system flow-rate following a change in load before an equilibrium condition is re-established. In such an arrangement, the surge tank is said to be 'riding' on the feed pump suction.

Suspension section: an arrangement for 'hanging' a fuel assembly inside a reactor core.

Swirler: a device for mixing the coolant to enable fully representative recordings of outlet temperatures and possible fission product release to be obtained. In the case of PFR fuel this duty is performed by the mixer–breeder.

Swirl vane separators: devices to separate steam from water which involve imparting a centrifugal motion to the mixture, so that the water is flung out to the walls of the separator to be collected, whilst the steam passes up the centre of the device.

Synchrotron: an accelerator in which charged particles follow a circular path in a magnetic field and are accelerated by synchronized electric impulses.

Tailings: mine or mill wastes consisting of crushed uranium ore from which the uranium has been extracted chemically.

Tails: the depleted uranium produced at an enrichment plant, typically containing only 0.25 per cent of ^{235}U.

TBP: tri-*n*-butyl phosphate—the solvent used as the extractant in the Purex process.

Teletherapy: treatment of tumours by radiations from outside the body.

Theoretical or ideal stage: a stage where equilibrium is not affected by chemical or physical influences.

Thermal diffusion: process in which a temperature gradient in a mixture of fluids tends to establish a concentration gradient. It has been used for isotope separation.

Thermal neutrons: neutrons in thermodynamic equilibrium (i.e. moving with the same mean kinetic energy) with their surroundings. At room temperature their mean energy is about 0.025 eV and their speed about 2.2 km/s.

Thermal reactor: a nuclear reactor in which the fission chain reaction is propogated mainly by thermal neutrons, and which therefore contains a moderator.

Thermite-type reaction: a strongly exothermic reaction between a metal and a metal oxide in which the oxide is reduced to the metal.

Thermocouple: a device for measuring temperature using the electrical potential produced by the thermoelectric effect between two different conductors.

Thermoluminescent dating: a further physical method of dating associated with the accumulated effect of radiation (α-particles, β-particles and γ-rays) emitted in the decay of the radioactive constituents in the material concerned (e.g. pottery). In crystalline material (e.g. quartz) a fraction of the energy associated with this radiation is permanently stored in the crystal lattice of the minerals in the material (as electrons trapped at regions of imperfection). Subsequent heating of the material allows this energy to be released in the form of light (thermo-uminescence), the quantity being related to the product of the time since the last heating (firing in the case of pottery) and the quantity of radioactivity present. Thus, from an assessment of the total radioactive content an estimate of the elapsed time can be determined.

Thermonuclear reaction: a nuclear fusion reaction brought about by very high temperatures.

Thermosetting resins: solid (plastic) compositions in which a chemical reaction takes place while they are being moulded under heat and pressure. The product is resistant to further applications of heat (up to charring point), e.g. phenol formaldehyde, urea formaldehyde resins.

Theta pinches: cylindrical plasmas constricted by external currents flowing in the θ-direction to produce solenoid magnetic fields.

Thorium: a naturally radioactive metal, the mineral sources of which are widely spread over the earth's surface. The main deposit from which thorium is industrially extracted is monazite sands.

Thorium cycle: a nuclear fuel cycle in which fertile thorium-232 is converted to fissile uranium-233.

THORP: the Windscale thermal oxide reprocessing plant.

Threshold energy characteristic: in the context of a nuclear reaction, the dependence upon incident neutron energy of an event, such as the probability of nuclear fission·in some nuclei, which only takes place above a certain energy—the threshold energy of the incident neutron.

Time-of-flight technique: a way of measuring the neutron spectrum in which the energy or speed of neutrons is determined by the time taken by the neutrons to traverse a known distance.

Tokamak: in fusion research, a toroidal magnetic trap whose poloidal field is generated by the current in the plasma. Tokamak is an acronym of the Russian words meaning toroidal magnetic chamber.

Toroidal: having the shape of a torus.

Toroidal magnetic field: magnetic field generated by current flowing in a toroidal solenoid.

Torus: a tube bent round in a ring with the ends joined up together to give the shape of a motor-tyre tube or an American doughnut.

Toxic: of poison.

Tracer: a small amount of easily-detectable material fixed to some substance whose movement one wishes to follow. If this material is radioactive it can easily be detected by devices such as γ-cameras or other nuclear radiation detectors.

Transient over-power: an accident in which the reactor power exceeds the normal safe upper limit, but full coolant flow continues throughout.

Transient, reactor: a change in power level and/or a temperature which may be accidental or deliberate, causing other reactor or plant parameters to change from their steady-state values.

Transition, nuclear: a change in the configuration of a nucleus, usually either a disintegration or a change in internal energy level, accompanied by emission of radiation.

Tri-normal: a solution of reagent containing three gram-equivalents per litre.

Transuranic elements: the elements of atomic number 93 and higher which have heavier and more complex nuclei than uranium. They can be made by a number of nuclear reactions, including prolonged neutron bombardment of uranium.

Trip: rapid automatic shutdown of a reactor caused when one of the operational characteristics of the system deviates beyond a present limit. Often applied to spurious shutdowns caused, for example, by an instrument malfunction.

Tritium: the isotope of hydrogen having an atomic mass number of 3.

Tritium unit: a proportion of tritium in hydrogen of one part in 10^{18}.

Troposphere: atmosphere up to region where temperature ceases to decrease with height.

Tube-and-shell units: a form of heat exchanger construction comprising a tube 'nest' supported within a casing referred to as the 'shell'. One fluid flows through the tubes, while the other passes over their outer surfaces, during which it is contained within the shell.

Tube-sheet: the thick cylindrical plate which is drilled to take the very large number, typically 5000, of tubes which go to make up a nuclear steam generator.

Tunable dye laser: a dye laser, consisting of an organic dye dissolved in a solvent, which fluoresces over a broad band of frequencies, enabling the output to be tuned.

Turnings: swarf produced by machining a piece of metal.

Two-phase coolant: a coolant which can exist in significant quantities in both the liquid phase and the gaseous phase.

Two-start, helical 'rip': a helical, longitudinal fin (here, specifically, there would be two such fins at any section) produced in the outer surface of a fuel clad. Used in conjunction with fuel elements having 'plain' cladding to space one element from another along their whole length within a fuel bundle assembly.

Unsealed source: radioactive material that is not encapsulated or otherwise sealed, and which forms a source of radiation. For example, radioactive material in use as a *tracer*.

Uranium: the heaviest element found in appreciable quantities in the earth's crust. Natural uranium contains 0.0055 per cent ^{234}U, 0.71 per cent ^{235}U, (by weight) the remainder being ^{238}U. Other isotopes of uranium are produced by irradiation in reactors.

Vacuum chamber: in fusion experiments, the vessel inside which the plasma is confined, so-called because it is first highly evacuated to remove impurities before the plasma is introduced. In actual confinement, the plasma is compressed by the magnetic field into a much smaller volume than the vacuum chamber, thus isolating it from the walls.

Valence: chemical unit of combining power.

Value function: a function representing the value of a quantity of separated material which depends on the number of separating elements required to produce it.

Venturi tube: a wasp-waisted tube used to measure the flow of fluids.

Vibro-compaction: mechanical vibration and compaction.

Vicor glass tube: a proprietory glass tubing (available in the USA).

Vipak: a vibratory compaction process.

Vitrification: the incorporation of radioactive waste oxides into glass (also known as glassification).

Void coefficient: the partial derivative of reactivity with respect to a void (i.e. the removal of the material) at a specified location within a reactor. It is equal to the reactivity coefficient of the material removed.

WAGR: Windscale advanced gas-cooled reactor.

Warm twisting operation: twisting at temperature of typially, 350–450 K.

Waste disposal: the consignment of radioactive wastes to areas or facilities from which there is no intent to retrieve them.

Waste storage: the consignment of radioactive wastes to areas or facilities from which there is an intent and capability to retrieve the waste for further treatment or disposal.

Waste transport: the movement of radioactive waste from one location to another.

Water reactors: nuclear reactors in which water (including heavy water) is the moderator and/or the coolant.

Weldment: a welded assembly.

Wet (saturated) steam: steam containing particles of unevaporated water.

Whole-body monitors: an assembly of large scintillation detectors, heavily shielded against background radiation, used to identify and measure the total gamma radiation emitted by the human body.

WIMS: the name of a family of computer codes used widely to predict the properties of thermal reactors. It is an acronym of Winfrith Improved Multigroup Scheme.

Xenon effect: the rapid but temporary *poisoning* of a reactor by the build-up of xenon-135 from the radioactive decay of the fission product iodine-135. Xenon-135 is a strong absorber of neutrons and until it (and its parent) has largely decayed away reactor start-up can be difficult.

X-rays: electromagnetic radiations having wavelengths much shorter (i.e. energy much higher) than those of visible light. X-rays with clearly defined energies are produced by atomic orbital electron transitions. X-rays produced by the interaction of high-energy electrons with matter have a continuous energy spectrum, and it is these that are generally used in medical X-ray machines (see *Bremsstrahlung*).

Yellow-cake: concentrated crude uranium oxide, the form in which most uranium is shipped from the mining areas to the fuel manufacturers.

Zeolites: one of a number of minerals consisting mainly of hydrous silicates of calcium, sodium, and aluminium, able to act as cation exchangers.

Zero-power reactors: reactors not requiring high power or large coolant flow, used to study reactor physics of various designs without the build-up of significant quantities of fission products.

Zircaloy: an alloy of zirconium and aluminium used for fuel cladding in water reactors.

Zooplankton: planktonic animals.

INDEX

acid
 hydrochloric 374
 leaching process 186, 374
actinides 289-91, 315, 401
 activity in nuclear waste 291-3
 in fast reactor 27-9
 production of in PWR 23-4
actinium 89 6
advanced gas-cooled reactor
  (AGR)
 assembly in 201
 cladding manufacture in 198
 elements 170-3
 fuel in 27
 pins 172
aerodynamic methods, of
  uranium enrichment 132-
  6
 jet nozzle process 132-5
ALARA (as low as is reasonably
  achievable) discharge
  procedures 231
alkali–feldspar deposits 62
Alligator River deposits 60
Almelo, cascade hall at 133
Alpine period 59
aluminium nitrate, anion-
  deficient 271
aluminium rotors, critical speed
  for 127
americium 5, 8, 23, 27, 45
ammonium di-uranate
  (ADU) 190
ammonium uranyl carbonate
  process (AUC) 190
anion-deficient aluminium
  nitrate 271
Antarctic Treaty 315
argillaceous (clay) rock
  formations 318
argon-41 296
Atomic Energy, Geneva
  Conferences of Peaceful
  Uses of 219
atomic route, of laser isotope
  separation 143-5
'Atoms for Peace' policy 219
Atucha 175
austenitic stainless steel 182
Australia 59
 Alligator River deposits 60
  Ranger vein 60
 Mary Kathleen deposits
   (Queensland) 63
 Yeelirrie deposits (Western
   Australia) 63
AVM (Atelier de Vitrification

Marcoule) 307
 process 308

'banana-peeling' technique, at
  Dounreay 235
barns 11
basket 240
Battelle Pacific Northwest
  Laboratories 356
Beaverlodge district, Canada 59,
  68
 vein deposits in 60
Becker jet nozzle 104, 132-5
Belgium 162, 300
 Eurochemic-Mol 43, 246, 354
Berkeley reactor 170
beryllium 172
beta-gamma activity, plutonium-
  contaminated solid wastes
  with 300-1
bioshield, at Elk River reactor
  (ERR) 360, 361
bismuth (^{214}Bi) 68
bitumen 299
BNFL CONPOR fuel 191, 193
Bohemia, vein uranium deposits
  in 59
boiling-water reactor (BWR) 18,
  173, 393
 fuel elements 179-80
boron 234
borosilicate glasses 303, 304,
  305, 324
bottles 221, 268
Brannerite 62, 73
breeding
 fusion 396
 gain 395
British Nuclear Energy Society
  Conference on Uranium
  Isotope Separation 137,
  140
British Tube Alloys Project 115
British Nuclear Fuels Ltd
  (BNFL) 210, 211, 267, 306
 CONPOR fuel 191, 193
Bugey 170
burn-up 19, 24, 29, 166-7
 high 393
Butex (dibutylcarbitol) 217, 219,
  226, 228, 254, 255, 257,
  258, 259, 266, 267, 400
 –TBP process 262

cadmium 234, 240

calandria 392
Calder Hall 169, 259, 264
Canada 85, 195, 219, 306
 Beaverlodge district 59
 CANDU 4, 175, 202-3, 392,
   393
 Chalk River, Nuclear Research
   Centre 216, 217
 Elliot Lake–Blind River 61,
   62, 68, 71, 73
CANDU (Canadian–Deuterium–
  Uranium) 4, 175, 392
 fuel assembly in 202-3
 fuel cycles 393
Cap de la Hague 43, 48, 247
Capenhurst plant 106, 107, 115,
  132
 cascade hall at 132
capture cross-sections, of
  plutonium isotopes 11, 10,
  24
carbide fuel 182
carbide, silicon coating 193
carbide, uranium (UC) 194, 195
carbonate leaching process 186
Carnallite 317
Carnotite, in Colorado 63
cascade 107-8
 basic theory 109-13
  ideal 111, 112
  jumped 111
  material hold-up and
   equilibrium time 112
  separative power 109-10
  squared-off 113
  valve function 110-11
 hall at Almelo 133
 hall at Capenhurst 133
 multistage 226
centrifugal contactor 249
centrifuges 106, 121-32
 Beams and 122, 124
 development in Holland and
   Germany 107
 flow in 127-32
 for liquor clarification 244
 maximum output rom 124
 plant extension at
   Capenhurst 107
 plasma 137
 stationary wall 132, 135-6
 Svedberg and 122
 Zippe and 124, 125
ceramics 305
chain reaction 370
Chalk River 216, 217, 258
Chapelcross 169, 259